Applied Maths for Engineering

Applied Maths for Engineering

Chris Potarzycki BSc Hons, Cert. Ed., IEng., MIIE
Havering College of Further and Higher Education

A member of the Hodder Headline Group
LONDON • SYDNEY • AUCKLAND

First published in Great Britain in 1999 by
Arnold, a member of the Hodder Headline Group,
338 Euston Road, London NW1 3BH

http://www.arnoldpublishers.com

British Library Cataloguing in Publication Data
A catalogue record for this book is available from the British Library

ISBN 0 340 72011 5

1 2 3 4 5 6 7 8 9 10

Commissioning Editor: Sian Jones
Production Editor: James Rabson
Production Controller: Priya Gohil
Cover Design: Richard Kwan

Typeset in 10/12pt Stone Serif by Academic & Technical, Bristol
Printed and bound in Great Britain by J W Arrowsmith Ltd, Bristol

What do you think about this book? Or any other Arnold title?
Please send your comments to feedback.arnold@hodder.co.uk

To Trish, for this was a joint effort

Grateful thanks to my friend and colleague,
Brian Stead, and to my students for putting me right so often

Contents

Preface

The words 'to engineer' mean to 'contrive' or 'bring about'. An engineer is a designer of artefacts that are intended to make our lives, safer, easier and more fulfilling. However, there are many that would argue that this is exactly what engineering technology does not do. The environmental and social impacts of engineering have brought us new problems. Nevertheless, the technology is here to stay and the problems that challenge us can only be solved by engineers.

In designing a system, an engineer takes the laws of science, selects mathematical methods and then applies them to the problems posed by the practical situation. These activities are the very essence of engineering.

Around the late 1600s Isaac Newton studied dynamic systems and established the laws that determine how things move and change over time. The maths that was around in Newton's time was not good enough for what he needed. In order to describe the systems he was studying, Newton needed a method for describing rates of change. Since such a method was not available, he invented one, *the differential calculus* – probably the most important mathematical invention of the millennium.

In complete contrast, around 1850, the mathematician George Boole developed a new kind of algebra that was concerned with logical thought. However, there was no practical use for it at the time. The Boolean algebra lay around for over half a century before computer engineers realized it was just what they needed for modelling digital information systems.

The two illustrations provide a clue to what mathematics is to engineers. It is, essentially, a man-made tool that can be used to model engineered systems. In some cases it may be invented with no clear purpose in mind, in other cases the mathematical technique is invented because it is needed to deal with a specific problem.

This book is not for students of mathematics. Its main purpose is to guide students of engineering towards the mathematical methods they are likely to need as they go about their careers. However, it would be disappointing if the diligent student failed to be impressed by the elegance of mathematical thinking and did not learn to appreciate the subject for its own sake.

Chris Potarzycki
April 1998

Introduction

This book is designed to meet the requirements of students studying for the Advanced GNVQ in Engineering. It covers the specifications laid down for all four of the mathematics Units: the Mandatory Unit, *Mathematics for Engineering*, the Optional Unit; *Further Mathematics* and the two Additional Units; *Additional Mathematics* and *Extended Mathematics*.

The content of the book also meets the specifications in two elements of the GNVQ Key Skills Unit, *Application of Number* at level III. Chapter 3 is designed with this Unit in mind. However, some additional data-gathering activity may be necessary for students to meet the requirements for all evidence indicators in *Application of Number*.

Eighteen assignments are included. Lists of performance criteria for the GNVQ appear at the head of each assignment. Each one is specifically designed to meet the requirements for evidence indicators and provides opportunities for grading in the Mandatory and Optional Units.

The book should be studied in the order that it is presented. Reinforcement is provided at several key points backed up by a number of revision exercises. However, the topic of Complex Numbers, which appears in Chapter 18, could be taken up at any point after Chapter 5. It would provide useful support to students who are taking *Electrical Principles* concurrently.

I have tried to bear the needs of distance learners in mind. However, distance learning is a lonely business so, for all but the most dedicated, some tutorial support is likely to be required.

Solutions to practice exercises and responses to short-answer questions are provided within the text. This will save the reader the inconvenience of thumbing to the back of the book and cross-referencing exercise numbers to find an answer. However, there is a disadvantage to this. The student should resist the inevitable temptation to look at the answers before attempting the question. The process of working through the practice exercises and attempting to answer questions is a vital part of developing mathematical skills. It is much more important than trying to fudge the right answer.

The Glossary can be used as a dictionary of important terms and as a quick reference source. Frequently, we don't quite remember something and need a short statement to jog the memory. However, if this does not work, page references are provided so the Glossary may be used as an index as well.

From experience, students often find the subject content of the first five Chapters to be the most difficult. If you are one of these students, be patient in your struggles, don't give up, it does get easier and more interesting as your basic skills improve.

chapter

Number

1.1 Types of number

Engineering quantities can be defined by different kinds of numbers. For example, the number of pulses per second sent out by a radar transmitter is a **discrete variable**. The numbers used to describe such a quantity can only be whole and positive, i.e. **natural numbers**. On the other hand, the length of a steel supporting beam is a **continuous variable**. The number that describes the length of a steel beam can be any positive number, depending on how accurate we wish to be and how accurate our instruments are.

What about describing a bank balance to the nearest £? My balance could be £235, −£51 or 0. So in describing the condition of a bank balance we could use a group of numbers which are whole, positive or negative and include zero. These are called **integers**. Now for some definitions.

Natural numbers $0, 1, 2, 3 \ldots$
Whole positive numbers including zero. These describe a quantity of discrete objects, or things that can be counted.

Integers $\ldots -2, -1, 0, 1, 2, 3 \ldots$
Whole positive or negative numbers including zero.

Directed numbers $+11, -256$
These are numbers with a positive or negative sign in front. Take two cars moving on a collision course. We could describe their velocities as being $+75$ and the other -75 miles per hour. An electric current that is reversing at regular intervals is alternating between positive and negative values. With positive numbers it is usual to omit the $+$ sign.

Modulus $11, 256$
A quantity that is described by a directed number such as -12 has two parts; in this case 12 is called the **modulus** (in engineering this is often called the **magnitude** or **amplitude** of the quantity) and '$-$' is the **sense**.

Rational numbers
This is a number that is arithmetically complete like the fraction 2/5 which, expressed as a decimal fraction, is 0.4. It is rational because its value is known to be exactly 0.4, no more and no less. The formal definition of a

rational number is: a number of the form m/n where m and n may be any integer, except n cannot be 0.

The last one is interesting. What is the value of m/n if $n = 0$?

∞. Infinity. It is a number which is mathematically **undefined**

There is one more kind of number.

Irrational numbers $1/7$, $\sqrt{2}$, π, e …
What about the fraction $1/7$? Expressed in decimal form, my calculator gives me 0.142857142. This is not the precise value. The calculator truncates (cuts-off) the value at the tenth digit. The true value of $1/7$ is unknown, the best we can do is write it as $0.142\,857\,142\ldots$ a non-terminating decimal.

The fact that the precise value of irrational numbers cannot be expressed is not too important to engineers. Why?

Because instruments which measure physical quantities
have a limited degree of accuracy

Real number system

The whole set of rational and irrational numbers is known as the **real** number system which can be represented graphically by the **real number line**.

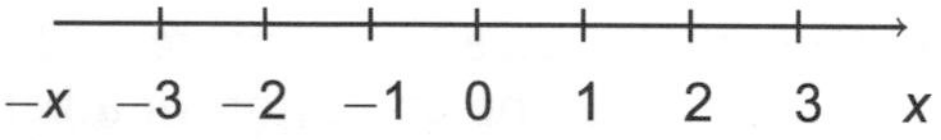

Figure 1.1 Real number line

Why does x appear at the ends of the line? Where will the line finish? Is $-\infty$ very large or very small?

- x is used as a symbol to represent a variable, i.e. it can be any number.
- The ends of the line will be ∞ and $-\infty$.
- $-\infty$ is a number which has the same modulus as ∞.

1.2 Laws of arithmetic

The arithmetical operations of $+$, $-$, $\times$ and $\div$ can only be carried out according to certain rules. Like any game invented by man, mathematics allows you to do anything as long as you don't break the rules.

The rules are **associative**, **commutative** and **distributive**. It is not necessary for you to remember the rules by heart, just learn to apply them. But before we look at some examples, remember the important rule about brackets.

In general, brackets mean that you must deal with the expression inside the brackets first.

The associative rule

Applies to addition and multiplication but it does not apply to subtraction and division.

$$a + (b + c) = (a + b) + c$$

and

$$a \times (b \times c) = (a \times b) \times c$$

Try some numbers in place of the variables a, b, c.

The commutative rule

Again, applies to addition and multiplication only. If you are not convinced try some numbers.

$$a + b = b + a$$

and

$$a \times b = b \times a$$

The distributive rule

Applies to multiplication over addition only. Note that this does not apply to addition over multiplication.

$$a \times (b + c) = (a \times b) + (a \times c)$$

Again, test the rule using some simple numbers.

Basically, what these rules mean is that:

- when adding or multiplying, the order of operations does not matter
- when addition and multiplication are mixed, the order does matter
- when subtracting and/or dividing, the order does matter
- brackets indicate the order of operations

1.3 Operations with directed numbers

There are only two important things to remember here:

- when multiplying and dividing like signs the result is always positive
- when multiplying and dividing unlike signs the result is always negative

With addition and subtraction you just collect up the minuses and collect up the pluses.

Before you do some exercises, a few more definitions.

Sum	$+$	The result of an addition of numbers
Difference	$-$	The result of a subtraction
Product	$\times$	The result of a multiplication
Quotient	$\div$	The result of a division.

A note about the use of × and ÷

In science and engineering it is customary to abbreviate the multiplication sign in the following ways:

$$a \times b \quad \text{looks better as} \quad a.b \quad \text{and even better as} \quad ab$$

Because it is easier to recognize, division is usually written as a fraction, so

$$a \div b \quad \text{looks better as} \quad \frac{a}{b}$$

Factor
This is a multiplier, so in the example ab above, a is one factor of ab and b is the other factor of ab.

Coefficient
In engineering, a factor is often called a coefficient. So with ab, a is said to be a coefficient of b. It is usual for a coefficient to be a constant value, i.e. a number such as $4x$ where 4 is a constant coefficient of a variable quantity x.

Reciprocal
When the result of a multiplication of two numbers is 1, then the two numbers are said to be reciprocals of one another. For example, 5 is the reciprocal of 0.2 and 0.2 is the reciprocal of 5. Try to think of a few more.

Term
An element of an expression. It can be a number, a symbol or a product of factors, for example:

$$5, \quad x, \quad 5x, \quad ax, \quad 2xy, \quad \frac{2xy}{w} \ldots \text{are all terms}$$

Expression
A collection of terms linked by + and −. For example

$$5x + 2y - 12(x - 1), \quad x^3 + 4x^2 - 16x + 31, \quad 12 + 3 - pq$$

Equation
An expression with an equals sign which mathematicians sometimes call an **equality**, e.g.

$$5x - 3 = 25, \quad 15 + 3 = y, \quad I_1 = 12 - I_2$$

Inequality
Examples are:

$$5x - 3 \neq 23, \quad 5x - 3 > 24, \quad 5x - 3 < 26$$

There are some other inequalities. Make a note of them.

$$\approx \quad \text{approximately equal to}$$
$$\sim \quad \text{of the order of magnitude}$$
$$\geq \quad \text{equal to or greater than}$$
$$\leq \quad \text{equal to or smaller than}$$
$$\propto \quad \text{proportional to}$$

Don't be put off by these laws and definitions, they are for reference. You will get to know them as we go along.

PRACTICE EXERCISE 1.1

This exercise gives you an opportunity of getting familiar with the arithmetical and statistical operations and the memory facility of your calculator.

1. Find the sum and differences of the following.
   (a) $107 - 104 + 63 - 48 + 137 + 50 - 149$
   (b) $45 - 764 + 418 - 382 + 1049 - 689 - 1000$
   (c) $-2 - 3 + 4 - (-5)$
   (d) $8 - (-6) + (-12 + 3) - (18 - 4 - 7)$

2. Evaluate the following expressions.
   (a) $12(-8) + 18 - 5(3 - 7)$

   (b) $134(8 - 6) - 12(-2 - 7) + \dfrac{25}{-5}$

   (c) $-3\left(\dfrac{-3072}{12}\right) + 79(18 \times 8) - (7 \times 5)$

   (d) $31\left[\dfrac{15(-6)(-7)}{67 - (-39)}\right] + 121(7 - 18)$

3. Find the product of 75 and 11.

4. A **mean** value is an average which is found by taking the sum of the values and dividing by the number of values. The following is a list of temperature readings in degrees celsius taken at regular intervals. Find the mean temperature.

   13.5, 12.0, 11.0, 9.5, 7.5, 8.0, 7.0, 6.0, 5.5, 4.0, 3.0, 3.5, 2.5, 1.5, 0.5, −0.5, −1.5, −2.5, −4.0, − 5.5, −6.5, −8.0, −9.5, −10.5, −11.0, −12.5, −13.5, −14.5, −15.5, −16.5, −17.0, −17.0, −16.5, −17.0, −16.5, −16.0, −15.5, −15.0, −14.0, −13.0, −11.5, −11.0, −9.5, −8.5, −7.5, −6.0, −5.0, −3.5, −2.5, −1.0, 0.5, 1.5, 2.5, 4.0, 4.5, 5.5, 6.5, 8.0, 9.5, 10.0, 11.0, 11.5, 12.0, 13.0, 12.5, 13.0, 13.5, 13.5, 14.0, 13.5, 13.5, 14.0

5. Which of the following equations is true?
   (a) $15 + 7 - 3 + 2 - 7 = 15 + 7 - (3 + 2 - 7)$
   (b) $58 - (27 - 12) + 6 = 58 - 27 - 12 + 6$
   (c) $115 - 2 + 13 - 46 + 1 = 115 - (2 + 13) - (46 - 1)$
   (d) $48 - 15 + 6 - 3 = 48 - (15 - 6 + 3)$

Solutions

1. (a) 56, (b) −1323, (c) 4, (d) −2;   2. (a) −58, (b) 371, (c) 12 109, (d) −1146.75;
3. 825;   4. $\bar{x} = -0.736111\ldots$, $n = 72$;   5. (d)

1.4 Number bases

Base 10

Probably for no other reason than the fact that the human species has evolved with ten digits, our number system is a base 10 system (denary). This simply means that we have ten different number symbols to represent values. Values greater than 10 must use combinations which repeat the ten symbols at higher levels. The levels are units, tens, hundreds, thousands and so on.

To understand number bases try to think about counting. Counting is a process of adding one to a previous value to make another value. $2 + 1$ makes another value which is represented by the symbol 3. $3 + 1$ makes 4, etc. When we get to 9 we run out of symbols so we record that fact with a 1 in the next decade and re-use our ten symbols 0 to 9. Think of the way a mileage recorder in a car works.

Base 2

Electronic devices are usually designed to count using base 2 (binary) numbers. This is simply because the two symbols 0 and 1 are easily represented by the electrical conditions of 'off' and 'on'. A binary count **trips** after 1 instead of 9. To compare the two number systems study the following count.

0	0
1	1
2	10
3	11
4	100
5	101
6	110
7	111
8	1000
9	1001
10	1010
11	1011
12	1100
13	1101
14	1110
15	1111

Notice how the ten symbols in the denary system and two symbols of the binary system are used to represent 16 unique states. In mathematics they usually represent 16 different values. In computers they can act as a code representing information, such as 16 letters of the alphabet or 16 different control instructions.

Arithmetical operations with binary numbers (*modulo-2 arithmetic*) follow similar, but not the same rules, as for denary numbers. You will learn modulo-2 arithmetic and binary–decimal conversion routines when you study microprocessor engineering.

Finally, it is worth noting that numbers can have any base. In computing, **octal** (base 8) and **hexadecimal** (base 16) numbers are used.

1.5 Denary to binary conversion

Denary to binary conversion is carried out by following the simple rule:

Divide repeatedly by 2 and note the remainder at each stage

EXAMPLE 1.1

Convert 171_{10} to its base 2 equivalent.

Divide	$1 \div 2$	$2 \div 2$	$5 \div 2$	$10 \div 2$	$21 \div 2$	$42 \div 2$	$85 \div 2$	$171 \div 2$
=	0	1	2	5	10	21	42	85
Remainder	1	0	1	0	1	0	1	1

$171_{10} = 1010\,1011_2$

So the remainders form the **weighted bits** of the binary number starting with the least significant bit (LSB).

1.6 Binary to denary conversion

In the denary system, the position of a digit can be indicated by the **exponent** of 10. The exponent is simply the power of a number. Take the following example

$$235 = 200 + 30 + 5 = (2 \times 10^2) + (3 \times 10^1) + (5 \times 10^0)$$

We can do exactly the same in the binary system using the exponent of 2. For example

$$101 = (1 \times 2^2) + (0 \times 2^1) + (1 \times 2^0) = 4 + 0 + 1 = 5$$

So the rule for conversion from binary to denary is:

Find the sum of the weighted bits as powers of 2, starting with 2^0 for the least significant bit (LSB).

EXAMPLE 1.2

Convert $1010\,1011_2$ into its denary equivalent.

Binary	1	0	1	0	1	0	1	1
=	1×2^7	0×2^6	1×2^5	0×2^4	1×2^3	0×2^2	1×2^1	1×2^0
=	128	0	32	0	8	0	2	1
=	171							

$1010\,1011_2 = 171_{10}$

And that is all there is to it. Practise with some numbers of your own.

1.7 Powers and roots of numbers

When dealing with very large (or very small) quantities we can use a form of shorthand by selecting a convenient **base** and raising it to a **power**. For example

$$10^7 = 10 \times 10 \times 10 \times 10 \times 10 \times 10 \times 10 = 10\,000\,000$$

How much simpler to use 10^7!

Numbers expressed in this way are called **orders of magnitude**. They are used in situations when we want a general idea about a quantity and accuracy is not important. For example, the National Grid at peak hours could be delivering 10^7 watts of power. The fact that the actual figure might be $10\,000\,100$ watts may be of little consequence.

Indices

The power of a number is often called an **index** (plural *indices*) or an **exponent**. It can be any value.

Bases

The base can be anything. In a denary system it is natural to favour 10 but in binary systems we can use 2. For example

$$2^8 = 2 \times 2 \times 2 \times 2 \times 2 \times 2 \times 2 \times 2 = 11111111_2 = 256_{10}$$

Note the method of indicating binary and denary values by an appropriate subscript.

Roots

The root of a number is given by a base raised to a **fractional index**. For instance, this means that

$$\sqrt[2]{4} = 4^{\frac{1}{2}} = 2 \quad \text{and} \quad \sqrt[3]{27} = 27^{\frac{1}{3}} = 3 \quad \text{and} \quad \sqrt[5]{1024} = 1024^{\frac{1}{5}} = 4$$

Try these on the calculator and be sure that you can evaluate any root.

But what does *root of a number* actually mean?

One way of looking at it is to say that, just like minus is the inverse of plus, and divide is the inverse of multiply, the root is the inverse of a power. To put it another way:

For numbers greater than 1: a power gives us a large number which has been multiplied by itself, while a root finds the small number which, when multiplied by itself, gives us the large number.

Quite a mouthful! Don't bother trying to memorize it.

Rules of indices

Rules of indices are very important. If you don't already know them, they take some getting used to. We will deal with them in two ways. For the moment we will look at them stated in words and consider numerical examples. Later we will express them algebraically, which is much neater.

Rule	Examples
Any base raised to a power of 0 is equal to 1	$7^0 = 1, 1^0 = 1, 2.53^0 = 1, \ldots$
Changing the sign of the power gives the reciprocal	2^{-3} and 2^3 are reciprocals
To multiply the same bases, sum their powers	$2^4 \times 2^3 = 2^7, 5^3 \times 5^{-5} = 5^{-2}$
To divide the same bases, take the difference of their powers	$6^3 \div 6 = 6^2, 12^{-2} \div 12^3 = 12^{-5}$
For a number raised to a power twice, multiply the powers	$(3^2)^4 = 3^8, (12^3)^{-5} = 12^{-15}$

If you find the rules confusing just consider the third example, multiplication, a little more closely.

$$2^4 \times 2^3$$

We are dealing with the product of 2 multiplied by itself four times and 2 multiplied by itself three times.

$$(2 \times 2 \times 2 \times 2) \times (2 \times 2 \times 2)$$

So how many times all together is 2 multiplied by itself?

$$2^7$$

Yes, 7 times!

1.8 Fractions, ratios and percentages

Fractions

A fraction is a part of a whole. It is the result of dividing a whole into a number of equal parts and then specifying some of those parts. For example: 3/7 specifies three parts of something divided into seven parts. The structure of a fraction is as follows:

$\dfrac{\text{top}}{\text{bottom}}$ **numerator** specifies the number of parts of interest
denominator the number of parts into which the whole is divided

Fractions can be:

proper $\frac{3}{7}$
vulgar $\frac{15}{8}$
mixed $1\frac{7}{8}$

Can you see how we might convert between vulgar and mixed fractions?

In the mixed fraction above, 1 is 8 parts of 8.
So we have 8 parts of 8 and 7 parts of 8.
That adds up to 15 parts of 8.

Be careful though. Mixed fractions are avoided in engineering situations because numbers like $3\frac{3}{4}$ might easily be mistaken for $3 \times \frac{3}{4}$. Using $\frac{15}{4}$ is much safer.

What about decimal fractions? Well, a fraction such as $\frac{3}{7}$ is really a division of 3 by 7. So what is the decimal value of $\frac{3}{7}$? According to my calculator it is:

$$0.428571428\ldots$$

Awful. Too many digits and not complete. Which, perhaps explains why proper fractions are often preferred to their decimal approximations.

Equivalent fractions

There are a great many number of ways (infinite, in fact) in which the same fraction can be expressed. Take the following example.

$$\tfrac{3}{7} = \tfrac{6}{14} = \tfrac{9}{21} = \tfrac{12}{28} = \cdots$$

$\frac{6}{14}$ is just $\frac{3}{7}$ multiplied by 2. Six parts out of 14 is the equivalent of $\frac{3}{7}$. We can multiply $\frac{3}{7}$ by any number and still get the same fraction except that $\frac{3}{7}$ is the smallest and therefore the most convenient; it is said to be in its **lowest terms**.

Notice that when we say $\frac{3}{7}$ multiplied by 2, we mean the *whole fraction*, both the numerator and the denominator were multiplied by 2. In this example, 2 is said to be a **common factor** of $\frac{6}{14}$.

Ratios

These are comparisons between two quantities. If we wish to build a 1 metre model of a ship that is 100 metres long, then all the dimensions must be reduced in the ratio 100:1. Fractions and ratios are just alternative ways of stating the same information.

If the number of workers in an engineering plant is 510, of which 34 are male and 476 female, then we can say that the ratio of male to female workers is 34:476 or we can say that the fraction of males is $\frac{34}{510}$ and the fraction of females is $\frac{476}{510}$. Express these fractions in their lowest terms.

$$\tfrac{1}{15} \quad \text{and} \quad \tfrac{14}{15}$$

Ratios, like fractions, are normally expressed in their lowest terms. Put the ratio 34:476 into its lowest terms.

$$1 : 14$$

Addition and subtraction of fractions

This can only be carried out with fractions of the same denominator. Addition of fractions with different denominators would be like adding metres to feet;

the result would be meaningless. Look at the following

$$\frac{5}{8}+\frac{3}{5}=\frac{25}{40}+\frac{24}{40}=\frac{25+24}{40}=\frac{49}{40}$$

And

$$\frac{5}{9}-\frac{7}{12}+\frac{1}{2}=\frac{20}{36}-\frac{21}{36}+\frac{18}{36}=\frac{20-21+18}{36}=\frac{17}{36}$$

We usually look for the *lowest* common denominator to keep the arithmetic simple, but what do we do with fractions like $\frac{13}{57}$ and $\frac{8}{13}$? Well, in desperation, multiply the denominators by each other and you have a denominator which is common even if it is not the lowest. How would you go about finding the sum and differences of whole numbers and fractions? The following should jog your memory.

$$3\tfrac{1}{4}+5\tfrac{1}{2}-17\tfrac{1}{8}=\frac{13}{4}+\frac{11}{2}-\frac{137}{8}$$

$$=\frac{26}{8}+\frac{44}{8}-\frac{137}{8}=\frac{26+44-137}{8}=-\frac{67}{8}$$

Multiplication of fractions

This is easy. Multiply the numerators, then multiply the denominators, then try and reduce the result to its lowest terms. For instance

$$\frac{2}{5}\times\frac{3}{7}=\frac{2\times3}{5\times7}=\frac{6}{35}$$

Multiplication of whole numbers and fractions is easy if you remember that a whole number is a fraction of 1. So 4 could be written as $\frac{4}{1}$ and 28 could be written as $\frac{28}{1}$. For example

$$3\times\frac{2}{5}\times\frac{2}{3}=\frac{12}{15}=\frac{4}{5}$$

Division

The rule here is to take the reciprocal of the number you are dividing by and then multiply. So for example

$$\frac{3}{8}\div\frac{5}{7}=\frac{3}{8}\times\frac{7}{5}=\frac{21}{40}$$

QUESTION
Evaluate $4\times\dfrac{8}{15}\div\left(\dfrac{3}{4}+\dfrac{5}{6}-\dfrac{2}{3}\right)\times\dfrac{7}{8}\div5$

ANSWER
$\dfrac{112}{275}$

Conversion and comparison

To convert a fraction into a decimal fraction simply divide the denominator into the numerator

$$\tfrac{3}{16}=0.1875$$

To go the other way, multiply the decimal by 10, 100 or whatever is necessary to make it a whole number, but remember what you do to the top you must do to the bottom so

$$0.875 = \frac{875}{1000} = \frac{175}{200} = \frac{35}{40} = \frac{7}{8}$$

This example reduced to its lowest terms very neatly. It is not always that easy.

When comparing fractions it is often difficult to judge which is bigger and which is smaller. Take $\frac{146}{216}$ and $\frac{27}{36}$. If you are good with numbers you may spot that the second is the larger of the two but how could you be sure?

Convert them into decimals.

$$\frac{146}{216} \approx 0.676 \quad \text{and} \quad \frac{27}{36} = 0.75$$

Reciprocals revisited

You should, by now, remember two things. The product of reciprocals is 1. Changing the sign of the power of a number gives its reciprocal. Take the number 5. It can be written as

$$\frac{5}{1}$$

Now, 0.2 is the reciprocal of 5 because $5 \times 0.2 = 1$. So let's look at 0.2. We can write it as a fraction over 1.

Then multiply by 10.

$$\frac{0.2}{1} = \frac{2}{10} = \frac{1}{5}$$

So the reciprocal of 5 is $\frac{1}{5}$. Exchanging the denominator with the numerator gives us the reciprocal. So it is just a matter of turning a fraction upside-down.

Now look at the **indicial** form of 5. We know that changing the sign of the power gives the reciprocal of the number. 5 is just 5 to the power of 1 so its reciprocal must be 5 to the power of minus 1.

$$5^1$$
$$5^{-1} = \frac{1}{5}$$

Now we can look at three rules of indices and test them. The product of reciprocals should be 1. Multiplying the same numbers together involves adding their powers. And anything to the power of zero is equal to 1.

$$5^1 \times 5^{-1} = 5^{1-1} = 5^0 = 1$$

Understanding reciprocals is not always easy at first, but knowing how to make use of them makes multiplication and division of fractions very easy because

multiplying by some number is the same as dividing by its reciprocal and vice versa.

Remember this. It is important!

Take some examples:

$$\tfrac{1}{2}(2+5) = \tfrac{7}{2} = 3.5$$

$$3 \div \tfrac{1}{12} = 36$$

$$\tfrac{1}{\pi}\left(\tfrac{13}{\sqrt{2}+36}\right) \div \tfrac{1}{18} = 1.991$$

Percentages

These are simply fractions of a 100, so that 37% really means $\tfrac{37}{100}$. To convert any fraction into a percentage first convert it into a decimal fraction then multiply by 100. For instance:

$$\tfrac{17}{225} = 0.075555\ldots \approx 7.56\%$$

To convert a percentage to a fraction simply write it as a fraction then reduce it to its lowest terms (if possible), for example

$$14\% = \tfrac{14}{100} = \tfrac{7}{50}$$

PRACTICE EXERCISE 1.2

1. Evaluate $\left(\tfrac{2}{3}\right)^3$

2. Evaluate the following.
   (a) $\sqrt[3]{571} \times 4^3$
   (b) $\dfrac{3^7 \times 3^2}{3^4}$
   (c) $(2^4)^4$
   (d) $\sqrt{2^{16}}$

3. $6\tfrac{2}{3} + \tfrac{1}{6} - \tfrac{5}{12}$ is equal to (note that $6\tfrac{2}{3} \neq 6(\tfrac{2}{3})$):
   (a) $\tfrac{5}{12}$
   (b) $\tfrac{20}{48}$
   (c) $\tfrac{-2}{12}$
   (d) $\tfrac{77}{12}$

4. Evaluate:
   (a) $\tfrac{3}{7}\left(\tfrac{4}{10} - \tfrac{21}{25} + \tfrac{2}{5}\right)$
   (b) $\tfrac{25}{137} - \tfrac{18}{61}$
   (c) $\tfrac{27(-4)}{2(-9)}$
   (d) $\tfrac{7(-3)\times 2(-5)}{6\times 5}$

5. If a population of fruit flies increases from 92 to 138 in a certain period, what percentage increase does this represent? What is the ratio of the new population to the original one?

Solutions

---

1. $\frac{8}{27}$;   2. (a) 531, (b) 243, (c) 65 536, (d) 256;   3. (d) $\frac{77}{12}$;
4. (a) $-\frac{6}{350}$, (b) $-\frac{941}{8357}$, (c) 6, (d) 7;   5. 1.5:1

1.9 Accuracy

Some of the exercises which you have done have produced results like 0.428 571 428...

If this was an amount of energy in joules or an electric current in amperes we would have to decide whether it is worth, or even correct, recording it using so many decimal places. Most ammeters will only display three or four figures. The resolution of a conventional ruler only allows you to record length to the nearest millimetre.

There is an important point here.

> **If your original data has limited accuracy then you have no right to present results which give the impression of an accuracy which was never there in the first place!**

Besides, calculations, even with the best modern calculators, introduce errors in the form of rounding and truncation so the result of any calculation must be less accurate than the original data.

> **It is important to work to a degree of accuracy which is appropriate to the engineering situation**

It is also important to be able to assess error, which reflects the degree of confidence we have in the results of calculations.

Decimal places

One way of indicating the level of accuracy is to quote a value to a certain number of decimal places (d.p.). 235.09 is accurate to 2 d.p., 0.5342 is accurate to 4 d.p.

Significant figures

Significant figures can be a more useful means of expressing accuracy in engineering. 253, 1.25, 2.00, 0.000 321 are all accurate to 3 significant figures (s.f.). Notice the last value consists of **non-significant zeros** while 2.00 has two **significant zeros**. Just writing 2 would imply an accuracy of 1 s.f.

Rounding

There are a number of different rules for rounding but the most popular is the **5-up rule**. If the number you wish to round off is 5 or above then round up, if the number is 4 or below then round down. Some examples:

$$0.543\,2 = 0.54 \qquad \text{to 2 decimal places}$$
$$0.543\,2 = 0.543 \qquad \text{to 3 significant figures}$$
$$127.893\,536\,7 = 127.894 \qquad \text{to 3 decimal places}$$
$$25.34 = 25.3 \qquad \text{to 3 significant figures}$$
$$0.000\,125\,55 = 0.000\,13 \qquad \text{to 5 decimal places}$$
$$0.000\,125\,55 = 0.000\,126 \qquad \text{to 3 significant figures}$$
$$0.098\,744 = 0.9874 \qquad \text{to 4 significant figures}$$
$$299\,792\,458 = ?$$

The last value seems to present a problem. There are no decimal places indicated and rightly so – what difference would fractions make to a number which is approximately 300 million! So we have to resort to significant figures but how? 300 is the value of the four most significant digits rounded to 3 places but it is nowhere near 300 million! Ponder over this for a while. We will come back to it shortly.

Error

Given a number like 1.54 you may rightly assume that the true value lies somewhere between 1.535 and 1.545. So you can confidently state that the true value of the number is 1.54 ± 0.005. Values like ± 0.005 are known as the **limits of confidence**.

± 0.005 is an indication of error which reflects your degree of confidence. An indication of error should always be given when dealing with **continuous data**.

Effect of calculations

Consider an example in which you have two numbers 69 and 33 to add. They are accurate to 2 s.f. which means the actual value of 69 lies between 69.5 and 68.5 and the actual value of 33 lies between 33.5 and 32.5.

Suppose we wish to add the numbers. The possibilities are

$$69.5 + 33.5 = 103 \quad \text{the extreme upper limit}$$
$$69.0 + 33.0 = 102 \quad \text{the nominal value}$$
$$68.5 + 32.5 = 101 \quad \text{the extreme lower limit}$$

So the simple process of addition has extended the limits of confidence from ± 0.5 to ± 1.

Suppose we wish to subtract the numbers. The possibilities are:

$$69.5 - 32.5 = 37 \qquad \text{one worst combination}$$
$$69.0 - 33.0 = 36 \qquad \text{the ideal}$$
$$68.5 - 33.5 = 35 \qquad \text{the other worst combination}$$

Again a subtraction has increased the possible error from ±0.5 to ±1. The general rule is:

with addition and subtraction the maximum error is the sum of the original errors.

Multiplication could produce the following results.

$$69.5 \times 33.5 = 2328.25$$

$$69.0 \times 33.0 = 2277$$

$$68.5 \times 32.5 = 2226.25$$

The three results only agree to the first significant figure because all three round to 2000. Note that the original data was accurate to 2 significant figures.

Division could produce the following.

$$69.5 \div 32.5 = 2.138461538\ldots$$

$$69.0 \div 33.0 = 2.090909091\ldots$$

$$68.5 \div 32.5 = 2.044776119\ldots$$

Again, with the original data correct to 2 s.f. the answers only agree to 1 s.f. The general rule is:

with multiplication and division the result is accurate to 1 significant figure less than the least significant value used.

Finally, when dealing with mixed operations the following should be applied.

When a calculation requires a variety of part calculations, do not be tempted to round off too early. Work to the maximum number of significant figures which your calculator allows, then round off your final result to a degree of accuracy that is appropriate to the problem.

Rough checks

There is, for all of us, a temptation to rely completely on what the calculator tells us. The trouble is that we can make mistakes in entering data and executing operations. You should develop a habit of rough checks *before you pick up your calculator.*

$$\text{Take} \quad \frac{31 \times 18}{1.7}$$

This is roughly 30 multiplied by 20 and divided by 2 which comes to 300. Now use the calculator to get the exact value – it is 328 correct to 3 s.f. This is not very far out from our rough check so the answer is likely to be correct.

When dealing with very large numbers the mental arithmetic starts to get difficult. There is a way around this. Earlier, we looked at orders of magnitude: number values expressed to the nearest power of ten. You can use orders of magnitude together with some crude rounding to get a rough idea of what an answer should be. For example:

$$\frac{2400 \times 819\,000}{753} \approx \frac{10^3 \times 10^6}{10^3} = 10^6$$

Roughly 1 million. The actual answer, correct to 3 s.f., is 2.61 million.

Although this is more than twice our rough estimate at least we are in the millions. Calculator errors with large numbers are likely to give results which are out by orders of magnitude, not just by a factor of 2 or 3.

1.10 Measures of accuracy

As you know, the accuracy of continuous data can never be exact and it is important that we are aware of the degree of error contained in an item of data. So, for example, 273 newtons is the approximate value of a force correct to 3 significant figures. And because of rounding conventions we can place a value on our limits of confidence. In this case it is $\pm 0.5\,\text{N}$.

Now, there are two ways of expressing accuracy: they are called **absolute** and **relative accuracy**.

Absolute accuracy

If a quantity, x, has an approximate value $\bar{x}$ then:

The absolute error in the approximation in $x = \bar{x} - x$

For instance, the limits of confidence of ± 0.5 newtons in the force of 273 N quoted above are, in fact, a measure of absolute error.

EXAMPLE 1.3

What is the absolute accuracy when a length of workpiece is measured to be 560 mm?

Here we have $560 \pm 0.5\,\text{mm}$ so the absolute error is $560 - 559.5 = 0.5$

Relative accuracy

If a quantity, x, has an approximate value $\bar{x}$ then:

The relative error in the approximation to $x = \dfrac{\bar{x} - x}{x}$

Multiplying the relative error by 100 gives **relative percentage error**.

EXAMPLE 1.4

Correct to 3 significant figures, what is the relative percentage error in the length of the workpiece used in the last example?

Relative percentage error $= \dfrac{560 - 559.5}{559.5} \times 100 = 0.0894\%$

Whether we use absolute or relative error as a measure of accuracy depends on the circumstances. For example in specifying the length of a span of a road bridge, it would be essential to know absolute error. The span either fits or does not fit, regardless of its overall length. On the other hand the temperature control of a furnace would need to operate within certain limits of relative error.

1.11   Standard index notation

Scientists who deal with very large quantities such as 40 400 000 000 000 000 metres, which is the distance to our nearest neighbouring star Alpha Centauri; or very small quantities such as 0.000 000 000 000 000 000 160 2 coulombs, which is the amount of electric charge carried by one electron, have to resort to a shorthand which makes the numbers manageable.

For example the distance to Alpha Centauri may be given as a simple order of magnitude. What is it?

$$10^{16} \, \text{m}$$

If we wished to be more accurate we might quote this distance as $4.0 \times 10^{16} \, \text{m}$ correct to 2 significant figures. What would be the distance stated correct to 3 s.f.?

$$4.04 \times 10^{16} \, \text{m}$$

This takes us back to our earlier problem of expressing 299 792 458 (the speed of light in m s^{-1}) in a sensible form. Write it down correct to 3 s.f. using scientific notation.

$$3.00 \times 10^8 \, \text{m s}^{-1}$$

Very small numbers

So much for writing very large values in a convenient form and using a suitable degree of accuracy, but how do we deal with very small numbers? Take a look at indices again. We said earlier that the reciprocal of 5 was $\frac{1}{5}$ which could be written as 5^{-1}. Check this with your calculator. Remember, a negative index indicates a fractional value.

Using base 10 we have:

$$1\,000\,000 = 10^6$$

$$100\,000 = 10^5$$

$$10\,000 = 10^4$$

$$1000 = 10^3$$

$$100 = 10^2$$

$$10 = 10^1 \qquad \text{Remember that any base to the power of 0}$$

$$1 = 10^0 \qquad \text{is always equal to 1}$$

$$0.1 = 10^{-1} \quad \text{The reciprocal of 10}$$
$$0.01 = 10^{-2} \quad \text{The reciprocal of 100}$$
$$0.001 = 10^{-3} \quad \text{etc...}$$
$$0.0001 = 10^{-4}$$
$$0.00001 = 10^{-5}$$
$$0.000001 = 10^{-6}$$

A convenient rule of thumb is:

A *positive* index shows the number of places the decimal point is shifted to the *right*.

A *negative* index shows the number of places the decimal point is shifted to the *left*.

What about numbers like 35 328 and 0.000 046 78? The procedure is to express the numbers in units and decimals and multiply by the right order of magnitude.

$$35\,328 = 3.5328 \times 10^{4} \quad \text{and} \quad 0.000\,046\,78 = 4.678 \times 10^{-5}$$

Notice how easy it is. Changing 35 328 to 3.5328 we have made the number 10 000 (10^4) smaller than it really is, so to compensate we must multiply it by 10^4. Similarly, 4.678 is 10^5 larger so we compensate by multiplying it by 10^{-5}. What remains is to express the values to an appropriate degree of accuracy. 3 s.f. gives us

$$3.53 \times 10^{4} \quad \text{and} \quad 4.68 \times 10^{-5}$$

So that is how we express numbers using **scientific notation**. Before going on to examples of arithmetic you need to know about a similar notation called **standard index notation** or **engineering notation**.

Engineering notation

This is more formal than the scientific form because the powers of ten can only be multiples of three. There is a very good reason for this: it allows us to use standard prefixes such as kilo, mega and micro. The following table should be self-explanatory.

Standard index notation

Order of magnitude	Prefix	Abbreviation	Example
10^{12}	tera	T	13.5 Tm
10^{9}	giga	G	12.1 GHz
10^{6}	mega	M	560 MJ
10^{3}	kilo	k	275 kV
10^{0}			1.36 N
10^{-3}	milli	m	55.4 mW
10^{-6}	micro	μ	7.68 μA
10^{-9}	nano	n	12.5 ns
10^{-12}	pico	p	360 pF

Note the convention of upper and lower case letters when prefixes are abbreviated. You must keep to this convention otherwise prefixes may be confused with units.

Standard index notation is used by engineers, hence the name engineering notation, and that is why calculators often have an 'Eng' key.

Note also that **standard index notation** or **engineering notation** is also called **preferred standard form**.

PRACTICE EXERCISE 1.3

With each of the following problems you should:

- first carry out a rough check and write down your estimate before using your calculator,
- use preferred standard form where it is appropriate,
- give the result to a sensible degree of accuracy.

This exercise will provide you with the opportunity of getting to know your calculator a little more. Particularly the 'Exp' and 'Eng' key operations.

1. Evaluate

   (a) $\dfrac{25.3 \times 2.10}{34.567}$

   (b) $13.5 \times 10^3 \times 71.6 \times 10^6$

   (c) $45.6 + 0.23 - 6.84 - 2.51 + 60.2$

   (d) $\dfrac{\pi(314 \times 10^{-3} + 9.82 \times 10^{-3})}{\sqrt{2}(97.1 - 162.78)}$

2. Write the following in standard engineering form correct to 3 s.f.

   (a) 250 000 000

   (b) −0.000 000 354 5

   (c) 0.012 000 29

   (d) 97.1×10^7

   (e) 2.389×10^{-2}

3. Express the following values placing standard prefixes before the units.

   (a) 0.005 68 metres

   (b) 765×10^{-9} seconds

   (c) 18.854×10^7 joules

   (d) 25.7×10^{-12} farads

   (e) $4\pi \times 10^{-7}$ henrys

4. Evaluate

   (a) $\sqrt{(13.1 \times 10^2) - (2.15 \times 10^2)}$

   (b) $\dfrac{\frac{1}{\sqrt{2}}(184 \times 10^{-6} - 36.4 \times 10^{-6})}{(79.823 \times 10^{-3} - 976.32 \times 10^{-3})^2}$

   (c) $37.5\sqrt{0.0402^2 - 0.01675^2}$

   (d) $\dfrac{1}{2\pi}\sqrt{\dfrac{1}{127 \times 10^{-6} \times 68 \times 10^{-12}} - \dfrac{43^2}{(68 \times 10^{-6})^2}}$

5. An experiment used to determine the modulus of elasticity of a sample of low carbon steel revealed that when stress was increased from $200\,MN\,m^{-2}$ to $460\,MN\,m^{-2}$ the strain changed from 625×10^{-6} to 1.784×10^{-3}. Given that Young's modulus of elasticity can be calculated using

$$E = \frac{\text{change in stress}}{\text{change in strain}}$$

determine the modulus of elasticity of the steel, calculate the error accumulation and state your limits of confidence.
Note: a change in a quantity can be found by taking the difference between its highest and lowest values.

6. In Question 4 of Practice Exercise 1.1 you calculated the mean temperature from a set of 72 values. If each temperature reading is accurate to $\pm 0.5°C$, calculate the error and state the limits of confidence in the mean value which you should have calculated to be $-0.736\,111 \ldots °C$.

7. 25.4 km expressed in standard form is:
   (a) $25\,000\,m$
   (b) $2.5 \times 10^4\,m$
   (c) $25 \times 10^3\,m$
   (d) $0.025 \times 10^6\,m$

8. Dividing watts by volts gives a result which is in amperes. Hence, 234.6 GW divided by 33.0 kV is:
   (a) $7.11\,mA$
   (b) $7.11\,A$
   (c) $7.11\,kA$
   (d) $7.11\,MA$

9. The nearest value to $\dfrac{0.034}{7 \times 10^{-3}} + \dfrac{371}{6 \times 10^{-4}}$ is
   (a) 618×10^3
   (b) 14.5×10^3
   (c) 626×10^3
   (d) 88.3×10^6

10. When calculating the electrical resistance of a circuit the answer given by the calculator is $835\,475.2345\,\Omega$. Which of the following is the best way of recording the value?
    (a) $835\,000\,\Omega$
    (b) $835 \times 10^3\,\Omega$
    (c) $835\,k\Omega$
    (d) $835.475\,234\,5\,k\Omega$

Solutions

1. (a) 1.5, (b) 970×10^9, (c) 96.7, (d) -11×10^{-3}
2. (a) 250×10^6, (b) -355×10^{-9}, (c) 12.0×10^{-3}, (d) 971×10^6, (e) 23.9×10^{-3}
3. (a) 5.68 mm, (b) 765 ns, (c) 189 MJ, (d) 25.7 pF, (e) 400π nH or 1.26 μH
4. (a) 33, (b) 130×10^{-6}, (c) 1.37, (d) 1.71×10^6
5. $224 \pm 1\,GN\,m^{-2}$
6. Accumulated error is $\pm 36°$ but division by integer value of 72 gives $\pm 0.5°C$
7. (c); 8. (d); 9. (a); 10. (c)

1.12 Dimensions and units

Some of the problems in the last exercise involved units like metres, newtons, seconds and ohms. It is important not to confuse units with dimensions. For example, distance which is usually abbreviated (*s*) is a dimension whose unit of measure is the metre (m), time (*t*) is a dimension measured in units of seconds (s), energy (*W*) is a dimension measured in joules (J) and so on.

This is not the time or place for you to start learning all the units and dimensions that exist in engineering. Those you will learn as you go along. But you do need to be aware of international standards for dimensions and units, the conventions which apply, what they actually mean and how to convert between 'popular' and the internationally agreed units.

SI units

These are internationally agreed standard units which are part of the *Système Internationale d'Unites*. They consist of six **fundamental** units and a vast number of **derived** units.

Fundamental SI units

Dimension	Unit	Symbol
length	metre	m
mass	kilogram	kg
time	second	s
electric current	ampere	A
temperature	kelvin	K
amount of substance	mole	m

All other units are derived from these six. For example, the unit of force

$$1 \text{ newton} = \frac{1 \text{ kilogram} \times 1 \text{ metre}}{1 \text{ second}^2}$$

The conventional abbreviation for this would be $N = \text{kg m s}^{-2}$. It is worth examining why units can be equated in this way.

Newton's second law of motion leads us to the equation

$$F = ma$$

where *F* is force in newtons, *m* is mass in kilograms and *a* is acceleration in m s^{-2}, which explains why we get newtons when we multiply kg m s^{-2}.

Note the use of a negative index to denote the reciprocal of seconds. Acceleration is given by

$$a = \frac{s}{t^2}$$

where *s* is distance in metres and *t* is time in seconds. So dividing metres by seconds squared gives acceleration. Can you see that the old fashioned way of representing the unit of acceleration as m/s^2 is the same as saying $\mathrm{m\,s^{-2}}$ (spoken 'metre seconds to the minus 2')? The use of negative indices has now been adopted for SI units and you should keep to it. Now for some examples of derived units.

Some examples of the derived SI units

Dimension	Unit	Symbol	Derivation
force	newton	N	$\mathrm{kg\,m\,s^{-2}}$
energy	joule	J	$\mathrm{kg\,m^2\,s^{-2}} = \mathrm{N\,m}$
power	watt	W	$\mathrm{J\,s}$
electric charge	coulomb	C	$\mathrm{A\,s}$
electric potential	volt	V	$\mathrm{J\,A^{-1}\,s^{-1}}$
magnetic field density	tesla	T	$\mathrm{N\,m^{-1}\,A^{-1}}$
frequency	hertz	Hz	$\mathrm{s^{-1}}$

One further convention to note is the use of upper and lower case letters. The rule is that if the unit name is adopted from the name of a person, it should be written as a capital letter when abbreviated.

QUESTION

In question 5 of Practice Exercise 1.3 you calculated the elasticity of a sample of steel. Given that stress = force per unit area ($\mathrm{N\,m^{-2}}$) and strain is the change in length divided by the original length ($\mathrm{m\,m^{-1}}$), carry out a **dimensional analysis** to determine the unit of elasticity.

ANSWER

$$\text{Elasticity} = \frac{\mathrm{N\,m^{-2}}}{\mathrm{m\,m^{-1}}} = \mathrm{N\,m^{-2}}$$

You may have noticed how easy it is to confuse m-for-milli with m-for-metre. We need to be alert to this and never use m-for-milli within a calculation – always write 10^{-3}. Only put m in front of a unit.

More information on SI units and dimensions can be found in Appendix I.

Dimensional analysis of this kind is a useful method for checking that an equation is correct. For example, what is 3 mm × 2 m? Putting both values into units of metres gives us

$$3 \times 10^{-3}\,\mathrm{m} \times 2\,\mathrm{m} = 6 \times 10^{-3}\,\mathrm{m^2}$$

Notice how multiplying m × m gives m^2. The way in which I have written this calculation is unusual although it is perfectly correct. 3 mm means literally three times the unit of a millimetre and 2 m means two times the unit of a metre. Normally we do not write the units into the calculation but we must always include them with the result.

Length, area and volume

Difficulty sometimes arises when we use standard notation. The things to remember are that area is measured in m^2 so all the numbers have to be squared, and volume is in m^3 so all the numbers have to be cubed.

The following are a couple of examples.

Given a plane measuring 2 mm by 78 mm its area is

$$2 \times 10^{-3} \times 78 \times 10^{-3} = 2 \times 78 \times (10^{-3})^2 = 156 \times 10^{-6}\,m^2 = 156\,mm^2$$

Given a solid measuring 2 mm by 78 mm by 12 mm, its volume is

$$2 \times 10^{-3} \times 78 \times 10^{-3} \times 12 \times 10^{-3} = 2 \times 78 \times 12 \times (10^{-3})^3$$
$$= 1872 \times 10^{-9}\,m^3$$
$$= 1872\,mm^3$$

The important point to note here is that the prefix milli becomes squared (or cubed) as well. A result like $1872\,mm^3$ really means $1872\,(mm)^3$.

Other units

In an ideal world we would all be using a standard set of units (such as SI) and no other. Unfortunately this does not happen. Many of the old Imperial units such as inches, feet and pounds are still in common usage. We measure temperature in degrees celsius and even fahrenheit, not in kelvin, and how many people use radians instead of degrees when measuring angles? The point here is that we must be prepared to convert units from one form to another.

Given a conversion factor the best way to go about carrying out the conversion is to write an equation and then to remember that *what you do to one side you must do to the other*.

Given that 1 inch = 25.4 mm it is easy to find the metric value of any number of inches simply by multiplying by the number of inches. To find the mm value of 3.45 inches we multiply by 3.45.

$$3.45 \times 1\,in = 3.45 \times 25.4\,mm = 87.63\,mm$$

But how would you convert mm to inches using the information given above? First, turn the equation around to put mm on the left-hand side.

$$25.4\,mm = 1\,in$$

Then divide by 25.4 so that you know what 1 mm is in inches.

$$1\,mm = 0.0394\,in$$

You are now in a position to multiply by the required number of millimetres to get the equivalent value in inches. To convert 12 mm we multiply by 12.

$$12\,mm = 12 \times 0.0394\,in = 0.472\,in$$

Some conversions are even easier. For example $\theta_K = \theta_C + 273$, where θ_K is temperature in kelvin and θ_C is temperature in °C. What is 10°C in kelvin?

$$\theta_K = 10 + 273 = 283\,K$$

How would you go about converting °C into kelvin? Rewrite the equation making θ_C the subject. This means taking 273 away from the right-hand side (and therefore taking it away from the left-hand side because you must balance the equation) and then writing the equation the other way around. Try it and find out how much 135 K is in °C.

$$\theta_C = \theta_K - 273 = 135 - 273 = -138°C$$

Other conversions are less easy but the same idea applies. First form an equation. Take the conversion of degrees fahrenheit to degrees celsius, where θ_C is the temperature in °C and θ_F is the temperature in °F. What is 50°F in celsius?

$$\theta_C = \tfrac{5}{9}(\theta_F - 32) = \tfrac{5}{9}(50 - 32) = 10°C$$

Now to transpose the equation into one which will allow you to convert from °C into °F. The RHS (right-hand side) is being multiplied by 5 so divide by 5 to make it 1. *Remember that dividing by something is the same as multiplying by its reciprocal.*

$$\tfrac{1}{5}\theta_C = \tfrac{1}{9}(\theta_F - 32)$$

Now the RHS is being divided by 9 so multiply by 9 to make this 1 as well.

$$\tfrac{9}{5}\theta_C = \tfrac{1}{1}(\theta_F-32) = \theta_F - 32$$

$\tfrac{1}{1}$ is just 1 and anything multiplied by 1 does not change. Now, the RHS is having 32 subtracted from it so let's add 32 to make it zero.

$$\tfrac{9}{5}\theta_C + 32 = \theta_F$$

Finally make θ_F the subject by turning the equation around.

$$\theta_F = \tfrac{9}{5}\theta_C + 32$$

Now you can test it by converting 10°C into °F.

$$\theta_F = \tfrac{9}{5}(10) + 32 = 50°F$$

Algebraic representation using symbols like θ_F and θ_C and transposing equations so that we can use them for a different purpose is the next main topic you will study. In the meantime a little practice with conversions.

PRACTICE EXERCISE 1.4

Given that:

1 mile = 1.61 km	1 lb = 0.454 kg = 16 ounces
1 mm^2 = 10^{-6} m^2	1 g = 10^{-3} kg
1 mm^3 = 10^{-9} m^3	1 gal = 4.55 l
1 cm^2 = 10^{-4} m^2	1 rad = 57.3°
1 cm^3 = 10^{-6} m^3	1 mile = 5280 ft
1 l = 10^{-3} m^3	1 yard = 3 ft

1. Convert 156 feet into metres
2. Convert 68 gallons into cubic metres
3. Express 60 miles per hour in km h^{-1} and m s^{-1}
4. Convert 4.2 square feet into m^2

5. Express 220 cm^2 in mm^2 and m^2
6. How much is 5.83×10^9 l in m^3?
7. Convert 5.367 radians into degrees
8. What is 98.4°F in °C and in kelvin
9. If a car travels 30 miles on 1 gallon of fuel how many litres per 100 km does it burn?
10. A design drawing gives the dimension of a part as 0.732 ± 0.0010 inches. Convert this to mm
11. Convert 150×10^3 kg into pounds
12. How many pounds and ounces is 2500 g?
13. The dimensions of a vessel are 12 ft, 15 ft and 9 ft. What is the volume of the vessel in litres?
14. The circumference of the earth at the equator is 4.01×10^7 m. How many miles is this?
15. If a runner covers 1 mile in four minutes how long would you expect her to take to run 1500 m?
16. Bending moment is given by the equation $M = EI/R$, where E is the modulus of elasticity in Nm^{-2}, I is the second moment of an area in m^4 and R the radius of the bend in m. What is the unit of M?
17. Hooke's law gives rise to the equation $T = \lambda x/l$, where T is the tension of a spring, λ is the coefficient of elasticity in newtons, l is the natural length of a spring in metres and x is the extension of the spring in metres. What is the unit of tension?
18. Given that $\omega = 2\pi f$, where 2π is the angular displacement of a circle in radians and f is frequency in s^{-1}, what is the unit of ω?
19. The electrical resistivity of a conductor can be calculated using $\rho = Ra/l$, where R is the resistance in ohms, a is cross-sectional area in m^2 and l is length in m. What is the unit of ρ?
20. The pascal is the SI unit of pressure. How is it derived?

Solutions

1. 5280 ft = 1610 m, 1 ft = 0.305 m, 156 ft = 47.6 m
2. 68 gal = 309 l, 309 l = 309×10^{-3} m^3
3. 60 mph = 96.6 km h^{-1} = 96.6×10^3 m h^{-1} = 26.8 m s^{-1}
4. 1 ft = 0.305 m, 1 ft^2 = 0.0930 m^2, 4.2 ft^2 = 0.391 m^2
5. 220×10^{-4} m^2 = $22\,000 \times 10^{-6}$ m^2 = 22 000 mm^2
6. 5.83×10^9 l = $5.83 \times 10^9 \times 10^{-3}$ m^3 = 5.83×10^6 m^3
7. 5.367 rad = 308°
8. 36.9°C, 309.9 K
9. 30 mpg = 48.3 km/4.55 l = 100 km/9.42 l = 9.42 l/100 km
10. 18.6 ± 0.025 mm
11. 1 kg = 2.20 lb, 150×10^3 = 330×10^3 lb
12. 2500 g = 2.5 kg = 5.51 lb = 5 lb 8 oz
13. 5280 ft = 1610 m, 1 ft = 0.305 m,
    $12 \times 15 \times 9 = 12 \times 0.305 \times 15 \times 0.305 \times 9 \times 0.305 = 46.0$ m^3 = 46×10^3 l
14. 4.01×10^7 m = 4.01×10^4 km, 1 km = 0.621 ml, 4.01×10^4 km = 24.9×10^3 ml
15. $1500/1610 \times 4 = 3.73$ min = 3 min 44 s
16. N m
17. N

18. rad s^{-1}
19. Ω m
20. N m^{-2}

1.13 Summary

In this chapter you have reviewed:

- different types of numbers and number bases
- laws governing the use of numbers
- some terminology that is essential for communicating mathematical ideas
- powers and roots
- fractions, ratios and percentages
- accuracy and error; very important in engineering
- engineering notation
- SI dimensions and units
- conversion of units

I would be surprised if you did not find some of this rather tedious. The basics of any subject are usually the most boring but unfortunately they are also the most important. So, beware; what follows depends on these basics.

You should find the next chapter on algebra more interesting.

chapter

2 Algebra

2.1 Introduction

The first chapter introduced numbers that are used to represent and calculate engineering quantities. It is a point of interest that our denary number symbols, 0, 1, 2, 3 and so on, are Arabic numerals. From the Middle Ages onward mathematicians universally found them to be the most convenient. Prior to that there were a variety of numerals in use, Babylonian, Egyptian, Greek and Roman; all rather complicated and difficult to use.

Anyway, this section introduces the use of algebra to generate and model data in common engineering situations. It is not by coincidence that the word algebra is also derived from the Arabic word that means *calculating with symbols*. The Arabs followed the Greeks and developed many of our modern mathematical techniques. Algebra was their main contribution.

Often in mathematics and engineering we are more interested in the relationship between quantities rather than their values. In such situations it is convenient just to represent a quantity with a letter of the alphabet (Greek or Roman).

In arithmetic we use numbers to make *particular* statements, whereas in algebra we use symbols to make *general* statements. If, to take a simple example, we wished to convert temperature values between kelvin and °C we might use a table of values such as the one below.

$\theta\,\mathrm{K}^{-1}$	$\theta°\mathrm{C}^{-1}$
286	13.0
301	28.0
308.9	35.9
446.5	173.5

This is clumsy, however. It involves keeping a list of data and does not allow conversions of data outside the given range. Far better is to have some general *rule* or *formula* that can be used to convert any value. For temperature conversion we have the following formula.

$$\theta_{\mathrm{K}} = \theta_{\mathrm{C}} + 273$$

In this equation θ_K and θ_C are called the **variables** and 273 is a **constant**. An equation like this can also be rearranged or **transposed** to suit our purpose. Do you remember how we rearranged this equation to allow us to convert from kelvin to °C?

General and particular solutions

An algebraic equation like $\theta_K = \theta_C + 273$ is a **general solution** to the problem of converting temperatures. We can use general solutions to arrive at **particular solutions** to problems, such as how much is 12°C in kelvin?

$$\theta_K = 12 + 273 = 285 \,\text{K}$$

So, a general solution to a problem is an **algebraic equation**. A particular solution to a problem is a **number**.

Equations of motion and energy

Let's look at a couple of algebraic equations which describe motion in a straight line.

$$v = u + at$$
$$s = ut + \tfrac{1}{2}at^2$$

In the first, v is velocity in metres per second at some time t, u is initial velocity when t is zero and a is acceleration.

In the second equation, s is distance travelled in metres. The equations are general solutions to problems that involve moving objects. They are **mathematical models** which allow us to predict what the velocity will be or how far an object has travelled after some specified time. They also allow us to explore relationships.

Another equation models kinetic energy E_k in terms of mass m and velocity v.

$$E_k = \tfrac{1}{2}mv^2$$

We shall return to these equations as we work through the chapter, so note where they are. You may need to refer back to them occasionally. After a while I hope you will get to remember them.

2.2   Transposition

First, some basic conventions, rules and techniques.

Constants and variables

In the equation $v = u + at$,

v is the dependent variable
t is the independent variable
a is a constant coefficient of t
u is a constant

In most equations involving time, *t* is the independent variable upon which the other variable depends. We say it *varies over time*. Usually, a variable can take any numerical value. But we need to be cautious. In practical situations certain values may not make sense and it may be that the range of a variable has to be restricted. For example if $y = \log x$ then we might add the statement 'with $x > 0$, because the log of 0 is undefined and logs of negative numbers do not exist.

Constants are a fixed value and will usually be a number, but they can also be a symbol. Often they are values like π or $\sqrt{2}$. In the equation of motion *u*, the initial velocity, is fixed at some value. If an object started from rest the value of *u* would be zero. Similarly, acceleration due to gravity is a constant value at $9.81\,\mathrm{m\,s^{-2}}$.

In mathematics, we tend to represent constants with the early letters of the alphabet, *a*, *b*, *c*, *d* etc. and variables with the later letters: *p*, *q*, *r*, *s*, *t*, *u*, *v* and the favourites *x*, *y* and *z*. The equations of motion show that this convention does not always apply in engineering!

Another convention is the sequence of terms in a product; some examples are

$$2xy, \quad \tfrac{1}{4}uv, \quad \omega L, \quad 2\pi fC$$

Can you see what it is?

> Numbers first, Greek letters next, Roman letters last,
> *usually* in ascending order.

Finally, the subject of a mathematical formula is normally written on the left-hand side. We will make use of the abbreviations LHS and RHS for convenience.

Products and quotients

Products and quotients are easy to deal with if you remember that **a number multiplied by its reciprocal is always 1**.

EXAMPLE 2.1

Take the equation $F = ma$, which is based on Newton's second law.

To make *a* the subject we must get rid of its coefficient *m*. Multiplying *m* by $1/m$ gives us 1 and $1 \times a$ is just *a*. So this is what we must do.

But the equation must be balanced so we multiply the LHS by $1/m$ as well.

$$\frac{1}{m}F = a$$

Now we tidy up. Remember that multiplying by something is the same as dividing by its reciprocal so $1/m \times F$ can be written as F/m, and since *a* is to be the subject, we turn the equation around:

$$\frac{F}{m} = a$$

$$a = \frac{F}{m}$$

EXAMPLE 2.2

Another example is the equation that is derived from Ohm's law of electricity.

$$I = \frac{V}{R}$$

In words, this tells us that current I is directly proportional to voltage V and inversely proportional to resistance R. To transpose for V we must get rid of R on the RHS by turning it into 1. Now V/R is the same as $V \times 1/R$ so how do we turn $1/R$ into 1?

Multiply by its reciprocal R

$$IR = V \qquad V = IR$$

EXAMPLE 2.3

Finally, what do we do if the term we want is the denominator? Take Young's modulus of elasticity.

$$E = \frac{\sigma}{\varepsilon}$$

To transpose for ε we must do something to the equation to turn ε into a numerator. An obvious way is to take reciprocals.

$$\frac{1}{E} = \frac{\varepsilon}{\sigma}$$

Now to get rid of $1/\sigma$ on the RHS we multiply by its reciprocal.

$$\frac{\sigma}{E} = \varepsilon \qquad \varepsilon = \frac{\sigma}{E}$$

A note about proportionality

When formulating physical laws, people like Newton and Ohm did not give us the equations we use today. Their contribution was to recognize a proportionality between variables. For example, Newton implied that the acceleration of an object was proportional to the force acting on the object. This can be stated mathematically as

$$a \propto F$$

This means that if you double the force you double the acceleration, or if you halve the force you reduce the acceleration by a half. To turn this into an equation we need a constant of proportionality. In the case of an object acted upon by a force it is the mass of the object. Common sense tells us that the greater the mass the less the acceleration. In other words acceleration of an object is inversely proportional to its mass. Double the mass and you halve the acceleration. The law now becomes an equation,

$$a = \frac{F}{m}$$

which transposes to the more familiar form $F = ma$.

Try to look for proportionality when you look at equations – it is a good way of understanding what they actually mean.

Sums and differences

EXAMPLE 2.4

Take an equation for z where the expression on the RHS consists of four terms linked as sums and differences, $z = 2x + 5y - 4 - 3x$. In mathematics, an expression like this is called a **polynomial** – meaning literally, many names. They are many names because they are different things. $2x$ and $5y$ cannot be added together because they are different. For a similar reason 4 cannot be taken away from $5y$. However, $2x$ and $-3x$ are **like terms**; there are two xs and minus three xs which can be collected up to make minus $1x$. Now our expression for z can be simplified.

$$z = 5y - 4 - x$$

Notice how we treated $2x$ and $3x$ as directed numbers.

Now, to transpose the equation for x. We want x on its own on one side. On the RHS we have $5y$, so if we subtract $5y$ that is one unwanted term out of the way. Remember, we must subtract $5y$ from *the whole equation*.

$$z - 5y = -4 - x$$

We still have -4 on the RHS so to get rid of that we can add 4 because the sum of 4 and -4 is zero.

$$z - 5y + 4 = -x$$

One problem remains. We were supposed to transpose for x, not $-x$. How can we turn $-x$ into x?

$$\text{Multiply by } -1$$

And tidy up.

$$x = 5y - z - 4$$

Let us do a couple more.

EXAMPLE 2.5

Transpose $v = u + at$ for u.

$$v - at = u$$

Subtract at to leave u on its own. Write the equation with the subject on the left.

$$u = v - at$$

EXAMPLE 2.6

Transpose $y = 24 + 3z - 5x - 10 + x$ for x.

This is not in its simplest form so collect up the like terms.

$$y = 3z - 4x + 14$$

Subtract $3z$ and 14.

$$y - 3z - 14 = -4x$$

Multiply by the reciprocal of -4 to make it 1 and leave just $1x$ on the RHS.

$$-\tfrac{1}{4}(y - 3z - 14) = x$$

Notice the use of brackets to show that we are multiplying the whole of the LHS by $-\frac{1}{4}$. Also notice how we multiplied by minus 1 at the same time to make the RHS positive.

We could leave it like that but since the original equation contained no brackets it is only right to leave things as we found them.

$$-\frac{y}{4}+\frac{3z}{4}+\frac{14}{4}=x$$

Finally, 4 is a common denominator so we can tidy up a little more.

$$x=\frac{3z-y+14}{4} \quad \text{or} \quad x=\tfrac{1}{4}(3z-y+14)$$

Brackets help us to deal with mixed operations of products, quotients, sums and differences. We shall look at them next but first you must practise with products, quotients, sums and differences.

A problem that often arises for students is that the result of a transposition can take different forms that are equally valid. Sometimes, as with the last example, it is a matter of simplification; how far do we go? Don't worry if your solution does not appear to be the same as the one I have provided. It may still be correct. You have to accept the fact that alternative algebraic solutions usually exist; something that does not occur with numerical solutions.

PRACTICE EXERCISE 2.1

1. Transpose $P=IV$ for V
2. Transpose $T=\dfrac{mv^2}{r}$ for r
3. Transpose $H=I^2Rt$ for R
4. Transpose $M=\dfrac{EI}{R}$ for I
5. Transpose $\dfrac{PV}{T}=mR$ for T
6. Transpose $y=3z^3+z^2-x+0.75$ for x
7. Transpose $R_0\alpha t=R_t-R_0$ for R_t
8. Transpose $2x-y+12=3x^2-10x-15$ for y
9. Transpose $C=7n+21$ for n
10. Transpose $A=2\pi r^2+2\pi rh$ for h and simplify

Solutions

1. $V=\dfrac{P}{I}$; 2. $r=\dfrac{mv^2}{T}$; 3. $R=\dfrac{H}{I^2t}$; 4. $I=\dfrac{MR}{E}$; 5. $T=\dfrac{PV}{mR}$;
6. $x=3z^3+z^2-y+0.75$; 7. $R_t=R_0(\alpha t+1)$; 8. $y=27+12x-3x^2$;
9. $n=\dfrac{C}{7}-3$; 10. $h=\dfrac{A}{2\pi r}-r$

Brackets

We have looked at how to deal with products and quotients and sums and differences. Brackets help us to deal with expressions in which all four are mixed. So what is the purpose of brackets?

Brackets turn a collection of terms into one thing. For example:

$$m(u + at) \quad \text{means that } u \text{ and } at \text{ are multiplied by } m$$
$$(u + at)^2 \quad \text{means that } u + at \text{ is all squared}$$

Expressions containing brackets tend to be shorter in length than expressions in which the brackets have been cleared out. Consequently clearing brackets is often called **expansion**.

Study the following examples and copy them out.

EXAMPLE 2.7

(a) $2(x - 4) = 2x - 8$

(b) $-\frac{2}{3}y(3y^2 - \frac{1}{6}y + 12) = -2y^3 + \frac{1}{9}y^2 - 8y$

(c) $(a + b)^2 = (a + b)(a + b) = a^2 + ab + ab + b^2 = a^2 + 2ab + b^2$

(d) $2(x - y)^2 = 2(x - y)(x - y) = 2(x^2 - xy - xy + y^2) = 2x^2 - 4xy + 2y^2$

(e) $\frac{1}{2}m(u + at)^2 = \frac{1}{2}m(u + at)(u + at) = \frac{1}{2}m(u^2 + atu + atu + a^2t^2)$

$$= \frac{1}{2}m(u^2 + 2atu + a^2t^2) = \frac{mu^2}{2} + amtu + \frac{ma^2t^2}{2}$$

With quotients, the fraction bar has the same function as brackets. Consider the following example.

$$\frac{8x^2 + 12x - 16}{4} = 2x^2 + 3x - 4$$

Notice that it can be written in two ways. Dividing by 4 is the same as multiplying by the reciprocal of 4. Which looks neater.

$$\frac{1}{4}(8x^2 + 12x - 16) = 2x^2 + 3x - 4$$

The most important thing about brackets is that when transposing equations or evaluating an expression you must *deal with the brackets first*.

EXAMPLE 2.8

Now look at the equation which links °F with °C. Use it to convert -15°C into °F.

$$C = \frac{5(F - 32)}{9}$$

The first step is to transpose for F. So multiply by 9.

$$9C = 5(F - 32)$$

The brackets are acting as a shield. We cannot get to F until we have cleared the brackets. There are two ways of doing this. We could expand the brackets by multiplying out by 5 which would lead to bigger numbers.

Or we can multiply by the reciprocal of 5 to get rid of it and thereby remove the need for brackets. The second option is better.

$$\tfrac{1}{5}(9C) = F - 32 \qquad F = \frac{9C}{5} + 32$$

so

$$F = \frac{9(-15)}{5} + 32 = 5°F$$

So much for clearing brackets, or **expansion**. The opposite to this process is **factorization**. The purpose of factorization is to make expressions more compact and easier to deal with. Factorization involves finding the common factor(s) of a set of terms. Let's start with an earlier example.

$$2x - 8$$

Both terms have the common factor 2, in other words they have both been multiplied by 2 so we can write 2 outside a set of brackets.

$$2(x - 4)$$

Which shows that $2x - 8$ has the factors 2 and $(x - 4)$.

Work through the following examples. Don't be put off if you find them difficult. They always are at first.

EXAMPLE 2.9

(a) $5y + 5 = 5(y + 1)$
(b) $35x^2 - 14x = 7x(5x - 2)$
(c) $pq + rq - ps - rs = q(p + r) - s(p + r) = (q - s)(p + r)$
(d) $\dfrac{R^2}{sT^2} + \dfrac{R^5}{sT} - \dfrac{R^4}{s^2T^2} = \dfrac{R^2}{sT}\left(\dfrac{1}{T} + R^3 - \dfrac{R^2}{sT}\right)$
(e) $ab - bd + ac - cd = a(b + c) - d(b + c) = (a - d)(b + c)$
(f) $7x + xz - 7y - yz = 7(x - y) + z(x - y) = (7 + z)(x - y)$

Roots and powers

There are two important things to remember about these,

roots and powers are inverse operations

and

treat roots and powers as brackets when they apply to more than one term.

EXAMPLE 2.10

Transpose $P = I^2R$ for I.

$$I^2 = \frac{P}{R}$$

Position, s, as a function of time can be calculated from another equation.

$$s = ut + \tfrac{1}{2}at^2$$

This can be transposed for t. Initial speed is zero so $ut = 0$ (anything multiplied by zero is zero) and we are left with a simpler equation for distance s.

$$s = \tfrac{1}{2}at^2$$

Multiply by 2 to make the $\tfrac{1}{2}$ on the RHS equal to 1.

$$2s = at^2$$

Divide by a to turn that into 1 on the RHS.

$$\frac{2s}{a} = t^2$$

Take the square root to 'un-square' t.

$$\sqrt{\frac{2s}{a}} = t$$

For neatness, write it the other way around.

$$t = \sqrt{\frac{2s}{a}}$$

We have found an algebraic expression for t which can be substituted into the energy equation.

We now have a model that we can put to practical use. However, you should always try to obtain the simplest model you can.

$$E_k = \frac{1}{2}\left\{m\left[u + a\sqrt{\frac{2s}{a}}\right]^2\right\}$$

u is zero so it can be removed.

$$E_k = \frac{1}{2}\left\{m\left[a\sqrt{\frac{2s}{a}}\right]^2\right\}$$

Let's square the terms in the inner brackets and see what we get.

Squaring a square root simply removes the square root.

$$E_k = \tfrac{1}{2}ma^2\frac{2s}{a}$$

The 2's cancel to 1.

$$E_k = ma^2\frac{s}{a}$$

a^2 divided by a is just a.

$$E_k = mas$$

Finally, acceleration is g, acceleration due to gravity and distance is s, the height of the shaft, h.

$$E_k = mgh$$

We obtain an elegant (and well known) general solution for the energy of an object moving under the force of gravity.

Now we can go for a numerical solution of the Cornish pasty problem. m is 0.1, g is approximately 10 and the height of the shaft is 300.

$$E_k = (0.1)(10)(300) \approx 300\,J$$

Note the indication of the low degree of accuracy. This kind of 'back of the envelope' calculation is typical in situations when we want a first approximation.

So how much is 300 J? Well, imagine the dent left by a 30 kg mass falling through a height of 1 m and you have the equivalent of the dent left by our pasty.

Now, some engineering related problems for you to do on your own. Work through them carefully and be patient. Above all, don't skip over anything that you do not understand. Seek guidance if necessary.

PRACTICE EXERCISE 2.3

With each of the problems that follow you should:

- first put the equation or expression into a suitable algebraic form
- substitute numerical values using preferred standard form
- evaluate to an appropriate degree of accuracy
- express the solution in terms of SI units

1. A rectangular box has a volume of $3.96 \times 10^3\,cm^3$. The length of the box is 27.5 cm, and its width is 15.15 cm. Determine the depth of the box.
2. If an object travels 20.0 mm in 1.00 ms, what is its speed in kilometres per hour?
3. A production machine weighing 50 000 lb and measuring 4 ft by 6 ft is to be installed on a shop floor which can withstand a loading of $25\,000\,kg\,m^{-2}$. Determine if it is safe to install the machine.
4. Given that $s = ut + \frac{1}{2}at^2$, find the stopping distance of the Eurostar train travelling at 120 miles per hour if it can brake at $7.50\,ms^{-2}$. Note: braking rate is deceleration, which is negative acceleration.
5. A formula for calculating the total electrical resistance in ohms (Ω) of two resistors that are connected in parallel is

$$R = \frac{R_1 R_2}{R_1 + R_2}\,\Omega$$

   A measurement indicates that a circuit consisting of two resistors connected in parallel has a total resistance of 836 kΩ and you know that one of the two resistors has a resistance of 1.27 MΩ. Calculate the value of the other resistor.
6. The ideal gas equation,

$$\frac{pV}{T} = mR$$

   relates pressure, p, volume, V, temperature, T, the mass, m, and the characteristic gas constant, R. At what temperature in °C will the pressure of 2.00 kg of a gas with a characteristic gas constant of $287\,J\,kg^{-1}K^{-1}$ be 1.10 MPa when occupying a volume of $0.153\,m^3$?
7. The density of a substance is given by its mass per unit volume, i.e.

$$\rho = \frac{m}{V}\,kg\,m^{-3}$$

A stainless steel cylindrical storage tank for storing sulphuric acid is to be welded to the deck of a ship. The outside dimensions of the tank are 4.50 m long with a diameter of 1.80 m. The thickness of the steel is to be 7.5 mm. The density of sulphuric acid is 1250 kg m^{-3} and the density of stainless steel is 3560 kg m^{-3}.

Formulate a suitable equation and hence calculate the total mass of the tank and its contents in tonnes.

Solutions

1. 95.0 mm;   2. 72 km h^{-1};   3. 10 200 kg m^{-2} so OK;   4. 190 m;   5. 2.4 MΩ;
6. 20°C;   7. 15 tonnes

That concludes this chapter on algebra. We will return to the topic later. Work through the revision problems which follow – they bring together all the important aspects of Number and Algebra and prepare you for an assignment.

PRACTICE EXERCISE 2.4 (REVISION)

1. The equivalent denary value of 10010111_2 is:
   (a) 5
   (b) 312
   (c) 151
   (d) 171
2. The equivalent binary value of 231_{10} is:
   (a) 1110 0111
   (b) 1100 0111
   (c) 1111 1100
   (d) 0001 1000
3. $(4.56 \times 10^2) + (6.725 \times 10^3)$ expressed in preferred standard form is:
   (a) 7.181×10^3
   (b) 71.81×10^2
   (c) 718.1×10^1
   (d) 0.7181×10^4
4. $18.6^{-\frac{1}{4}}$ expressed to three significant figures is:
   (a) -4.56
   (b) 1.00
   (c) 2.08
   (d) 0.482
5. The length of a steel bolt measured to the nearest 0.1 mm is 12.6 mm. The relative accuracy of the measurement is:
   (a) 0.1 mm
   (b) 0.05 mm
   (c) 0.398%
   (d) 0.397%
6. The distance between two points is measured in three sections, 123 ± 0.5 m, 112 ± 0.5 m, 265 ± 0.5 m. The total distance between the two points is:

(a) 500 m exactly

(b) $500 \pm 1.5\,\text{m}$

(c) $500 \pm 0.5\,\text{m}$

(d) $500 \pm 0.125\,\text{m}$

7. If the angular velocity of a pendulum is given by $\omega = \sqrt{g/l}\,\text{rad s}^{-1}$, then

(a) $l = g - \omega^2$

(b) $l = \dfrac{g}{\omega^2}$

(c) $l = \dfrac{g}{\sqrt{\omega}}$

(d) $l = \omega g^2$

8. Clear the brackets of the following expression,

$$-4(2m - 5)$$

9. Factorize the following expression,

$$9b - 9$$

10. Factorize the following expression,

$$\frac{Q^2}{td^2} + \frac{Q^5}{td} + \frac{Q^3}{t^2 d^2}$$

11. The strain energy, U, of a material under stress is given by

$$U = \frac{f^2 V}{2E}$$

Make f the subject of the formula.

12. Transpose the following equation for f,

$$\frac{D^2}{d^2} = \frac{f + p}{f - p}$$

Solutions

1. (c); 2. (a); 3. (a); 4. (d); 5. (c); 6. (b); 7. (b); 8. $-8m + 20$; 9. $9(b - 1)$;

10. $\dfrac{Q^2}{td}\left(\dfrac{1}{d} + Q^3 + \dfrac{Q}{td}\right)$; 11. $f = \sqrt{\dfrac{2EU}{V}}$; 12. $f = \dfrac{p(D^2 + d^2)}{D^2 - d^2}$

Now tackle Assignment I.

Assignment 1

This is a 'desk-top' assignment that is designed to assess what you have learned so far. It is not a project but a simple test in two sections. Section A consists of a set of multiple choice questions. Answer each question and see if your answer matches one of the four options given. Do not be tempted to work 'backwards'; looking at the answers and seeing if they suit the question. Don't try to guess! Stretch yourself and work as quickly as you can.

Section B consists of four short-answer questions and one problem. Answer each as fully as you can. Show what you know. In this section you can make mistakes and still gain marks. Mistakes do not gain negative marks! So proceed with confidence. Use the marks as a guide for how long you should spend on each question; roughly two minutes a mark is about right.

The recommended time for you to complete this assignment is $1\frac{1}{2}$ hours. You should do this assignment under normal exam conditions with no notes or books to help you; just a calculator and writing materials.

The main purpose of this assignment is a diagnostic one. The result will indicate your strengths and weaknesses and show whether you need to go back and review some of the topics you have covered or whether you are ready to continue. A score below 40 marks indicates that you probably need to take some remedial action.

Good luck

Section A

This section contains 16 multiple choice questions. Each question is worth 2 marks.

Answer each question by writing down the letter that represents the **one** option you think is correct.

QUESTION 1

The binary equivalent of 79_{10} is:

(a) 1111 0010
(b) 0100 1111
(c) 1001 1110
(d) 0100 0101

QUESTION 2

A strain gauge instrument has a digital display which indicates a force of 153 N. The relative accuracy of this value is:

(a) 0.327%
(b) 0.328%
(c) 0.500%
(d) 0.500 N

QUESTION 3

$(342.2 \times 10^{-9} \div 12.6 \times 10^{-3})$ expressed in **preferred standard** form is:

(a) 27.159×10^{-6}
(b) 2.7159×10^{-5}
(c) $27\,159 \times 10^{-9}$
(d) 2715.9×10^{-8}

QUESTION 4

The nearest value to $\dfrac{0.063}{8 \times 10^{-3}} + \dfrac{123}{5 \times 10^{-4}}$ is:

(a) 246×10^{3}
(b) 0.8121
(c) 30.75×10^{6}
(d) 24.6×10^{3}

QUESTION 5

$(3^{1.6} - 9^{-2.5})$ equals:

(a) -237.2
(b) 5.795
(c) 531 441
(d) 237.2

QUESTION 6

Given that $3(6t + 2y) = 0$:

(a) $18t + 2y = 0$
(b) $18t = -6y$
(c) $6t + 2y = -3$
(d) $9t + 5y = 0$

QUESTION 7

Given that $I = V/(R + r)$ when $V = 12\,\text{V}$, $R = 100\,\text{k}\Omega$ and $I = 92.3\,\mu\text{A}$:

(a) $r = 15\,\Omega$
(b) $r = 30\,\Omega$
(c) $r = 15\,\text{k}\Omega$
(d) $r = 30\,\text{k}\Omega$

QUESTION 8

$\dfrac{1}{\pi}\left(1\tfrac{1}{9} + \dfrac{2}{6} - \dfrac{4}{3}\right)$ evaluates to:

(a) $\dfrac{2}{3\pi}$
(b) $\dfrac{1}{9\pi}$
(c) $\dfrac{\pi}{9}$
(d) $\dfrac{4}{9\pi}$

QUESTION 9

If the number of engineering students is 543 out of a total number of college students of 3258, the ratio of engineering students to non-engineering students is:

(a) $1:6$
(b) $543:3258$
(c) $181:1086$
(d) $1:5$

QUESTION 10

Given $\dfrac{32 \times 10^3 \times 7235.9}{0.00312}$ a suitable rough check would be:

(a) $\dfrac{10^4 \times 10^4}{10^{-3}} \approx 10^{11}$

(b) $\dfrac{10^4 \times 10^4}{10^{-3}} \approx 10^5$

(c) $\dfrac{10^3 \times 10^4}{10^{-3}} \approx 10^{10}$

(d) $\dfrac{10^3 \times 10^3}{10^{-2}} \approx 10^7$

QUESTION 11

The following numbers from a set of data were added,

$$(68 \pm 0.5) + (220 \pm 5) - (77.0 \pm 0.05) + (2.1 \pm 0.05)$$

The limits of confidence in the result are:

(a) ± 5
(b) ± 5.55
(c) ± 5.6
(d) ± 0.25

QUESTION 12

The tolerance value of a resistor represents the limits of confidence in its value. If a resistor of 33 kΩ is quoted as having a tolerance of $\pm 5\%$ its highest possible value is:

(a) 31.350 kΩ
(b) 33 005 Ω
(c) 38 kΩ
(d) 34.65 kΩ

QUESTION 13

Given that 1 mile is 1.61 km, 140 miles per hour is:

(a) $62.6 \, \text{m s}^{-1}$
(b) $3.76 \, \text{km s}^{-1}$
(c) $24.2 \, \text{m s}^{-1}$
(d) $225 \, \text{m s}^{-1}$

QUESTION 14

If $E = 4v^2 + 981$

(a) $v = \sqrt{\dfrac{981E}{4}}$
(b) $v^2 = 4(E - 981)$
(c) $v = \frac{1}{2}\sqrt{E - 981}$
(d) $v = \frac{1}{4}\sqrt{E - 981}$

QUESTION 15

Given that $12x - y - 4 = 7y - 2x + 8$

(a) $y = \dfrac{7x - 3}{2}$
(b) $y = 2x - 1$
(c) $x = \dfrac{4y + 6}{7}$
(d) $x = 28y + 21$

QUESTION 16

If $I = \dfrac{240}{R + 12}$

(a) $R = \dfrac{240 + 12}{I}$

(b) $R = \dfrac{240 - 12}{I}$

(c) $R = 240I - 12$

(d) $R = \dfrac{240}{I} - 12$

Turn to Section B

Section B

This section contains five questions and carries a total of 28 marks. Answer each question in full. Remember to show all steps in your working.

QUESTION 17

Remove the brackets and simplify the following expression,

$$\tfrac{1}{2}(h - 2) + \tfrac{1}{4}(4h - 12)$$

2 marks

QUESTION 18

Factorize the following expression,

$$ab + nb - ap - np$$

4 marks

QUESTION 19

Temperatures in degrees fahrenheit (F) and in degrees celsius (C) are connected by the formula,

$$F = \tfrac{9}{5}C + 32$$

(a) Convert 92°C into degrees fahrenheit.
(b) Convert 57°F into degrees celsius.

6 marks

QUESTION 20

Transpose the following equation for p,

$$\frac{D^2}{d^2} = \frac{f + p}{f - p}$$

8 marks

QUESTION 21

A cylindrical oil storage tank is fitted with a sight glass. The tank is standing upright on one of its ends, its inside diameter is 5.82 m and its height inside is 1.98 m. The sight glass indicates the oil level to be 1.25 m from the bottom of the tank. Formulate suitable equations and calculate:

• the contents of the tank as a percentage of its maximum capacity,
• the contents of the tank in litres.

Devise a means of checking both results.

Note: to help you plan your solutions you should start by sketching a diagram which illustrates the problem situation.

8 marks

End

Total marks 60

chapter

3

Use of number

Statistics is an area of mathematics that is concerned with the collection, representation, and analysis of data. A statistical analysis is likely to involve a search for patterns, so graphical techniques (graphs, pie-charts, histograms) are widely used. The goal of nearly every statistical analysis is to find the value of two important quantities:

a measure of a **central tendency**
a measure of **dispersion**

You will learn what these mean, how to evaluate them and how to judge their usefulness by working through a simple example.

In statistics we deal with two types of data, **discrete** and **continuous**. How did we define these?

Discrete data are those that can be counted
Continuous data are those that can be measured

Think about the following, which is continuous or discrete data?

- accident figures,
- temperatures,
- failures of electronic components,
- screw thread sizes,
- lengths of telephone calls.

Accidents are discrete events. A half an accident has no meaning.

Temperature is measured on a continuous scale.

Failures, like accidents, can only be counted.

In practice, screws are manufactured to certain standard pitch sizes which would make them discrete but, since the actual size depends on the level of precision, screw sizes are continuous variables.

Time is a continuous variable but telephone calls are usually priced by discrete intervals, nearest minute, ten seconds or similar.

Table 3.1. Discrete data

49	96	67	78	78	66	100	45	89	42
45	89	61	75	74	62	86	39	85	34
74	58	70	59	85	30	69	56	57	68
69	56	18	79	79	26	67	51	53	67

Our chosen example relates to a set of discrete data, which is shown in Table 3.1.

Frequency distribution

The data in Table 3.1 are a collection of marks gained by a student in Engineering. The marks are unsorted so it is difficult to detect any pattern or **central tendency**. In other words, just looking at the raw data, it is difficult to say how well this student has done. The first step of the analysis is to arrange the data into **rank order** (Table 3.2).

This looks a little better but it needs more sorting. The next step is to **group** the data into **classes**. Let's set a **class width** of 10. Now we are in a position to draw up a table of **frequency distribution** (Table 3.3).

The **class boundaries** selected (10.5, 20.5 …) ensure that an actual item of data does not coincide with a boundary. They guarantee that the data fall into definite classes.

The number of classes to use is a decision that depends on the circumstances: how many items of data, whether discrete or continuous and the desired degree of accuracy. If in doubt, remember that the wider the class interval the

Table 3.2. Data sorted into rank order

18	26	30	33	34	39	42	45	45	49
51	53	56	57	58	59	61	62	66	67
67	67	68	69	70	74	74	75	78	78
79	79	80	85	85	86	86	89	96	100

Table 3.3. Frequency distribution

Class	Frequency (f)
0 −10.5	0
10.5–20.5	1
20.5–30.5	2
30.5–40.5	3
40.5–50.5	4
50.5–60.5	6
60.5–70.5	9
70.5–80.5	8
80.5–90.5	5
90.5–100.5	2
Total frequency	40

less accurate the analysis. On the other hand, there is no point in grouping if the class interval is too narrow.

Histogram

This is the most common method of representing a frequency distribution (see Fig. 3.1).

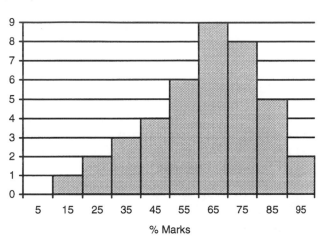

Figure 3.1 Histogram

Notice how each column is marked with the mid-point value of each class. This is a typical value that is representative of that class. The histogram now gives me an impression of overall performance. This student's marks are clustered around the 65% mark.

Frequency curve

Interpolation between the mid-point values of a histogram leads to a frequency curve (see Fig. 3.2).

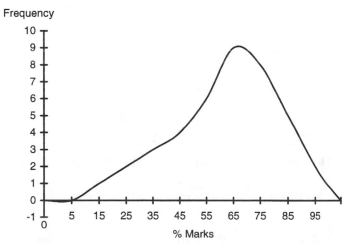

Figure 3.2 Frequency curve

Table 3.4. Cumulative frequency distribution

Class	Frequency	Cumulative frequency
≤ 10.5	0	0
≤ 20.5	1	1
≤ 30.5	2	3
≤ 40.5	3	6
≤ 50.5	4	10
≤ 60.5	6	16
≤ 70.5	9	25
≤ 80.5	8	33
≤ 90.5	5	38
≤ 100.5	2	40

Note the approximate 'bell' shape of the curve. The bell-shape is what statisticians normally look for. It indicates a pattern of concentration around a central value and is called a **normal distribution**. In this case we are not dealing with a normal distribution but one that is approximately so.

Cumulative frequency

The cumulative frequency of a class is the total frequency up to and including the upper limit in that class (Table 3.4).

Notice how the total cumulative frequency corresponds to the total number of items of data.

A cumulative frequency curve is called an **ogive** because of its resemblance to the architectural feature that goes by that name.

It provides us with a means of obtaining direct information such as **percentile values**. Percentiles divide the total sample into 100 equal parts. Those percentiles which divide the sample into four equal parts (25th, 50th and 75th) are called **quartiles**. The quartiles are indicated in Fig. 3.3. They show, for example, that a quarter of the sample of assessments was marked below 50.

Measures of a central tendency

These are an average of some kind and an average is a single value that is representative of the data sample. There are three types of average:

The Mode – the most commonly occurring value(s)

The Median $= \dfrac{n+1}{2}$ the middle value of the set of values

The Mean, $\bar{x} = \dfrac{\sum x}{n}$ the sum of the samples divided by the number of samples

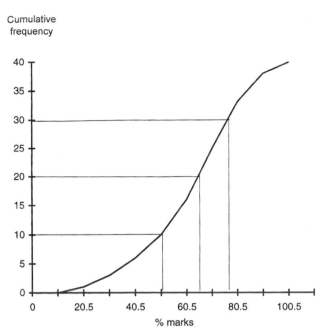

Figure 3.3 Cumulative frequency curve with quartiles marked

EXAMPLE 3.1   Find the mode of the student's marks shown in Table 3.1.

The easiest way to find this is to have the data arranged in rank order. Table 3.2 shows that 67 occurs most often so this is the mode. If 74 occurred three times as well, it would make the distribution **bi-modal**, in which the modes were 67 and 74.

Now, for large amounts of data the mode may be difficult to find, especially if there are too many data to arrange in rank order. In this case it is possible to estimate the mode from the histogram in the way shown in Fig. 3.4.

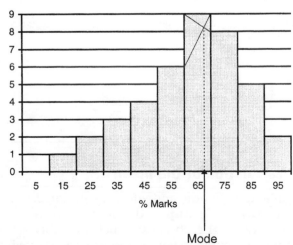

Figure 3.4 Estimating the mode

EXAMPLE 3.2   Find the median of the same set of assessment marks.

The number of values in the sample $n = 40$ so

$$\text{median} = \frac{40 + 1}{2} = 20.5$$

This is the middle value between the 20th and 21st sample.

Table 3.2 shows this to be 67.

There is also a graphical method of finding the median. We can estimate it from the cumulative frequency curve. Look at Fig. 3.3 again; the middle value corresponds to the 50th percentile (or 2nd quartile), it is approximately 66. Finding the 50th percentile from an ogive to obtain the median is a less accurate but quicker method when dealing with large amounts of data.

EXAMPLE 3.3   Find the mean of the assessment marks.

One approach with a large set of data is to use a frequency distribution table. Taking the values from Table 3.3 gives Table 3.5.

Table 3.5. Frequency distribution table using data from Table 3.3

Marks	x	f	fx
0.5 –10.5	5.5	0	0
10.5–20.5	15.5	1	15.5
20.5–30.5	25.5	2	51.0
30.5–40.5	35.5	3	106.5
40.5–50.5	45.5	4	182.0
50.5–60.5	55.5	6	333.0
60.5–70.5	65.5	9	589.5
70.5–80.5	75.5	8	604.0
80.5–90.5	85.5	5	427.5
90.5–100.5	95.5	2	191.0
		$\sum f = 40$	$\sum fx = 2500$

$\sum$ is the upper-case Greek letter 'sigma', which has the mathematical meaning 'sum of . . .'.

Hence,

$$\bar{x} = \frac{\sum x}{n} = \frac{\sum fx}{\sum f} = \frac{2500}{40} = 62.5$$

A note about calculation of the mean

In the last example, to save time and effort, I calculated the mean (full name is **arithmetic mean**) of a **grouped** distribution. In order to do that I had to take the median value of each class, which probably introduced an error. To calculate the mean accurately you would use un-grouped data, i.e. the raw data as listed in Table 3.2.

Accurate calculation of the mean can be laborious but not if you use the STATS or SD mode on your calculator. For practice, find the exact value of the mean of the assessment marks we have been analysing. If you don't know how to do this, look it up in your calculator handbook; it's not difficult. And, yes, you do need a handbook to go with your calculator! It is essential.

I make:

$$n = 40, \qquad \sum x = 2536, \qquad \bar{x} = 63.4$$

So using the grouped distribution introduced an error of -0.9. However, rounding to the nearest integer makes no difference in this particular case. Our student's mean assessment mark is 63%.

Comparison of mode, median and mean

For the set of data we have been analysing we now have:

$$\text{mode} = 67\%$$
$$\text{median} = 67\%$$
$$\text{mean} = 63\%$$

The fact that 67 is both the median and the mode might tempt us to opt for 67% as being the central tendency in the assessment data. This would lead us to award a Merit grade (65% and above is normally considered to be worth a Merit). If we just look at the mean our student would only get a Pass. Which would be fair?

	Advantages	Disadvantages
Mode	Typical item of data. Not affected by extreme values. Easily understood.	Can be more than one mode. Can be no mode at all. Does not use all the data.
Median	Not affected by extreme values. Easily understood.	Takes no account of the distribution of data.
Mean	Mathematically sound. Simple to calculate. Widely accepted.	Affected by extreme values which may not be representative.

So apart from being affected by extreme values, the mean appears to be the most useful average.

The effect of extreme values is worth considering. Suppose a quarterly telephone bill includes 100 local calls of between 1 and 5 minutes in length and one call of 60 minutes to Australia. In trying to assess the typical cost of a quarterly bill would it be sensible to include the 60 minute call? The effects of extreme values can be measured and there are methods for ensuring that the mean is not affected by them.

Measures of dispersion

There are four measures of dispersion (also called scatter or variability). For the moment let's just examine two of them.

Range

The difference between the highest and lowest values in a set of data.

Interquartile range

The range over the middle 50% of the data sample.

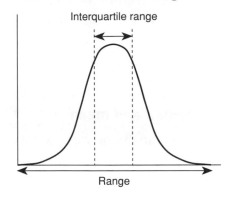

Interquartile range

Range

EXAMPLE 3.4

What is the range of the assessment marks shown in Table 3.2?

$$100 - 18 = 82$$

EXAMPLE 3.5

What is the interquartile range of the same data?

The quartiles shown in Fig. 3.3 are 50 and 77 approximately,
so the interquartile range $= 77 - 50 = 27$

The range is clearly a very crude measure. In this case it shows us that there is a wide spread in the assessment marks and judging by the **skew** of the distribution curve shown in Fig. 3.2 the low marks, 18 and 26, are not representative of the main trend.

Now, the interquartile range is a little more interesting. If we decide that we wish to discount extreme values we can calculate the mean over the interquartile range. Taking the values between 50 and 77 in Table 3.2 and entering them into the calculator, I get:

$$n = 18, \qquad \sum x = 1154, \qquad \bar{x} = 64.1\%$$

This only allows our student a Pass. The mean over the interquartile range has ignored the effects of low marks like the 18 and 26%, but at the same time it has discounted the effects of 12 of the highest marks. Do we have a fair measure of the central tendency of the marks? I shall leave you to decide whether this student deserves a Pass or a Merit. But before you decide, take another look at the skew of the distribution curve in Fig. 3.2.

Mean deviation and **standard deviation** are two other important measures of dispersion but we do not need to go into them here.

3.2   Area, volume and mass

Here, we will review some of the basic techniques for calculating volumes of regular solids and introduce the idea of using surface areas to help determine the volume of regular and regular composite solid objects.

Regular plane shapes

The surface area of a plane is measured in two dimensions. It is the result of multiplying one length by another. A good way to think about it is to think of *sweeping*. For example, taking the width of a rectangle and multiplying by its height is equivalent to sweeping the length through a second dimension. The proper name for this sweeping action is **integration**. But more about that later.

The line *ab* swept through the second dimension *bc* 'paints' the area *abcd*.

In Table 3.6, notice how, with the exception of the circle, all the area calculations are variations on the calculation of the area of a rectangle. This is the approach we take with surface areas of composite planes.

Regular composite planes

The total area of a composite plane is simply the sum of the areas of the parts that make up the composite plane.

Figure 3.5 shows a section through the centre of a circular water cylinder. Calculate the area of the section in square metres.

$$\text{Total area} = \text{area of semicircle} + \text{area of large rectangle}$$
$$+ \, 2(\text{area of quarter circles}) + \text{area of small rectangle}$$
$$= \tfrac{1}{2}\pi r_1^2 + (2r_1 \times 1.15) + 2(\tfrac{1}{4}\pi r_2^2) + 0.4r_2$$
$$\doteq \tfrac{1}{2}\pi r_1^2 + 2.3r_1 + \tfrac{1}{2}\pi r_2^2 + 0.4r_2$$
$$= \tfrac{1}{2}\pi (0.25)^2 + 2.3(0.25) + \tfrac{1}{2}\pi (0.05)^2 + 0.4(0.05)$$
$$= 0.697\,\text{m}^2$$

Regular volumes

The volume of a solid is a measure in three dimensions and the same way that we think of sweeping a line as describing a plane area in two dimensions. Then we can think of the sweep of a plane as describing a volume in three dimensions.

Table 3.6. Regular plane shapes

Surface	Comment	Area
Rectangle	A 'square' is a special case	$A = ab$
Parallelogram	b is height. A parallelogram is just a skewed rectangle of sides a and b	$A = ab$
Triangle	This is half a rectangle	$A = \frac{1}{2}ab$
Trapezium	h is the height and $\frac{1}{2}(a+b)$ is average width	$A = \frac{1}{2}(a+b)h$
Circle	If you are interested, for *any circle*, π is the number you get when you divide the circumference by the diameter	$A = \pi r^2$

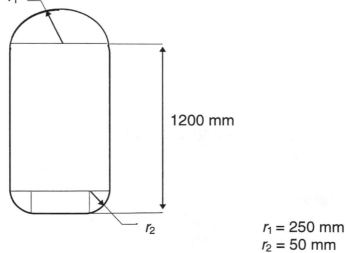

1200 mm

$r_1 = 250$ mm
$r_2 = 50$ mm

Figure 3.5 Section through a circular water cylinder

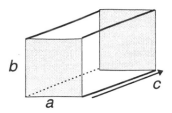

You should be able to see that sweeping the plane *ab* through a distance *c* describes the volume of a **cuboid**.

A cuboid is one of a group of solids called **prisms** that have end planes which are equal and parallel. The end planes of prisms may be any regular polygon, including triangles, parallelograms, trapezia, pentagons, hexagons and so on.

What do we call the prism that has parallel end planes which are circular?

A cylinder

So to determine the volume of a prism is easy.

**Calculate the area of one of the parallel end planes
and multiply by the distance between the end planes.**

Not all regular solids are prisms of course. There are: **pyramids**, **cones**, **spheres**, **hemispheres** and the bottom parts of these, called **frusta**. (A frustum is any part of a solid figure cut off by a plane parallel to its base.) To refresh your memory, sketch the shapes of these solids and note the formulae listed in Table 3.7.

Table 3.7. Shapes and their volumes

Solid	Volume
Sphere	$V = \frac{4}{3}\pi r^3$
Cone	$V = \frac{1}{3}\pi r^2 h$
Pyramid	$V = \frac{1}{3}abh$
Frustum of a cone or pyramid	$V = \frac{1}{3}d(a + \sqrt{aA} + A)$
Hemisphere	$V = \frac{2}{3}\pi r^3$
Frustum of a hemisphere	$V = \frac{\pi h}{6}(h^2 + 3r^2 + 3R^2)$

Theorem of Pappus

The first theorem of Pappus allows us to calculate volumes of solids which are neither prisms or the solids shown in the table above. It tells us that:

volume = area × distance of path of centroid

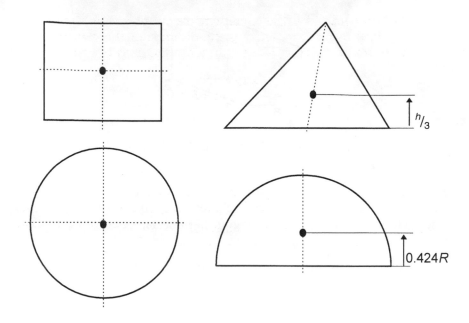

Figure 3.6 The centroids of four types of plane

The theorem is easy to understand once you have seen it used and when you know what a centroid is (see Fig. 3.6).

> **A centroid of a plane is its geometric centre: the point at which external forces are focused so it is the centre of gravity.**

We can now put the theorem to use by finding the volume of the ring shown in Fig. 3.7.

$$\text{Width} = \text{difference between inner radius and outer radius}$$

$$= \frac{1.35}{2} - \frac{1.2}{2} = 0.075\,\text{m}$$

$$\text{Area} = \text{height} \times \text{width} = 0.2 \times 0.075 = 0.015\,\text{m}^2$$

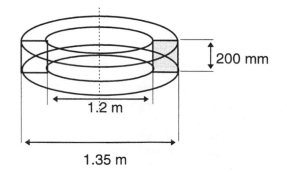

Figure 3.7 To find the volume of this ring

Distance of path of centroid is the circumference swept out by the centroid, which can be found using $2\pi r$. Here r is the distance of the centroid of the area from the centre of the ring. It is the inner radius plus half the width of the ring

$$r = \frac{1.2}{2} + \frac{0.075}{2} = 0.6375\,\text{m}$$

Now

$$\text{volume} = \text{area} \times \text{distance of path of centroid}$$

$$= 0.015 \times 2\pi(0.6375) = 0.060\,\text{m}^3$$

Again, you should recognize the principle of *sweeping*. We took the rectangular cross-section area and swept it through the circumference $(2\pi r)$ which was the length of the path of the centroid of the area. The delicate bit was finding the distance of the centroid from the axis of rotation.

Regular composite volumes

Let's return to the water cylinder example shown in Fig. 3.5. Here it is again:

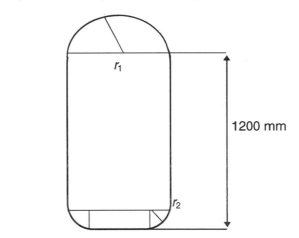

$r_1 = 250$ mm
$r_2 = 50$ mm

1200 mm

Now calculate the volume of the water cylinder.

Total volume = volume of hemisphere + volume of large cylinder
+ volume of curved quarter circle
+ volume of small cylinder in base

Before we can go ahead with the calculation we need to clarify one or two things.

• The volume of the small cylinder is a function of its radius which is $250 - 50 = 200$ mm and its height which is 50 mm. So its volume is $\pi(0.2)^2(0.05) = 0.002\pi\,\text{m}^3$.
• The volume of the quarter circle section is the area of a quarter circle times the path of the centroid of the quarter circle.
• The position of the centroid is $0.424r_2 + 0.2 = 0.221$ m from the centre of the cylinder.

Volumes of: hemisphere large cylinder curved quarter-circle small cylinder

$$\text{Total volume} = \tfrac{1}{2}[\tfrac{4}{3}\pi(r_1)^3] + \pi(r_1)^2 h + [\tfrac{1}{4}\pi(r_2)^2 \times 2\pi(0.221)] + 0.002\pi$$

$$= \pi[\tfrac{2}{3}(0.25)^3 + (0.25)^2(1.15) + \tfrac{1}{2}(0.05)^2\pi(0.221) + 0.002]$$

$$= 0.268\,\text{m}^3$$

Mass

According to Newton, mass is a measure of how difficult it is to accelerate an object. We see this if we transpose the force equation. Mass is, therefore, a measure of **inertia**.

$$F = ma \qquad m = \frac{F}{a}$$

It is logical that a mass of 10 kg is ten times more difficult to accelerate than a mass of 1 kg.

So what is the **weight** of an object? What we often mistakenly call weight is in fact mass! *Weight* is the gravitational force exerted on an object; it is the product of mass and acceleration due to gravity.

$$\text{Weight} = F = mg$$

Our habit of quoting weight in kilograms is a scientific nonsense. In talking about weight we are referring to force for which we use the unit of mass!

For example, in describing the 'weight' of a bag of sugar as being 1 kg, we are really talking about its mass being 1 kg. So what actually is the weight of a bag of sugar on Earth? Approximately 10 newtons.

$$\text{Weight} = 1(9.81) \approx 10\,\text{N}$$

There are two other things which you should know: the **density** of a substance is defined as mass per unit volume,

$$\rho = \frac{m}{V}\ \text{kg m}^{-3}$$

and **pressure** is force per unit area in units of pascals

$$P = \frac{F}{A}\ \text{Pa}$$

Now for a last look at the water cylinder. Water has a density of 1000 kg m^{-3}. Neglecting the metal cylinder and pipework, we wish to know what pressure the tank exerts on the floor when it is full.

The radius of the circular base in contact with the floor is 0.2 m, $V = 0.268\,\text{m}^3$, $\rho = 1000\,\text{kg m}^{-3}$ and $g = 9.81\,\text{m s}^{-2}$

$$P = \frac{F}{A} = \frac{mg}{A} = \frac{V\rho g}{\pi r^2}$$

$$= \frac{0.268 \times 1000 \times 9.81}{\pi(0.2)^2}$$

$$= 21\,\text{kPa}$$

21 kPa is 21 thousand newtons per square metre or just over 2 tonnes per square metre.

You are now ready to tackle a rather interesting assignment which will require you to draw on information from this chapter and exercise your skills with number and algebra.

Assignment II

Problems with mud

Problem situation

Figure II.1 shows the Mud System of a drilling rig. The mud is mixed in two identical mud pits and pumped down a $1''$ diameter hole, through the middle of the drill pipe, to the bottom of the well. The mud returns under pressure on the outside of the drill pipe.

The mud is used for four reasons. It lubricates the drilling bit, it cleans the well by carrying away fragments of formation from the drilling bit and it prevents the formation from collapsing prior to the casing being installed. The level of mud in the pit also acts as a barometer.

A sudden rise in the level of mud in the pits indicates that a high pressure formation of gas or oil has been reached. This gives warning of a blow-out and allows the blow-out preventer to be activated. On the other hand, if the level in the pits falls suddenly, this indicates that mud is being lost to a low pressure formation so its density needs to be reduced.

The normal level of mud in the pits should be of sufficient quantity to fill the well when the drill-pipe is removed for a bit change, i.e. it must be of sufficient volume to displace the drill-pipe. There should also be a further 5% in reserve in the pits in case of a sudden loss of mud into the formation.

The density of the mud is critical. It must be greater than the density of the surrounding formation in order for the well not to collapse. At the same time the mud must not be too dense so that it flows into the formation and becomes lost.

The composition of the mud is a dense mineral called barytes mixed in a solution of salt water.

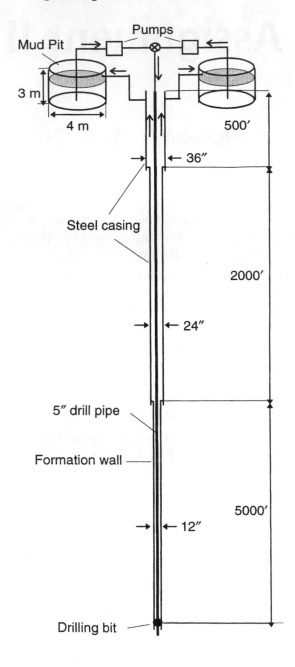

Pumps

Mud Pit

3 m

4 m

500′

36″

Steel casing

2000′

24″

5″ drill pipe

Formation wall

5000′

12″

Drilling bit

Figure II.1

PROBLEM 1

You, as Drilling Engineer on board the North Sea rig Ekomist II, have been given the well plan shown in Fig. II.1. The geologists have told you that the expected average formation density is $3000 \, kg \, m^{-3}$. So you have decided that the mud must have a density a little above that; say, $3050 \, kg \, m^{-3}$. The density of barytes is $3700 \, kg \, m^{-3}$ and the density of the salt water is $1100 \, kg \, m^{-3}$. By solving a pair of simultaneous equations you have determined that the mixture of barytes to water is in the ratio of $3:1$.

Your job now is to calculate the amount of barytes required to mix sufficient mud to fill the well and allow an extra 5% in the pits. The volume of mud contained in the pumps and surface pipes is small enough to be neglected.

Your supplier is Imperial Chemicals Ltd, who can supply the barytes in sacks weighing 50 kg at £2.40 per sack or in plastic shrink-wrapped bundles of 1 cubic metre at a cost of £185.00 a bundle. You must calculate which is the cheaper option.

Finally, you must advise the toolpusher the correct height of mud in the pits when drilling under stable conditions. Remember, this height acts like a barometer, it indicates whether it is safe to continue drilling.

You are working alternate shifts with another drilling engineer. You must leave her with a brief but clear report which provides her with a detailed record of your calculations including:

- your choice of appropriate units,
- correct use of mathematical terms,
- the use of algebraic expressions to get data,
- evidence of checking procedures,
- justification that your results make sense within the situation.

PROBLEM 2

One of the many products sold by Imperial Chemicals is the barytes used in the drilling industry. The Company buys in the barytes in rock form, pulverizes, purifies and packages it for resale. One of the Company's accountants is unhappy with the latest cost/revenue figures. He suspects that there may be something wrong with the amount of barytes being sold, so he has asked the works manager to carry out checks.

The works manager decides to start with the bagging machine to ensure that it is correctly calibrated. He has delegated the task to you, his assistant, asking you to make a statistical analysis and report back to him. You have taken random samples of the 50 kg sacks over a period of two days and had them weighed in the standards room to an accuracy of ± 0.005 kg.

The following are the data you have gathered:

50.02	49.87	50.30	50.08	50.23	50.11	50.18	50.12	50.16	50.01
49.95	50.15	50.80	50.15	50.28	50.13	50.26	49.64	49.87	50.20
50.07	50.18	49.94	50.09	50.30	50.49	50.26	50.20	50.21	50.14
50.05	50.06	50.17	50.22	50.14	50.17	50.23	50.09	49.95	50.00
50.31	50.25	50.15	50.26	50.32	49.83	50.10	50.12	50.07	50.11

50.23	50.10	50.12	50.41	49.97	50.21	50.18	50.29	50.00	49.76
50.22	50.66	50.10	50.44	50.25	50.11	50.47	50.16	49.92	50.13
50.18	50.02	49.99	50.39	50.13	50.09	50.16	50.41	50.17	50.21
49.97	50.37	50.09	50.27	50.22	50.39	50.17	50.26	50.21	50.19
50.09	50.05	49.81	50.32	50.26	50.04	50.29	50.51	50.56	50.16

Your analysis will involve sorting the data, presenting it in at least two suitable graphs, finding the mean, median, mode and interquartile range and using your findings to make a recommendation to the works manager by suggesting a percentage correction to the calibration of the bagging machine if necessary.

You want to impress the manager, so you intend to show him that you:

• use mathematical terms correctly,
• carry out calculations correctly,
• use checking procedures to confirm results of calculations,
• check that results make sense.

Note: You will find that there is a little too much data here to sort into rank order. In which case you will need to use a histogram to find the mode and an ogive to find the median.

Start with a tally chart to form a grouped distribution. Form 13 groups with a class width of 0.1 kg, starting with the lower boundary of 49.595 for the first group and finishing with the upper class boundary of 50.895 for the last group.

A general note about assignments

The work of carrying out a project-sized assignment involves a number of stages which need to be planned, monitored and reviewed. A structured approach is needed in order that time is not wasted through mistakes or omissions. In fact, sometimes, the most difficult step is simply knowing where to start! How often have you simply sat on a job, wasting time because the fuzziness of the situation makes the task seem too difficult to tackle. We have all done it. A structured approach helps prevent this kind of gridlock and makes more efficient use of time and resources. There are four themes which underlie a structured approach to project management:

• planning
• information seeking and handling
• evaluation
• outcomes.

PLANNING THE ASSIGNMENT

Write a task list. Start by stating four or five obvious steps. Write a second and third version in which you include the more detailed steps. Against each task put down what resources/information you think might be needed and estimate how long it will take. To help you monitor your plan, against each item of your task list put an empty box which you can return to and enter the actual time spent and/or add comments about what changes are necessary and why. They will help you manage your time.

INFORMATION SEEKING AND HANDLING

Identify, find and manipulate data which leads to solutions. For example, in the drilling problem you will need conversion factors to put imperial units into SI units. You will also need an equation to convert mass to weight. And much more.

This theme covers the main body of work in the assignment and tests your technical skills.

EVALUATION

At key stages you should consider your solutions. Make rough checks. Do the results you are getting make sense within the problem situation? Is the accuracy of your calculation appropriate to the accuracy of the data provided? Was it necessary to modify your original plan? If so, what did you do and why?

OUTCOMES

The outcome of an assignment is the final product. In a commercial situation this is what you are paid for so it must be correct. The outcomes of your solutions to the mud problems should be final recommendations stated in quantitative terms. Correct terminology, symbols and units should be used throughout your solutions. Your report should be presented neatly. Your findings should be stated briefly and in the clearest terms. Keep to the language of engineering mathematics.

Finally, there is nothing worse than when a challenge becomes a chore. Involve yourself in the project and enjoy it!

chapter

More algebra

4.1 Linear equations

These equations are called linear because they obey the law of the straight line which has the following general form:

$$y = mx + c$$

where

y is the dependent variable
x is the independent variable
m is a constant coefficient which gives the gradient of the line
c is a constant which is the intercept on the y-axis

So in any particular equation linking two variables, m and c would be numbers. The following are particular examples; for each one, check that you can identify the values of m and c.

$y = 2x + 5$	$m = 2,$	$c = 5$
$p = 12.7q - 92.4$	$m = 12.7,$	$c = -92.4$
$t = 8 - 3s$	$m = -3,$	$c = 8$
$v = 55 + 3t$	$m = 3,$	$c = 55$
$F = 5a$	$m = 5,$	$c = 0$
$V = 47I$	$m = 47,$	$c = 0$
$y = \frac{1}{3} - \frac{x}{7}$	$m = -\frac{1}{7},$	$c = \frac{1}{3}$

So, can m and c be any number?

Yes, m and c can be any number.

What do you notice about the powers of the variables?

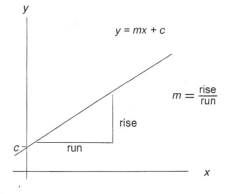

Figure 4.1 Graph of a linear equation

A linear equation is also called an *equation of the first degree* because the highest power of the variable is 1.

The reason that $y = mx + c$ is the straight line law is that for any variable (and constant) the graph of the equation would *always* be a straight line. One in which m is the gradient or slope of the line and c is the intercept on the y-axis (see Fig. 4.1).

PRACTICE EXERCISE 4.1

Plot a graph of $y = 3 - 2x$ by choosing three convenient values for x. Measure c and m.

You should have found that y is 3 when x is 0 and that a fall in the value of y is a 'negative rise' so m is -2.

Enough about graphs for the moment. Let's stick to algebra.

Solutions of linear equations in one unknown

This is a simple process. If a linear equation has just one unknown term, we transpose (if necessary) to make that term the subject and then evaluate.

Solve:

$$\frac{2}{x+1} = \frac{3}{2x-3} \qquad\qquad 4x - 6 = 3x + 3, \qquad x = 9$$

$$3V = \frac{V+12}{220} \qquad\qquad 660V = V + 12, \qquad 659V = 12, \qquad V = 18.2 \times 10^{-3}$$

$$0 = 15 - 25t \qquad\qquad 25t = 15, \qquad t = \tfrac{3}{5}$$

4.2   Simultaneous linear equations

Often, we do not have the luxury of dealing with a problem where only one thing is unknown. There is now a whole field of mathematics called 'Linear Programming' which deals with practical problems involving many unknown variables. Let's start by considering situations with two unknowns.

Linear equations in two unknowns

Cast your mind back to the Drilling Engineers problem in the *Mud* assignment. Given that the density of barytes is $3700\,\mathrm{kg\,m^{-3}}$ and the density of seawater is $1100\,\mathrm{kg\,m^{-3}}$ the engineer needs to calculate what mix of barytes and water will give the required mud density of $3050\,\mathrm{kg\,m^{-3}}$.

In other words he needs to calculate two things to give the required mud density. So having identified the unknowns we assign a symbol to them.

- the number of parts of barytes
- the number of parts of water

Let x be the number of parts of barytes, and let y be the number of parts of water.

Now we need to formulate equations. We know that x parts of barytes plus y parts of water will give us a density of $3050\,\mathrm{kg\,m^{-3}}$.

$$3700x + 1100y = 3050$$

This needs to be checked. The LHS consists of a density multiplied by a number, x, plus another density multiplied by another number, y. So its total is in units of $\mathrm{kg\,m^{-3}}$. The RHS is a density as well, so both sides of the equation are in $\mathrm{kg\,m^{-3}}$. A dimensional analysis of this kind is an important check to make sure your equation makes sense. The next thing is to simplify the equation if you can. Here the numbers are inconveniently large so let's divide the equation by 100.

$$37x + 11y = 30.5$$

Notice that this is a linear equation; the highest power of the variables is 1. However, it cannot be solved because we have two unknowns so it is necessary to formulate a second equation linking x and y.

We know that $1\,\mathrm{m^{-3}}$ of mud will contain $x\,\mathrm{m^{-3}}$ of baryte plus $y\,\mathrm{m^{-3}}$ of water, so we can write an equation which equates $\mathrm{m^{-3}}$.

$$x + y = 1$$

This cannot be simplified any more, so we will just call the two equations A and B and write them together for clarity.

$$\text{A} \qquad 37x + 11y = 30.5$$

$$\text{B} \qquad x + y = 1$$

Because we have managed to write two equations with the same two unknowns we now have a means of solving them.

Solution by substitution

Take the simplest equation and transpose, making one of the variables the subject. So here, take equation B and transpose for x.

$$x = 1 - y$$

Now substitute for x in the other equation.

$$37(1 - y) + 11y = 30.5$$

Expand the brackets.

$$37 - 37y + 11y = 30.5$$

Collect-up terms.

$$37 - 26y = 30.5$$

Now we can transpose and solve for y.

$$26y = 37 - 30.5 = 6.5 \qquad y = 0.25$$

Having found a value for y, we can substitute and solve for x. Pick the easier equation, B.

$$x + 0.25 = 1 \qquad x = 0.75$$

x is the proportion of baryte and y is the proportion of water, so the ratio of baryte to water is

$$0.75 : 0.25 \quad \text{or} \quad 3 : 1$$

PRACTICE EXERCISE 4.2

Solve the following pairs of equations by substitution.

(a) $4x - 3y = 1$ (b) $3x + 2y = 7$ (c) $2x - 3y = 5$

     $x + 3y = 19$ $x + y = 3$ $x - 2y = 2$

Solutions

(a) $x = 4, y = 5$;   (b) $x = 1, y = 2$;   (c) $x = 4, y = 1$

Checks: (a) $4 + 3(5) = 19$;   (b) $1 + 2 = 3$;   (c) $4 - 2(1) = 2$

Since we were dealing with integers, the solutions to Practice Exercise 4.2 were fairly straightforward. However, this is not always the case so it is easy to make errors. Checking by back-substitution is essential in every case.

74 Applied Maths for Engineering

Solution by elimination

This involves equalizing the coefficients of one of the variables. At first, it may seem a more complicated method but it can avoid some of the manipulation that is necessary with the substitution method. First, line up the equations.

$$A \qquad 37x + 11y = 30.5$$

$$B \qquad x + y = 1$$

Now, if we multiply equation B by 11, we will equalize the coefficients of y.

$$A \qquad 37x + 11y = 30.5$$

$$B \qquad 11x + 11y = 11$$

Subtracting equation B from equation A will eliminate the y term since $11y - 11y = 0$,

$$A \qquad 37x + 11y = 30.5$$

$$B \qquad 11x + 11y = 11$$

$$A - B \qquad 26x = 19.5$$

This is easy to deal with, dividing by 26 gives us the value of x.

$$x = 0.75$$

We are now at a similar point as we were with the substitution method. We have the value of x so we can substitute for x to find the value of y.

$$0.75 + y = 1$$

$$y = 0.25$$

PRACTICE EXERCISE 4.3

For practice, solve the following pairs of equations by elimination and substitution.

(a) $2x + 3y = 23$ (b) $2x - 2y = -22$ (c) $5x - 3y = 7$

     $x + 2y = 13$ $22x + 16y = -128$ $-3x + 7y = 1$

(d) A mesh analysis of an electric circuit gives rise to the following pair of equations,

$$470I_1 + 330I_2 = 6.96$$

$$22I_1 - 680I_2 = -2.456$$

where I_1 and I_2 are two electric currents. Calculate the values of these currents correct to three significant figures.

Solutions

(a) $x = 7, y = 3$;   (b) $x = -8, y = 3$;   (c) $x = 2, y = 1$;   (d) $I_1 = 12\,\text{mA}, I_2 = 4\,\text{mA}$

4.3 Quadratic equations

This is the name given to a group of non-linear equations which have the following general form:

$$y = ax^2 + bx + c$$

Here, a, b and c are constants and c is, once more, the **intercept on the** y-axis. The graph of a quadratic equation is not a straight line but a *parabola*, so neither a nor b give the gradient because the gradient is no longer a constant.

The following are some particular examples. In each case identify the values of the constants a, b and c.

$$y = x^2 + x + 1 \qquad\qquad a = 1, \qquad b = 1, \qquad c = 1$$

$$y = 5x^2 \qquad\qquad a = 5, \qquad b = 0, \qquad c = 0$$

$$y = \tfrac{1}{2}x^2 - 37 \qquad\qquad a = \tfrac{1}{2}, \qquad b = 0, \qquad c = -37$$

$$35 - \frac{3x}{2} - \frac{x^2}{3} = 0 \qquad\qquad a = -\tfrac{1}{3}, \qquad b = -\tfrac{3}{2}, \qquad c = 35$$

$$s = 2t + \tfrac{1}{2}t^2 \qquad\qquad a = \tfrac{1}{2}, \qquad b = 2, \qquad c = 0$$

$$A = \pi r^2 \qquad\qquad a = \pi, \qquad b = 0, \qquad c = 0$$

The same question arises, can a, b and c be any number?

a, b and c can be any number *except* a *cannot be zero*

If a is zero we have a linear, not a quadratic.

You can also see that, here, the highest power of the variable is 2. So,

quadratic equations are called *equations of the second degree* because the highest power of any variable is 2.

Quadratic expressions

Before we set about solving quadratic equations let's take another look at factorization and expansion of brackets. Quadratics can appear in disguise. Expand the brackets in each of the following expressions.

$$(x + 2)(x - 5) = x^2 - 3x - 10$$

$$(t - 12)(t + 12) = t^2 - 144$$

$$x(x + 3) = x^2 + 3x$$

$$(x + a)(x + b) = x^2 + ax + bx + ab$$

$$(x - 3)(x - 3) = x^2 - 6x + 9$$

$$(x+3)(x+3) = x^2 + 6x + 9$$

$$(x+4)^2 = x^2 + 8x + 16$$

Each expression is a product of factors which yields a quadratic term. There are two special cases which take the following general forms:

The perfect square

$$(a+b)^2 = (a+b)(a+b) = a^2 + 2ab + b^2$$

$$(a-b)^2 = (a-b)(a-b) = a^2 - 2ab + b^2$$

The difference of two squares

$$(a+b)(a-b) = a^2 - b^2$$

Recognizing that you are dealing with a perfect square or a difference of two squares can be useful, as you will see. In the meantime, which of the above examples is a perfect square or a difference of two squares?

$t^2 - 144$ is a difference of two squares

$x^2 - 6x + 9$ is a perfect square

$x^2 + 6x + 9$ is a perfect square

$x^2 + 8x + 16$ is a perfect square

Now, you should remember that the opposite to expansion is factorization. If you recognize that an expression is a perfect square or a difference of two squares, factorization can be done just by recognition. Factorize the following by recognition and explain.

$x^2 - 9 = (x+3)(x-3)$ Difference of two squares – middle term missing

$x^2 - 10x + 25 = (x-5)(x-5)$ Perfect square – middle term negative

$x^2 + 16x + 64 = (x+8)(x+8)$ Perfect square – middle term positive

If you are not dealing with the special cases, it is still possible to factorize but it does start to become more difficult. Carefully work through the following factorizations:

$$x^2 - 3x = x(x-3)$$

$$4x^2 + 12x = 4x(x+3)$$

$$x^2 - 3x + 2 = (x-1)(x-2)$$

$$t^2 - 2t - 8 = (t+2)(t-4)$$

$z^2 - 6z + 9 = (z-3)(z-3)$ Perfect square

$$2x^2 + 5x - 3 = (x+3)(2x-1)$$

$$4x^2 - 25y^2 = (2x + 5y)(2x - 5y) \quad \text{Difference of two squares}$$

$$9x^2 + 9x - 28 = (3x - 4)(3x + 7)$$

The solution of a quadratic equation

The solution of a quadratic equation is the value(s) of x that satisfy the equation in exactly the same way as it does with linear equations. Take,

$$y = 3 - 2x$$

of which you plotted the graph earlier. As it stands, the equation cannot be solved because it has two unknowns.

However, if we have a value for y we can solve for x. Let $y = 0$.

$$0 = 3 - 2x$$

$$x = 1.5$$

The solution is easy, but what does it mean? Well, if you look at the graph of the equation, you will see that 1.5 is the value of x when y equals zero, i.e. it is the point where the graph cuts the x-axis.

Now, give the same treatment to a quadratic. To solve:

$$y = 2x^2 - 18$$

Make

$$y = 0,$$

$$2x^2 - 18 = 0$$

$$2x^2 = 18$$

$$x^2 = 9$$

$$x = \sqrt{9}$$

At this point you might, understandably, say that $\sqrt{9}$ is 3 so the solution is $x = 3$. This is not correct; the solution may be 3 but it is not the only value of x that satisfies the equation.

What other number, when squared, gives us 9?

$$-3$$

So the solution to $2x^2 - 18 = 0$ is

$$x = 3, -3$$

How can we be certain that this is the solution?

Check by back-substitution

Substituting 3 for x.

$$2(3)^2 - 18 = 0$$

Substituting -3 for x

$$2(-3)^2 - 18 = 0$$

3 and -3 are the roots which both satisfy the equation. So a quadratic equation may have two solutions. We can summarize.

> **The solution of a quadratic equation may consist of two numbers which *satisfy* the equation. They are called the *roots of the equation*.**

PRACTICE EXERCISE 4.4

If area $A = \pi r^2$, what would be the radius of a circle of area, say $10\,\text{m}^2$?

Solution

Substituting for A gives me

$$\pi r^2 = 10 \quad \text{so} \quad r = \sqrt{\frac{10}{\pi}} = \pm 1.78\,\text{m}$$

The solution, 1.78 and -1.78, is mathematically correct because both solutions satisfy the equation. However, the solution to the practical problem is clearly 1.78 m because a radius of -1.78 makes no sense: it is an irrational solution.

Solution by factors

The previous example involved a quadratic with one x-term, so it was possible to solve directly by taking roots. This is not usually possible. Take the following equation.

$$x^2 + 4x + 3 = 0$$

This cannot be solved simply by transposing for x and taking roots. Try it. It does not work because we have an x^2 as well as an x term. Let's explore the solution-by-factors method. The LHS factorizes quite easily.

$$(x + 1)(x + 3) = 0$$

Now we have an equation which tells us that something in brackets multiplied by something else in brackets equals zero. This is interesting because we know that any multiplication which gives a zero result must involve at least one factor which is zero. Armed with this important information we are now able to make two statements.

either $x + 1 = 0$

or $x + 3 = 0$

Transposing both gives two values of x.

either $x = -1$

or $x = -3$

so $x = -1, -3$

-1 and -3 are the roots of $x^2 + 4x + 3 = 0$. Check by back substitution to see if the roots do in fact satisfy the equation.

So a solution by factors is straightforward as long as you are able to find the factors of a quadratic expression.

PRACTICE EXERCISE 4.5

Solve the following by factors.

(a) $x^2 - 3x = -2$; (b) $x^2 = 7 - 6x$; (c) $28 = 9x^2 + 9x$
(d) $z^2 - 6z + 3 = -6$; (e) $x^2 = 3x$; (f) $4x^2 + 12x = 0$
(g) $30x^2 + 60x = 240$; (h) $28p^2 + 24 = 58p$; (i) $x^2 + 16x + 64 = 0$

Solutions

(a) $x = 1, 2$; (b) $x = 1, -7$; (c) $x = 1.33, -2.33$; (d) $z = 3$ repeated; (e) $x = 0, 3$;
(f) $x = 0, -3$; (g) $x = 2, -4$; (h) $p = 1.5, 0.571$; (i) $x = -8$ repeated

Examples (d) and (i) gave you just one value for the unknown which appears to contradict what was said earlier; that quadratic equations have two solutions. Here we have special cases of equations which have **repeated roots**.

So solution-by-factors is fine if you can find the factors of an expression. What happens if the factors are difficult to find? What happens if they are impossible to find as with, say, $260w^2 - 713w = 1$?

Solution by formula

In such examples, where factorization is difficult or downright impossible, as it is in most practical situations, we can resort to solution-by-formula. (You may be interested in seeing how the formula for solving a quadratic equation comes about. In which case, look at 'Completing the square' in Appendix IV.)

For a quadratic equation which has the general form:

$$ax^2 + bx + c = 0$$

the solution is:

$$x = \frac{-b \pm \sqrt{b^2 - 4ac}}{2a}$$

In the formula:

- a is the coefficient of x^2 in the original equation
- b is the coefficient of x in the original equation
- c is the constant term in the original equation

Remember that a, b and c can be any number except that a cannot be zero. I will show you an example and then you must practise.

EXAMPLE 4.1

Solve $7x^2 + 8x - 6 = -4$.

First, we must rearrange to put it into a standard form. This gives us:

$$7x^2 + 8x - 2 = 0$$

Hence

$$a = 7 \qquad b = 8 \qquad c = -2$$

We substitute these values into the formula.

$$x = \frac{-b \pm \sqrt{b^2 - 4ac}}{2a} = \frac{-8 \pm \sqrt{8^2 - 4(7)(-2)}}{2(7)}$$

Evaluate the sum inside the square root and the denominator.

$$x = \frac{-8 \pm \sqrt{120}}{14}$$

14 is a common denominator so we can write a sum and difference of two fractions.

$$x = \frac{-4}{7} \pm \frac{\sqrt{120}}{14}$$

$$= -0.571 \pm 0.782$$

Complete the evaluation which gives two values for x.

$$x = 0.211, -1.353$$

PRACTICE EXERCISE 4.6

Find, correct to 3 s.f., the roots of the following by the formula method but do not spend too long trying to solve (g).

(a) $3x^2 - x - 5 = 0$; (b) $2x^2 - 5.3x + 1.25 = 0$
(c) $3x^2 - 6x - 11 = 0$; (d) $x^2 - x + \frac{1}{4} = \frac{1}{9}$
(e) $v(v+4) + 2v(v+3) = 5$; (f) $2w^2 + 16w = 17$
(g) $3x^2 + 12x = -15$

Solutions

(a) $x = 1.47, -1.14$; (b) $x = 2.39, 0.263$; (c) $x = 3.16, -1.16$;
(d) $x = 0.834, 0.167$; (e) $v = 0.442, -3.78$; (f) $w = 0.950, -8.95$

The solution to the last example is impossible to find because it is not possible in the **real** number system to find the square root of a negative number. Try squaring something to give you a negative result and you will see why. In such a situation we say that the equation has **complex roots** because it is only possible to solve it using **complex numbers**; but this will be for another time.

The last thing to mention about the solution-by-formula method is that this is the method most widely used. If you are able to spot the factors of a quadratic expression, fine; solve by factors. But do not waste time searching for factors, *the formula method works in every case*.

PRACTICE EXERCISE 4.7

(a) The electrical resistance of a conductor is given by: $R_t = R_0(1 + \alpha T)$, where R_0 is the resistance at $0°C$, T is temperature and α is the temperature coefficient of the material. Find α in $\Omega\,°C^{-1}$ if $R_t = 2.37\,\Omega$ at $T = 25°C$ and $R_0 = 2.20\,\Omega$.

(b) Find the values of x and y if $2x + 3y = 8$, $7x + 8y = 33$.

(c) Solve $33I_1 - 56I_2 - 52 = 0$, $12I_1 + 10I_2 - 56.56 = 0$.

(d) Solve $x^2 - 36 = 0$.

(e) The volume of the frustum of a cone is given by $v = \frac{1}{3}\pi h(R^2 + rR + r^2)$. If the height h is $4\,m$, the major radius R is $3\,m$ and the volume is $79.6\,m^3$, calculate the minor radius r.

(f) The equation of motion in a straight line is $s = ut + \frac{1}{2}at^2$. What is the time t that it takes to travel a distance s of $41.4\,m$, if the initial velocity u is $23.4\,m\,s^{-1}$ and the acceleration a is $5.61\,m\,s^{-2}$.

Solutions

(a) $3.09\,m\Omega\,°C^{-1}$;   (b) $x = 7, y = -2$;   (c) $I_1 = 3.68, I_2 = 1.24$;
(d) $x = 6, -6$;   (e) $2\,m$;   (f) $-9.84\,s, 1.5\,s$

4.4 Exponentials and logarithms

Exponential equations have the following general form:

$$y = Ab^{kx}$$

x, the independent variable, is the exponent of b (index or power of b).

y is the dependent variable as usual.

b is the base of x.

A is the intercept on the y-axis (the value of y when x is zero), i.e. a constant.

k is the constant coefficient of x which affects the gradient of the graph of y against x.

A quick look at the graphs of exponentials indicates that they can be **growth** or **decay** curves (see Fig. 4.2).

From the graphs, we can conclude that:

With an exponential growth, k is positive.

With an exponential decay, k is negative.

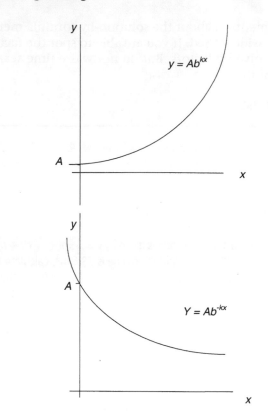

Figure 4.2 Growth (upper) and decay (lower) curves

In order to deal with exponentials we need to be competent in dealing with **indices** and **logarithms**.

Indices revisited

Earlier, we looked at indices in a numerical form. Now we look at the laws in algebraic form, which is the general form.

Laws of indices

$$a^m \times a^n = a^{m+n}$$

$$\frac{a^m}{a^n} = a^{m-n}$$

$$(a^m)^n = a^{mn}$$

$$\sqrt[n]{a^m} = a^{\frac{m}{n}}$$

Properties of indices

There are also some special properties which it is important to remember.

Since $\quad \dfrac{a^m}{a^m} = a^{m-m} = a^0$

then $\qquad a^0 = 1$

and $\qquad a^{-m} = \dfrac{1}{a^m}$

For practice, copy out the following and make sure you are satisfied how they are done.

$$2^{2m}2^{3m} = 2^{5m}$$

$$b^{-5t}b^{3t} = b^{-2t}$$

$$4^{2x}4^{-4x} = 4^{-2x}$$

$$e^{-5t}\,e^{3t} = e^{-2t}$$

$$(e^{-3t})^2 = e^{-6t}$$

$$(1 + e^{2t})^2 = (1 + e^{2t})(1 + e^{2t}) = 1 + 2\,e^{2t} + e^{4t}$$

$$\frac{1}{b^n} = b^{-n}$$

$$\frac{1}{e^{-3t}} = e^{3t}$$

$$\frac{a^{3x}}{a^{5x}} = a^{-2x}$$

$$\frac{5\,e^{3t}}{10e^t} = \tfrac{1}{2}e^{2t}$$

$$\frac{6b^{3n}b^{5n}}{2b^{-4n}} = 3b^{12n}$$

The base e

When dealing with exponentials we can take the exponent of any base: $a, b, 2, 10, e, \ldots$ However, in engineering we nearly always use the base e, which is a number:

$$e = 2.718281828\ldots$$

e is simply called 'the exponential'. The reason we use it is because it has the special property that when it is raised to a power its rate of change is equal to its value.

Try this with your calculator. If

$$y = e^x$$

when $x = 2$,

$$y = 7.39$$

The point is, that not only is y equal to 7.39 but when $x = 2$, the rate of change of y is also equal to 7.39. We will look at this later when we start calculus. For the moment accept that e is a number with a special property, so that is why it is the base most often chosen when dealing with exponentials.

So we can also write the general form of the exponential equation using e instead of b as the base.

$$y = A\,e^{kx}$$

PRACTICE EXERCISE 4.8

Plot, against the same axes, the graphs of $y = 2e^t$ and $y = 2e^{-t}$ from $t = -3$ to $t = 3$.

You should notice two things, the rate at which y changes is not constant. With the growth curve it increases at an ever increasing rate and with the decay curve it decreases at an ever decreasing rate.

Exponentials provide us with important models of physical change. For example, you will have probably heard that populations tend to grow exponentially and radioactivity decays exponentially. Since most things change over time, you will often encounter t as the exponent of e, not x.

There is one important variation of the general form of an exponential equation.

You can think of this as the growth-toward-a-target-value-exponential.

$$y = A(1 - e^{-kx})$$

This is the equation of a curve that grows at an ever decreasing rate. Let's see what it looks like.

PRACTICE EXERCISE 4.9

The rise in voltage across a charging capacitor is described by the following equation.

$$v = V\left(1 - e^{-\frac{t}{CR}}\right)$$

Plot a graph of this voltage from $t = 0$ to $t = 5$ s, at 1 s intervals, given that $C = 100 \times 10^{-6}$, $R = 10 \times 10^3$ and $V = 100$.

Your solution should look something like the sketch in Fig. 4.3.

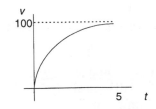

Figure 4.3 Curve of a 'growth toward a target value exponential'

The exercise was made easy for you by the choice of values of C and R so that $CR = 1$. The important thing to note, for the moment, is the slowing down of the rate at which the voltage increases. At first it rises very rapidly but after five seconds it is hardly rising at all.

PRACTICE EXERCISE 4.10

1. Evaluate the following correct to 3 s.f.
   (a) $y = -3e^5$; (b) $y = 31e^{-4.1}$; (c) $v = V e^{-\frac{2}{3}}$; (d) $y = 20(1 - e^{-250 \times 10^{-3}})$
2. Newton's law of cooling is modelled by $\theta_t = \theta_0\, e^{-kt}$. Calculate θ_t, the temperature at time t, given that:
   θ_0, the initial temperature, is 100°C
   k, the temperature change constant, is 0.002
   t, the time, is 60 s
3. The electric current in a circuit with inductance behaves according to the relationship

$$ i = I\left(1 - e^{-\frac{tR}{L}}\right) $$

   where i is the circuit current in amperes at instant t seconds, I is the maximum current in amperes, L is the inductance in henrys and R is the resistance in ohms. Given that: $I = 12\,A$, $L = 270\,mH$ and $R = 1.2\,k\Omega$, calculate the instantaneous current when $t = 800\,\mu s$.
4. The amplitude of the swing of a pendulum, A, when damped by friction and air resistance decreases exponentially over time. If A_0 is the initial amplitude and k is the damping constant, formulate a suitable equation and hence calculate the amplitude of a pendulum with a damping constant of 0.1, 15 s after its starting amplitude 15 mm.

Solutions

1. (a) −445, (b) 0.514, (c) 0.513V, (d) 4.42; **2.** 88.7°C; **3.** 11.7 A; **4.** 3.35 mm

Logarithms

Start by thinking of logarithms as powers. If N is a **number** and b is a **base** then x is the **power** to which the base is raised to give the number.

$$ N = b^x $$

Alternatively, **log** to the **base** b of the **number** N is x.

$$ \log_b N = x $$

Logs can be confusing to start with but don't be put off. Things start to become clear when you deal with numerical examples.

Study the following carefully.

If	Then	In log form
$10 = 10^1$	1 is log to the base 10 of 10	$\log_{10} 10 = 1$
$100 = 10^2$	2 is log to the base 10 of 100	$\log_{10} 100 = 2$
$1000 = 10^3$	3 is log to the base 10 of 1000	$\log_{10} 1000 = 3$
$8 = 2^3$	3 is log to the base 2 of 8	$\log_2 8 = 3$
$125 = 5^3$	3 is log to the base 5 of 125	$\log_5 125 = 3$
$1 = e^0$	0 is log to the base e of 1	$\log_e 1 = 0$
$0.001 = 10^{-3}$	−3 is log to the base 10 of 0.001	$\log_{10} 0.001 = -3$
$0.25 = 2^{-2}$	−2 is log to the base 2 of 0.25	$\log_2 0.25 = -2$

Let's see what things look like in reverse:

If	Then
$\log_{10} 10\,000 = 4$	$10\,000 = 10^4$
$\log_{10} 0.0001 = -4$	$0.0001 = 10^{-4}$
$\log_2 16 = 4$	$16 = 2^4$
$\log_e 7.39 = 2$	$7.39 = e^2$
$\log_e 0.135 = -2$	$0.135 = e^{-2}$

Notice that the log of 0 is 1, the logs of numbers greater than 1 are positive and the logs of numbers less than 1 are negative. This is consistent with indices where any base to a power of 1 is zero, any base to a positive power is greater than 1 and any base to a negative power is less then 1.

So if logs are derived from indices are the laws of logs the same? Yes.

Laws of logs

$$\log(mn) = \log m + \log n$$

$$\log\left(\frac{m}{n}\right) = \log m - \log n$$

$$\log m^n = n \log m$$

Properties of logs

$$\log_b 1 = 0$$

$$\log_b b = 1$$

These are clear if you remember that, with indices, any base raised to the power of zero is 1 and any base raised to the power of 1 is just the base.

Copy out the following and study the results carefully.

$$\log(xy) = \log x + \log y$$

$$\log(xy)^n = n(\log x + \log y)$$

$$\log\left(\frac{x^m}{y^n}\right) = m \log x - n \log y$$

$$\log_{10}(371 \times 23.9) = 2.569 + 1.378 = 3.95$$

$$\log_e\left(\frac{0.073 \times 745.7}{84^2}\right) = -2.617 + 6.614 - 2(4.431) = -4.87$$

For the last two you will have used your calculator to evaluate base 10 and base e logs. This would have meant pressing either the log or lg keys for base 10 and the ln or $\log_e$ key for base e, depending on which calculator you have.

Bases

As with indices, where you can take the exponent of any base, with logs you can also use any base. However, we tend to stick to base 10 (which are called **common logs**) or base e (which come by various names: **natural**, **Naperian** or **hyperbolic logs**).

Common logs are used especially in communication engineering (the decibel is obtained by taking common logs) while natural logs are more widely applied to exponential problems. Faced with a choice a mathematician or engineer would probably work with natural logs.

Solution of exponential equations

We now look at what logs are used for. In the past, before calculators and slide-rules, long multiplication and division was made easier with logs. Remember that the multiplication of numbers becomes an addition of the logs of those numbers, and a division becomes a subtraction. That is now history but we still need logs for certain engineering situations and in the solution of exponential equations.

Remember, an exponential equation has the general form

$$y = Ab^{kx}$$

where b is often the exponential, e. Now look at x. This is a variable that is the **exponent** of b, in other words the unknown term is a power. Contrast this with a **polynomial**

$$y = ax^2 + bx + c$$

In this case x is not a power. The powers are constants. Now we know how to solve linear and quadratic polynomials, how do we deal with exponentials? How do we transpose for x when it is, or forms part of, an exponent? The answer is to take logs.

EXAMPLE 4.2

Solve $y = Ab^{kx}$.

First take logs of both sides.

$$\log y = \log(Ab^{kx})$$

Apply the laws of logs.

Remember that the multiplication of Ab becomes an addition of the log of A and the log of b. Also the exponent of b, kx, becomes a multiplication factor of the log of b.

$$\log y = \log A + kx \log b$$

Now we are able to transpose for x in the usual manner. Subtract $\log A$.

$$\log y - \log A = kx \log b$$

Divide by $\log b$.

$$\frac{\log y - \log A}{\log b} = kx$$

Multiply by the reciprocal of k.

$$x = \frac{1}{k} \left(\frac{\log y - \log A}{\log b} \right)$$

This is a solution in algebraic form. It enables us to calculate the value of x for any combination of values of y, k, A and b.

Let's go through a numerical example.

EXAMPLE 4.3

If $R = R_0 \, e^{\alpha\theta}$, calculate the value of θ when $\alpha = 120 \times 10^{-6}$, $R_0 = 5000$, and $R = 5637$.

Keep the equation in algebraic form; avoid putting in numbers until you have an algebraic solution. Take logs.

$$\log R = \log R_0 + \alpha\theta \log e$$

Now, we can be a bit clever here. If we choose to take logs to the base of e, on the RHS we will get $\alpha\theta(\log_e e)$. But the log of a number to the same base is just 1 so $\log_e e = 1$.

$$\log_e R - \log_e R_0 = \alpha\theta$$

Note, we are now committed to using natural logs. Multiply by the reciprocal of α.

$$\theta = \frac{1}{\alpha}(\log_e R - \log_e R_0)$$

Finally substitute the given values and use your calculator with care.

$$\theta = \frac{1}{120 \times 10^{-6}}(\log_e 5637 - \log_e 5000) = 999$$

There are situations where we are required to take **antilogs**. This involves taking the exponent of the base of the log which you have used. It is very easy with a calculator.

The antilog of a common log is obtained by pressing the 10^x key and the antilog of a natural log is obtained by pressing the e^x key.

EXAMPLE 4.4

The overall voltage gain of an audio amplifier is given by

$$G = 20\log_{10}\frac{V_{out}}{V_{in}}\,\text{dB}$$

Given that an amplifier has a gain of 70 dB, calculate the output voltage V_{out} when the input voltage $V_{in} = 12\,\text{mV}$.

We need to transpose for V_{out}, so we start by dividing by 20.

$$\frac{G}{20} = \log_{10}\left(\frac{V_{out}}{V_{in}}\right)$$

To get to V_{out} we need to antilog.

$$\text{antilog}_{10}\left(\frac{G}{20}\right) = \frac{V_{out}}{V_{in}}$$

On your calculator the antilog will be 10 to the power of G-over-20.

$$10^{\frac{G}{20}} = \frac{V_{out}}{V_{in}}$$

Finally, multiply by V_{in}.

$$V_{out} = V_{in}\,10^{\frac{G}{20}}$$

This yields an expression for output voltage. Substituting values leads to a numerical solution.

$$V_{out} = 12 \times 10^{-3}(10)^{\frac{70}{20}} = 37.9\,\text{V}$$

Notice how we avoided putting in numbers until the transposition was complete. This prevents the accumulation of errors that comes about with part calculations. Wherever possible, deal with the algebra first. Then do the evaluation.

PRACTICE EXERCISE 4.11

1. Express the following as logs:
   (a) $5^{-2} = 0.04$; (b) $a^1 = a$; (c) $10^{-3} = 0.001$; (d) $e^2 = 7.39$
2. Find the value of x:
   (a) $\log_{10} x = 2$; (b) $\log_7 x = 0$; (c) $\log_x 81 = 4$; (d) $\log_n n = x$
3. Find the value of x:
   (a) $0.9^x = 2.176$; (b) $21.9^{(3-x)} = 7.334$; (c) $2.79^{(x-1)} = 4.377^x$

4. Given that $L = 644 \times 10^{-6}\left[\log_e\left(\frac{d}{r}\right) + \frac{1}{4}\right]$ find L when $d = 30.5$ and $r = 0.15$.

5. The instantaneous charge, q, on a capacitor is given by the equation $q = Qe^{-\frac{t}{CR}}$ coulombs. Calculate the charge on a capacitor after a time $t = 150\,\text{ms}$ if the initial charge Q was $1.82\,\mu\text{C}$, $C = 4.7\,\mu\text{F}$ and $R = 820\,\text{k}\Omega$.
6. The formula for a particular chemical decomposition is $C = A(1 - e^{-0.12t})$ where t is time in seconds. Calculate the time it takes the compound to decompose to a value of $C = 0.114$, given that $A = 8.9$.

Solutions

---

1. (a) $\log_5 0.04 = 2$;   (b) $\log_a a = 1$;   (c) $\log_{10} 0.001 = -3$;   (d) $\log_e 7.39 = 2$
2. (a) 100;   (b) 1;   (c) 3;   (d) 1
3. (a) -7.38;   (b) 2.354;   (c) -2.278
4. 3.58×10^{-3}
5. $1.75\,\mu C$
6. $107\,ms$

This concludes some important algebra. In your next assignment you will demonstrate what you have learned about solving:

- simultaneous linear equations,
- quadratic equations,
- exponential and logarithmic equations.

Assignment III

Three engineering problems

This is an 'open-book' assignment, which means that you should work through it with access to notes and textbooks. There is no time limit; just work at your own pace and finish when you feel you have solved all three problems to your satisfaction.

In planning your solutions to problems 1 and 2 you should start by stating relationships between variables in words.

Transposing equations into a suitable form and then substituting values will form part of the information handling process of this assignment.

Evaluate your solutions by carrying out checking procedures and making sure your results make practical sense. Your evaluation of problem 2 should involve some discussion supported by sketches.

The product of this assignment should be three correct solutions neatly presented quoting appropriate dimensions and their units.

Do your best!

1. A SIMPLY SUPPORTED BEAM

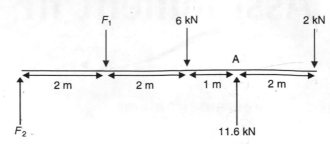

In vertical equilibrium the downward forces are equal to the upward forces so you can write an equation showing these forces are equal. Since the clockwise and anti-clockwise moments about A are also equal you can write another equation. Then, with two equations, you are in a position to find the magnitudes of the two forces F_1 and F_2.

Evaluate your results by carrying out a suitable check.

2. MAKING A TEMPLATE

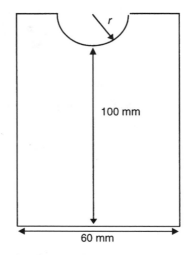

The template shown is to be designed to have a total area of $6500\,\text{mm}^2$. Formulate an equation for the area of the template then use it to calculate the required radius of the semi-circular cut-out.

Evaluate your solution(s) by considering if they suit the practical situation.

3. THE CURRENT IN AN INDUCTOR

The behaviour of electric current in an inductive circuit is described by the equation

$$i = \frac{V}{R}\left(1 - e^{-\frac{tR}{L}}\right)$$

Rearrange this equation to obtain an expression for time t, then calculate the time in seconds that it would take the current i to reach $150\,\text{mA}$ given that: $V = 240\,\text{V}$, $R = 1.2\,\text{k}\Omega$, $L = 8\,\text{H}$.

chapter

5

Trigonometry

Trigonometry was developed to deal with problems involving triangles. Knowing the relationships between the sides and angles of a right-angled triangle provides us with ways of calculating dimensions of sides, angles and areas that are unknown. For this, we rely heavily on the **trigonometric ratios** and **Pythagoras' theorem**. Today, the use of trigonometry extends well beyond the analysis of triangles. In engineering we use trigonometry to model two-dimensional quantities such as **vectors** and **phasors**; but first, some revision.

5.1   Triangles

The class of triangle that is easiest to deal with is the **right-angled triangle**. Its general form and the standard method of representing its sides and angles are shown in Fig. 5.1.

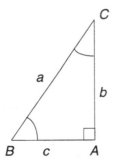

Figure 5.1 Right-angled triangle

Upper case letters represent the angles and the corresponding lower case letters represent the sides opposite. We know a great deal about such triangles. Pythagoras' theorem tells us that:

**the square of the hypotenuse is equal to the
sum of the squares on the other two sides.**

This theorem can be expressed as an equation for the triangle shown in Fig. 5.1

$$a^2 = b^2 + c^2$$

We also know that:

The sum of the angles inside any triangle is 180°

The equation for this is:

$$A + B + C = 180$$

About area:

area $= \frac{1}{2} \times$ base $\times$ height

Or, for our standard right-angled (RA) triangle:

$$\text{area} = \tfrac{1}{2}bc$$

The **trigonometric ratios:**

The *sine* of an angle is the ratio of its *opposite* side to the *hypotenuse*.

The *cosine* of an angle is the ratio of its *adjacent* side to the *hypotenuse*.

The *tangent* of an angle is the ratio of its *opposite* side to its *adjacent* side.

Some people find the mnemonic **SohCahToa** useful for remembering these. The algebraic equations expressing these ratios for angle B of the standard RA triangle are:

$$\sin B = \frac{b}{a} \qquad \cos B = \frac{c}{a} \qquad \tan B = \frac{b}{c}$$

For angle C the equations are:

$$\sin C = \frac{c}{a} \qquad \cos C = \frac{b}{a} \qquad \tan C = \frac{c}{b}$$

Make a dimensional analysis of these ratios. What are the units of sin, cos and tan?

> In units of SI the ratios are of metres over metres
> so each one is just a number

In certain circumstances the **reciprocal ratios** are useful.

The cosecant, $\qquad \operatorname{cosec} \theta = \dfrac{1}{\sin \theta}$

The secant, $\qquad \sec \theta = \dfrac{1}{\cos \theta}$

The cotangent, $\qquad \cot \theta = \dfrac{1}{\tan \theta}$

So, for instance, the reciprocal ratios for angle B of the standard RA triangle are:

$$\operatorname{cosec} B = \frac{a}{b} \qquad \sec B = \frac{a}{c} \qquad \cot B = \frac{c}{b}$$

PRACTICE EXERCISE 5.1

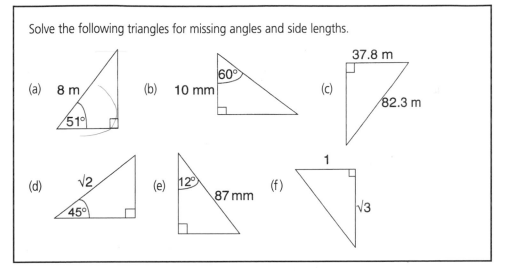

Solve the following triangles for missing angles and side lengths.

(a) 8 m, 51°

(b) 60°, 10 mm

(c) 37.8 m, 82.3 m

(d) √2, 45°

(e) 12°, 87 mm

(f) 1, √3

Solutions

(a) 39°, 6.22 m, 5.03 m;   (b) 30°, 20 mm, 17.3 mm;   (c) 73.1 m, 62.7°, 27.3°;
(d) 45°, 1, 1;   (e) 78°, 18.1 mm, 85.1 mm;   (f) 2, 30°, 60°

The Practice Exercise should have drawn your attention to two things. The inverse of a trig. ratio is an angle. So if:

$$\sin \theta = x \qquad\qquad \cos \theta = x \qquad\qquad \tan \theta = x$$

$$\theta = \arcsin x \qquad\qquad \theta = \arccos x \qquad\qquad \theta = \arctan x$$

Because of the way most calculator keys are labelled: arcsin, arccos and arctan are frequently written as $\sin^{-1}$, $\cos^{-1}$ and $\tan^{-1}$. *Be careful; $\sin^{-1}$ does not mean the reciprocal of sin.* What is the reciprocal of a sine called?

cosecant

The second thing you might have noticed are the ratios of the sides of the triangles in (b), (d) and (f). Like the famous '3-4-5' triangle, they are worth remembering, so here they are again:

Trigonometric identities

There are a number of useful identities but we will look at just two of them for now.

$$\tan A \equiv \frac{\sin A}{\cos A}$$

and

$$\sin^2 A + \cos^2 A \equiv 1$$

The reason these are called **identities** is that unlike equalities they are true for any value of the variable, which in this case is A. Identities are written with a ($\equiv$) sign instead of a ($=$). However, most engineers tend to treat identities as equalities, which is rather sloppy.

PRACTICE EXERCISE 5.2

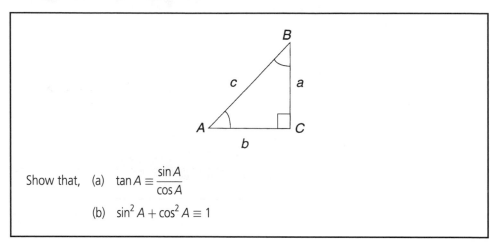

Show that, (a) $\tan A \equiv \dfrac{\sin A}{\cos A}$

(b) $\sin^2 A + \cos^2 A \equiv 1$

Solutions

(a) $\dfrac{\sin A}{\cos A} = \dfrac{\frac{a}{c}}{\frac{b}{c}} = \dfrac{a}{b} = \tan A$

(b) $\sin^2 A + \cos^2 A = \dfrac{a^2}{c^2} + \dfrac{b^2}{c^2} = \dfrac{a^2 + b^2}{c^2} = \dfrac{c^2}{c^2} = 1$

Scalene triangles

Right-angled triangles are a special case of a class of triangle known as **scalene** (non-equal sided). The solution of triangles which do not have a right angle is possible from our extensive knowledge of the properties of RA triangles.

The Sine Rule

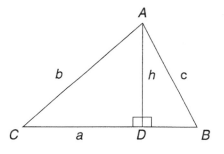

For the triangle shown:

$$\sin C = \frac{h}{b} \quad \text{and} \quad \sin B = \frac{h}{c}$$

$$\therefore \qquad h = b \sin C \quad \text{and} \quad h = c \sin B$$

$$\therefore \quad b \sin C = c \sin B$$

$$\frac{b}{\sin B} = \frac{c}{\sin C}$$

Dropping a perpendicular h from angle A, divided triangle ABC into two RA triangles. This allowed us to formulate the equation shown. A similar treatment would lead to an equation with $\sin A$. The general rule is therefore,

$$\frac{a}{\sin A} = \frac{b}{\sin B} = \frac{c}{\sin C}$$

The Sine Rule allows us to solve any triangle providing we know one side and any two angles.

EXAMPLE 5.1

Solve the following triangle,

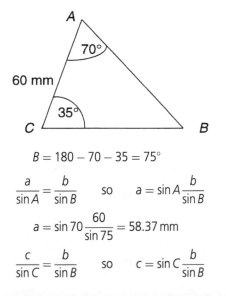

$$B = 180 - 70 - 35 = 75°$$

$$\frac{a}{\sin A} = \frac{b}{\sin B} \quad \text{so} \quad a = \sin A \frac{b}{\sin B}$$

$$a = \sin 70 \frac{60}{\sin 75} = 58.37 \, \text{mm}$$

$$\frac{c}{\sin C} = \frac{b}{\sin B} \quad \text{so} \quad c = \sin C \frac{b}{\sin B}$$

$$c = \sin 35 \frac{60}{\sin 75} = 35.62 \text{ mm}$$

Answer $B = 75°,$ $a = 58.4 \text{ mm},$ $c = 35.6 \text{ mm}$

With problems of this kind, we can apply a very simple checking procedure. Can you see what it is?

Longest side opposite to the largest angle,
shortest side opposite the smallest angle and so on.

The limitations of the Sine Rule are that, if:

- two sides and an angle opposite are known, the solution may be ambiguous
- two sides and an angle between are known, a solution is not possible

EXAMPLE 5.2

Solve:

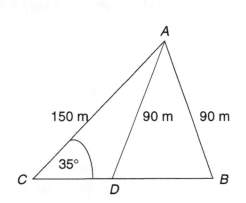

To understand what is happening you must, for the moment, accept that there are two angles between 0 and 360° which have the same sine. $\sin 72.9° = 0.9560$ and $\sin 107.1° = 0.9560$, try this with your calculator. So the solution in this example involves a second possible triangle ADC.

Possible solutions

$$\frac{b}{\sin B} = \frac{c}{\sin C} \quad \text{so} \quad \frac{\sin B}{b} = \frac{\sin C}{c}$$

$$\text{and} \quad \sin B = b\frac{\sin C}{c} \quad \text{so} \quad B = \arcsin\left(b\frac{\sin C}{c}\right)$$

First possibility:

$$B = \arcsin\left(150\frac{\sin 35}{90}\right) = 72.9°$$

$$A = 180 - 35 - 72.9 = 72.1°$$

$$\frac{a}{\sin A} = \frac{c}{\sin C} \quad \text{so} \quad a = \sin A\frac{c}{\sin C}$$

$$a = \sin 72.1\frac{90}{\sin 35} = 149.3 \text{ m}$$

1st solution: $a = 149\,\text{m}$, $A = 72.1°$, $B = 72.9°$

Second possibility:

$$D = \arcsin\left(150\frac{\sin 35}{90}\right) = 107.1°$$

$$A = 180 - 35 - 107.1 = 37.9°$$

$$\frac{a}{\sin A} = \frac{c}{\sin C} \qquad \text{so} \qquad a = \sin A \frac{c}{\sin C}$$

$$a = \sin 37.9 \frac{90}{\sin 35} = 96.39\,\text{m}$$

2nd solution: $a = 96.4\,\text{m}$, $A = 37.9°$, $D = 107.1°$

This kind of ambiguity is an interesting reminder that maths is artificial. In our desire to apply trigonometric ratios to all angles (those greater than 90), we create problems which do not have a unique mathematical solution. However, in an engineering situation you would recognize which of the two triangles makes sense. So, while, mathematically there are two perfectly good solutions, practically, only one of them might do.

EXAMPLE 5.3

Solve:

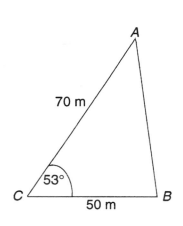

Attempted solution

$$\frac{a}{\sin A} = \frac{b}{\sin B} \qquad \text{Two unknowns.}$$

$$\frac{a}{\sin A} = \frac{c}{\sin C} \qquad \text{Two unknowns.}$$

$$\frac{b}{\sin B} = \frac{c}{\sin C} \qquad \text{Still two unknowns!}$$

Here, a solution is not possible.

The Cosine Rule

The Cosine Rule gets us out of trouble when the Sine Rule fails. The derivation of the Cosine Rule is similar to that of the Sine Rule because it depends on the properties of RA triangles, but we will not go into that here. The rule is:

$$a^2 = b^2 + c^2 - 2bc \cos A$$

$$b^2 = a^2 + c^2 - 2ac \cos B$$

$$c^2 = a^2 + b^2 - 2ab \cos C$$

These may seem a little overwhelming at first but there is a neat pattern which is not difficult to remember.

Armed with the Sine and the Cosine Rules we are in a position to solve almost any triangle. We try the Sine Rule first (because it is easier of course) and if that fails we use the Cosine Rule. The Cosine Rule works when we know:

• two sides and the angle between,
• three sides.

Note that we cannot solve a triangle for which we only know all three angles. Why is that?

The number of triangles with all three angles equal is infinite

PRACTICE EXERCISE 5.3

Solve the following triangles whose angles and sides are given as A, B, C and a, b, c. You will need to familiarize yourself with the degrees, minutes, seconds keys of your calculator. Remember that there are 60 seconds to the minute of a degree and 60 minutes to one degree.

(a) $A = 36°$, $B = 77°$, $b = 2.5$ m; (b) $B = 115°4'$, $C = 11°17'$, $c = 516.2$ mm
(c) $A = 77°3'5''$, $C = 21°2'56''$, $a = 9.80$ m; (d) $a = 7$ m, $c = 11$ m, $C = 22°7'$
(e) $b = 8.16$ m, $c = 7.14$ m, $A = 37°18'10''$; (f) $a = 312$ mm, $b = 527$ mm, $c = 700$ mm

Solutions

(a) $C = 67°$, $a = 1.51$ m, $c = 2.36$ m; (b) $A = 54°39'$, $a = 2126$ mm, $b = 2390$ mm
(c) $B = 81°54'$, $b = 9.96$ m, $c = 3.61$ m; (d) $A = 13°51'36''$, $B = 144°1'12''$, $b = 17.2$ m
(e) $a = 4.99$ m, $B = 82°19'1''$, $C = 60°22'48''$
(f) $A = 24°41'1''$, $B = 44°51'9''$, $C = 110°28'12''$

Minutes and **seconds** of a degree are important. You must be aware of them. Whilst most computers and calculators are programmed to work in decimals of a degree, mechanical engineers and navigators still use minutes and seconds. I remember an incident when the captain of a ship read the digital display of a navigation system to be something like: 12.3521S 10.4755E and charted it as 12°35'21''S 10°47'55''E, putting his position several miles away from where the ship actually was. A serious error but easily made!

Pitch circle diameter

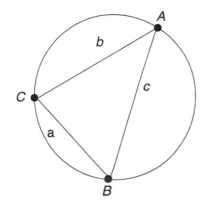

The **Sine Rule** is also useful for finding the diameter of a circle which circum-scribes a triangle using:

$$D = \frac{a}{\sin A} = \frac{b}{\sin B} = \frac{c}{\sin C}$$

Area of a triangle

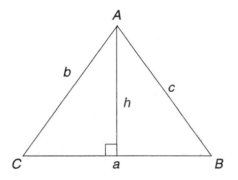

For the triangle shown, you should remember that the area can be calculated from

$$\text{area} = \tfrac{1}{2}ah$$

But we also have

$$\sin C = \frac{h}{b} \qquad \text{so} \qquad h = b\sin C$$

Substituting for h in the equation for area, we get:

$$\text{area} = \tfrac{1}{2}ab\sin C$$

So the sine ratio has proved itself useful once more. It provides us with a means of calculating the area of a triangle if we know two sides and the angle between.

PRACTICE EXERCISE 5.4

Find the areas of the following triangles:

(a) $A = 80°$, $b = 4\,\text{m}$, $c = 3\,\text{m}$
(b) $C = 40°$, $a = 80\,\text{mm}$, $b = 50\,\text{mm}$
(c) $B = 110°$, $a = 50\,\text{mm}$, $c = 140\,\text{mm}$
(d) $A = 145°$, $b = 30\,\text{mm}$, $c = 40\,\text{mm}$
(e) Find the pitch circle diameter of the three circles shown.

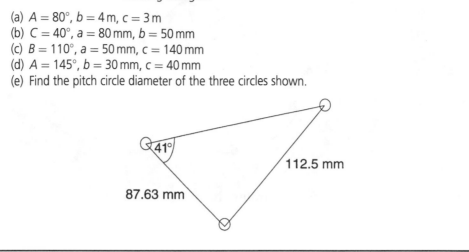

Solutions

(a) $5.91\,\text{m}^2$; (b) $1286\,\text{mm}^2$; (c) $3289\,\text{mm}^2$; (d) $344\,\text{mm}^2$; (e) $172\,\text{mm}$

The radian

The use of the **degree** as the unit of **angular displacement** is familiar to us. But the degree is not derived from SI units. The derived unit of angular displacement is called the **radian** which is defined in the following way:

> **An angle of 1 radian is subtended by the length of an arc of a circle radius *r* when the length of the arc is equal to *r*.**

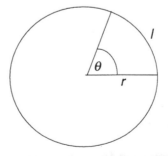

This is a bit of a mouthful; better to abbreviate it as an equation.

$$\theta = \frac{l}{r} \text{ radians}$$

Since l and r are both lengths in metres the unit of the radian is just a ratio of two lengths. How does the radian relate to the degree?

That's easy; within a complete circle, what is the maximum length of an arc?

$$2\pi r$$

So what is the angle in radians subtended by an arc which is the whole circumference of a circle?

$$\theta = \frac{2\pi r}{r} = 2\pi \text{ rad}$$

So the arc of a semicircle must subtend an angle of π radians. Compare this to the degrees subtended by a semicircle; it is 180. We can apply as a proportion

$$\frac{\theta°}{180} = \frac{\theta}{\pi}$$

In words: the ratio of an angle in degrees to 180 is the same as the ratio of an angle in radians to π.

PRACTICE EXERCISE 5.5

Convert the following:

(a) 60°; (b) 90°; (c) 30°; (d) 270°; (e) 360°; (f) 120°; (g) 1 rad; (h) 3.142 rad;
(i) $\frac{\pi}{4}$ rad; (j) 0.5171 rad; (k) 1.5632 rad; (l) $\frac{\pi}{6}$ rad

Solutions

(a) $\frac{\pi}{3}$; (b) $\frac{\pi}{2}$; (c) $\frac{\pi}{6}$; (d) $\frac{3\pi}{2}$; (e) 2π; (f) $\frac{2\pi}{3}$; (g) 57.3°; (h) 180°; (i) 45°;
(j) 29.6°; (k) 89.6°; (l) 30°

5.2 Angles greater than 90°

We must now look at angles which go beyond the confines of triangles. Take for example a rotating machine. One revolution represents an angular displacement of 2π rad or 360°. Many revolutions would correspond with angular displacements much greater than 2π.

2 revolutions	4π rad	720°
$\frac{1}{2}$ revolution	π rad	180°
10 revolutions	20π rad	3600°
1.25 revolutions	2.5π rev	450°
25.75 revolutions	51.5π rad	9270°
$\frac{3}{4}$ revolution	$\frac{3}{2}\pi$ rad	270°

The point needs to be made that in engineering systems we often deal with situations involving a change that is **periodic**. Something that is periodic undergoes a cycle of change which is repeated in equal time. Once a machine has turned through one revolution it has completed its full cycle of change. The next revolution is a second cycle and so on. An alternating current is also

periodic. It rises to a maximum positive, falls to zero, rises to a maximum negative, falls to zero and then repeats that cycle of change. The time it takes to complete such a cycle is called the **period**, and the **angular displacement** of a
period is always 2π rad (360°). The speed at which a periodic function changes
is called **angular velocity** (ω) which is measured in rad s^{-1}. Trigonometry is
a useful tool in modelling the behaviour of physical quantities that are
periodic because, for the most part, the pattern of change is **sinusoidal**.
Sinusoids have values that are proportional to the sine (or cosine) of angular
displacement.

Try to think of a few examples of things that have **periodicity**:

position of a pendulum, extension of a vibrating spring,
relative position of a satellite, displacement of a wave,
speed and position of an engine piston, alternating voltage, etc.

In devising a mathematical model for particular situations (pendulum, spring,
etc.) we like to start with a general model. Such a general model of **periodic
functions** which are **sinusoids** is sometimes called a rotating **complexor**.

Figure 5.2 shows the complexor, r. It is rotated to describe a circle about
the origin, O, of a set of x- and y-axes. By convention, the rotation is always

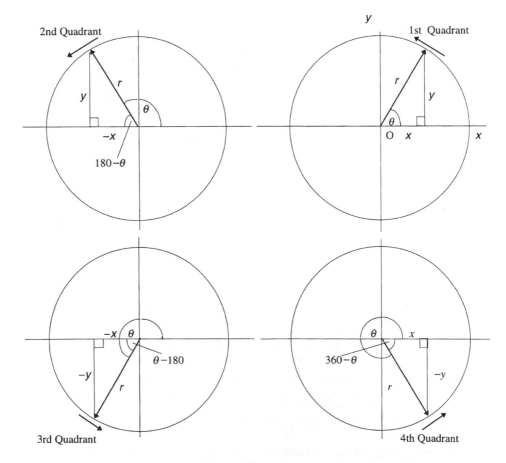

Figure 5.2 Trigonometric ratios in four quadrants

anti-clockwise starting in the 1st quadrant of the circle. Also, by convention, angle θ is always referenced to the positive x-axis. y and x are the sides opposite and adjacent to the angle of interest. They are sometimes referred to as components of r. So y is a vertical component of r and x is a horizontal component of r. Note that y and x can be positive or negative but r cannot be negative, as that would not make sense. Now look at the trigonometric ratios of angles that lie within each of the four quadrants in turn:

In the 1st quadrant

$$\sin\theta = \frac{+y}{+r}, \ (+\text{ve}) \qquad \cos\theta = \frac{+x}{+r}, \ (+\text{ve}) \qquad \tan\theta = \frac{+y}{+x}, \ (+\text{ve})$$

In the 2nd quadrant

$$\sin(180-\theta) = \frac{+y}{+r}, \ (+\text{ve}) \qquad \cos(180-\theta) = \frac{-x}{+r}, \ (-\text{ve})$$

$$\tan(180-\theta) = \frac{+y}{-x}, \ (-\text{ve})$$

In the 3rd quadrant

$$\sin(\theta-180) = \frac{-y}{+r}, \ (-\text{ve}) \qquad \cos(\theta-180) = \frac{-x}{+r}, \ (-\text{ve})$$

$$\tan(\theta-180) = \frac{-y}{-x}, \ (+\text{ve})$$

In the 4th quadrant

$$\sin(360-\theta) = \frac{-y}{+r}, \ (-\text{ve}) \qquad \cos(360-\theta) = \frac{+x}{+r}, \ (+\text{ve})$$

$$\tan(360-\theta) = \frac{-y}{+x}, \ (-\text{ve})$$

Now, the angles in the 2nd, 3rd and 4th quadrants correspond to the large angle θ which is swept out by the rotation of r, so each of the ratios above can be used to represent any angular displacement, even the displacement of many rotations.

Let's summarize.

- All angles lie in one of the four quadrants.
- For angles lying in the 1st quadrant the sines, cosines and tangents are all positive.
- For angles extending into the 2nd quadrant only the sines are positive.
- For angles extending into the 3rd quadrant only the tangents are positive.
- For angles extending into the 4th quadrant only the cosines are positive.

Figure 5.3 summarizes the situation in a concise form. Try to memorize it. Its purpose will become clear when you have done a few examples.

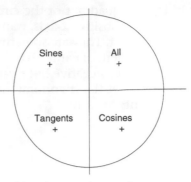

Figure 5.3 Quadrants with positive sines, cosines and tangents

PRACTICE EXERCISE 5.6

Evaluate the following to 4 d.p. They are quite straightforward, your calculator does all the work. Just be careful to put it into the correct mode; degrees or radians.

(a) $\sin 135°$; (b) $\cos 98°$; (c) $\tan 141°$; (d) $\sin 235°$; (e) $\cos 197°$; (f) $\tan 255°$;

(g) $\sin 334°$; (h) $\cos 279°$; (i) $\tan 300°$; (j) $\sin 0.4571$; (k) $\cos\frac{\pi}{3}$; (l) $\tan\frac{2\pi}{3}$;

(m) $\sin 4.265$; (n) $\cos 6$; (o) $\sin 5.912$; (p) $\sin 370°$; (q) $\cos 820°$; (r) $\sin 90°$;
(s) $\cos 90°$; (t) $\tan 90°$

Solutions

(a) 0.7071; (b) −0.1392; (c) −0.8098; (d) −0.8192; (e) −0.9563; (f) 3.7321;
(g) −0.4384; (h) 0.1564; (i) −1.7321; (j) 0.4413; (k) 0.5000; (l) −1.732;
(m) −0.9016; (n) 0.9602; (o) −0.3627; (p) 0.1736; (q) −0.1736; (r) 1.0000;
(s) 0; (t) ∞

So it is important always to make sure that your calculator is set to the correct mode. First you must recognize whether the angle you are dealing with is a radian or a degree value. Radians are written as numbers. Degrees always carry the (°) sign.

Practice Exercise 5.6 shows that the calculator can cope with all angles and give us the correct trig. ratio. That is not the case when going the other way and this is where we need Fig. 5.3. Think back to the ambiguous case when using the Sine Rule. We ended up with two possible triangles because, between 0 and 360°, there are two angles with the same sine. 72.9° lies in the first quadrant and its sine is 0.9560; and 107.1°, which is in the second quadrant, has the same sine, 0.9560.

This is the flaw in the system:

> **any trigonometric ratio can represent two possible angles in the range of 0 to 360°**

In fact how many possible angles can a trig. ratio represent if we go beyond 360°?

An infinite number

We need not worry about this infinite possibility. The angles which we are interested in are always those between 0 and 360°. Now, work carefully through the next set of exercises.

PRACTICE EXERCISE 5.7

Find, in degrees, all the possible angles between 0 and 360° of:

(a) arcsin 0.5;   (b) arcsin −0.5;   (c) arccos 0.5;   (d) arccos −0.5;   (e) arctan 0.5

My advice is not to enter the sign of the trig. ratio into the calculator. Just enter the sine, cosine or tangent as a positive number. Then look at the sign and decide on the possible quadrants:

- in the 2nd quadrant take the angle away from 180°
- in the 3rd quadrant add the angle to 180°
- in the 4th quadrant take the angle away from 360°

Solutions

(a) 270°, 150°;   (b) 210°, 330°;   (c) 60°, 300°;   (d) 120°, 240°;   (e) 26.6°, 206.6°

Try a few more.

PRACTICE EXERCISE 5.8

Evaluate, in degrees, the possible angles represented by the following:

(a) arcsin −1;   (b) arccos −0.9660;   (c) arctan 1.73;   (d) arctan −57.29;   (e) arccos 0;
(f) arccos 1;   (g) arcsin −0.8660;   (h) arctan ∞;   (i) arctan 2;   (j) arcsin 0.0349;
(k) arccos 0.0175;   (l) arctan −0.0875

Solutions

(a) 270°;   (b) 165°, 195°;   (c) 60°, 240°;   (d) 91°, 271°;   (e) 90°, 270°;   (f) 0°;
(g) 240°, 300°;   (h) 90°, 270°;   (i) 63°, 243°;   (j) 2°, 178°;   (k) 89°, 271°;
(l) 175°, 355°

This can be hard work but to learn it properly you need to practise with many examples so stay with it.

Now try the same again, remembering that 180° is equivalent to π rad.

PRACTICE EXERCISE 5.9

Evaluate in radians, correct to 4 d.p.

(a) arcsin −1;   (b) arccos −0.4813;   (c) arctan −1.7642;   (d) arcsin 0.7137;
(e) arccos 0;   (f) arccos 0.8660;   (g) arctan 99;   (h) arccos 0.707;   (i) arccos 1;
(j) arctan 0;   (k) arcsin −0.7880;   (l) arctan −2.9042

Solutions

(a) $\frac{3\pi}{2}$ rad;   (b) 2.0729 rad, 4.213 rad;   (c) 2.0865 rad, 5.2281 rad;

(d) 0.7948 rad, 2.3468 rad;   (e) $\frac{\pi}{2}$ rad, $\frac{3\pi}{2}$ rad;   (f) 0.5236 rad, 5.7595 rad;

(g) 1.5607 rad, 4.7023 rad;   (h) 0.7860 rad, 5.4980 rad;   (i) 0 rad;   (j) 0 rad, π rad;
(k) 4.0491 rad, 5.3756 rad;   (l) 1.9024 rad, 5.0440 rad

Just one more thing about angles and their trig. ratios.

Negative angles

Instead of specifying angles as lying between 0 and 360° (0 and 2π) there is an alternative and, perhaps, easier method. We can say that all the principal angles lie between +180° and −180°; equivalent to π and $-\pi$ (see Fig. 5.4).

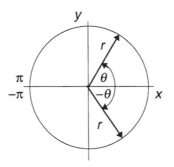

Figure 5.4 Negative angles

Angles below the x-axis are considered to be formed by a negative sweep of r and those above, by a positive sweep of r.

PRACTICE EXERCISE 5.10

Evaluate the angles between 180° and −180° of the following:

(a) arccos 0;   (b) arcsin 0.5;   (c) arcsin −0.8661;   (d) arctan −1

Evaluate the angles between π and $-\pi$ of the following:

(e) arcsin 1;   (f) arccos −0.5;   (g) arcsin 0.8661;   (h) arctan 1

Solutions

(a) 90°, −90°;   (b) 30°, 150°;   (c) −60°, −120°;   (d) 135°, −45°;   (e) $\frac{\pi}{2}$ rad;

(f) 2.0944 rad, −2.0944 rad;   (g) 1.0473 rad, 2.0942 rad;   (h) 0.7854 rad, −2.3562 rad

Practice Exercise 5.10 illustrates the fact that angles greater than 180° can be specified in two ways. 270° is the same as −90° and $\frac{3\pi}{2}$ is equivalent to $-\frac{\pi}{2}$.

Position of a point in a plane

In specifying the position of a point in a two-dimensional plane we need something more than just the real numbers. Real numbers are restricted to the one-dimensional number line. So we need two number lines. There are two such number lines called the **Cartesian coordinates** (named after the French mathematician Descartes); they are quite familiar, and we usually refer to them as the x- and y-axes.

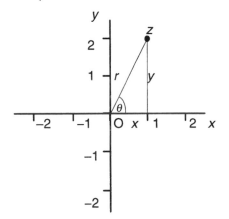

Figure 5.5 Cartesian coordinates and polar coordinates

Figure 5.5 illustrates how we use the Cartesian system of coordinates to describe the position of a point z in the x–y plane. The x-component of z is 1 and the y-component of z is 2, so we can specify the position of z as $(1, 2)$. The convention requires us to specify the x-component followed by the y-component, so the general form of Cartesian representation is:

$$(x, y)$$

Now there is an alternative method of representing the position of z. It is called **polar representation**. The x and y-components of z form two sides of an RA triangle with a hypotenuse that is labelled r in Fig. 5.5. It is possible to specify the position of z by stating the length of r and the angular displacement θ. Polar representation has the general form;

$$r \angle \theta$$

where r is called the **modulus** and θ is called the **argument**.

EXAMPLE 5.4

Refer to Fig. 5.5 and evaluate the position of point z in polar form.

The length of r can be found using Pythagoras' theorem.

$$r = \sqrt{x^2 + y^2} = \sqrt{1^2 + 2^2} = 2.24$$

and angle θ can be found using the tangent ratio (although any of the trig. ratios would do).

$$\tan\theta = \frac{y}{x}, \qquad \theta = \arctan\frac{y}{x} = \arctan\frac{2}{1} = 63°$$

So the position of z in polar form is $2.24 \angle 63°$.

EXAMPLE 5.5

Convert $5 \angle 53.1°$ into Cartesian form.

x is the side adjacent to θ so the cosine ratio can be used.

$$\cos\theta = \frac{x}{r}, \qquad x = r\cos\theta = 5\cos 53.1 = 3.00$$

y is the side opposite so we use the sine ratio.

$$\sin\theta = \frac{y}{r}, \qquad y = r\sin\theta = 5\sin 53.1 = 4.00$$

So $5 \angle 53.1°$ in Cartesian form is $(3, 4)$.

Summary of the relationship between Cartesian and polar representation.

In Cartesian form a point is represented as (x, y) where

$$x = r\cos\theta, \qquad y = r\sin\theta$$

In polar form a point is represented as $r \angle \theta$ where

$$r = \sqrt{x^2 + y^2}, \qquad \theta = \arctan\frac{y}{x}$$

PRACTICE EXERCISE 5.11

(a) Specify, in Cartesian form, the positions of the points z_1 to z_8 shown in Fig. 5.6.
(b) Convert to polar form the positions of the points in (a).
(c) Convert the following into Cartesian form:

$$5 \angle 20° \qquad 3 \angle -30° \qquad 2.6 \angle -134° \qquad 3.7 \angle 97°$$

$$2 \angle -170° \qquad 4.5 \qquad 1.8 \angle 180° \qquad 1 \angle -90°$$

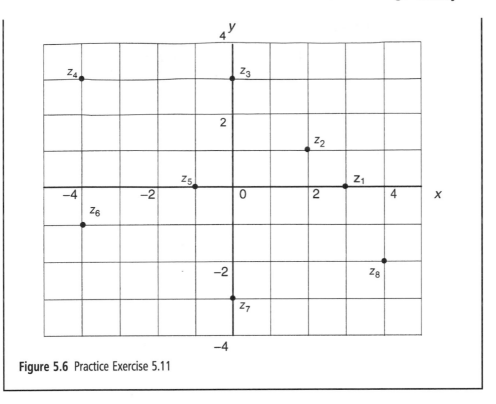

Figure 5.6 Practice Exercise 5.11

Solutions

(a) $z_1 = (3, 0)$ $z_2 = (2, 1)$ $z_3 = (0, 3)$ $z_4 = (-4, 3)$
    $z_5 = (-1, 0)$ $z_6 = (-4, -1)$ $z_7 = (0, -3)$ $z_8 = (4, -2)$

(b) $z_1 = 3 \angle 0$ $z_2 = \sqrt{5} \angle 26.6°$ $z_3 = 3 \angle 90°$ $z_4 = 5 \angle 143°$
    $z_5 = 1 \angle 180°$ $z_6 = 4.12 \angle -166°$ $z_7 = 3 \angle -90$ $z_8 = 4.90 \angle -26.6°$

(c) $(4.70, 1.71)$ $(2.60, -1.50)$ $(-1.81, -1.87)$ $(-0.451, 3.67)$
    $(-1.97, -0.347)$ $(4.5, 0)$ $(-1.8, 0)$ $(0, -1)$

Now for some practical applications of all this trigonometry.

5.3 Vectors and phasors

Scalars

Engineering quantities which can be specified in one dimension only, i.e. in terms of a modulus alone, are called **scalars**. Think of some examples of scalar quantities:

Energy, temperature, mass, volume, speed, etc.

Now, specifying the position of a point in a plane was a mathematical exercise which illustrated the need to use two-dimensional representation; polar or Cartesian. In mechanical engineering we apply the technique to quantities which have **magnitude** (modulus) and **direction** (argument). Examples of such quantities are: velocity, force, and acceleration. They are called **vectors**. In electrical engineering we have currents and voltages which have **amplitude** (modulus) and **phase** (argument). These are called **phasors**.

Vectors

Vectors are not represented as a point in a plane but as a line whose length represents magnitude and whose angle (relative to some reference point) is the direction. For example, the force of gravity (weight) of 30 kN can be described as having a modulus of 30k and an argument of $-90°$. Drawn to scale it would look as follows:

30 kN

Adding and subtracting vectors

When an object is subject to more than one force, the resultant of the forces is found by the addition of vectors. This addition can be carried out according to two rules:

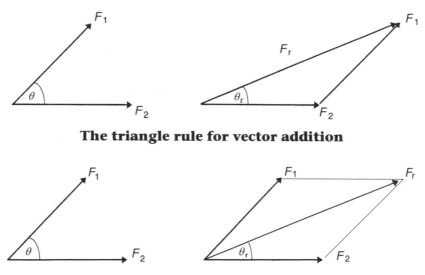

The triangle rule for vector addition

The parallelogram rule for vector addition

In both cases F_r and θ_r are the resultant modulus and argument. θ is usually referenced to the positive direction of the x-axis of the set of Cartesian coordinates.

So how are these rules applied? There are three approaches.

- We can, with the help of graph paper, protractor and compasses draw scale diagrams and measure the resultant.
- We can take an analytical approach and apply our knowledge of scalene triangles to find the resultant; using the sine rule or cosine rule, or both, if necessary.

• We can resolve the vectors by adding their Cartesian components.

The graphical method is not practical enough. It takes too long and the results are not accurate enough. However, a sketch drawn roughly to scale is always a good rough check.

Carefully copy out and study the next two examples.

EXAMPLE 5.6

A vector has a modulus of 12 and an argument of 20°, while another vector has a modulus of 5 and an argument of 80°. Find the resultant of the two vectors by applying the triangle rule.

The first step is to sketch the vectors:

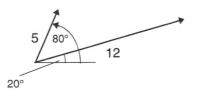

Then to form a triangle:

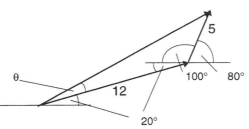

We have a triangle in which we know two sides and the angle between. The cosine rule can be applied. Let the unknown side be a:

$$a^2 = b^2 + c^2 - 2bc \cos A$$

$$a = \sqrt{12^2 + 5^2 - 2(12)(5) \cos 120°}$$

$$= 15.1$$

The sine rule now allows us to find the angle θ:

$$\frac{\sin \theta}{5} = \frac{\sin 120}{15.1}, \qquad \theta = \arcsin\left(\frac{5 \sin 120}{15.1}\right) = 16.6°$$

This is the angle of θ; with reference to the x-axis the angle is $16.6 + 20 = 36.6°$. So the resultant vector has a modulus 15.1 and argument of 36.6°.

EXAMPLE 5.7

Find the resultant of the two vectors in the previous example by resolving Cartesian components.

The two vectors are in the polar form, $r \angle \theta$. We have $12 \angle 20°$ and $5 \angle 80°$.

For vector 1, $x_1 = r \cos \theta = 12 \cos 20° = 11.28$
For vector 2, $x_2 = 5 \cos 80° = 0.8682$

The resultant vector will have an x-component that is a sum of these, so

Resultant vector $x_r = 11.28 + 0.8682 = 12.15$

For vector 1, $y_1 = r \sin\theta = 12 \sin 20° = 4.10$
For vector 2, $y_2 = 5 \sin 80° = 4.92$

Again, the resultant will have a y-component which is a sum of these, so:

Resultant vector $y_r = 4.1 + 4.92 = 9.02$

We now have the Cartesian components of the resultant vector. It is always sensible to sketch them:

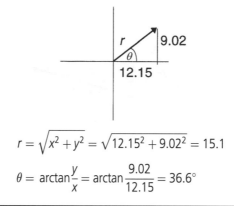

$$r = \sqrt{x^2 + y^2} = \sqrt{12.15^2 + 9.02^2} = 15.1$$

$$\theta = \arctan\frac{y}{x} = \arctan\frac{9.02}{12.15} = 36.6°$$

Phasors

The technique for finding the resultant of phasors is the same as it is for finding the resultant of vectors. Electrical engineers tend, however, to prefer resolving the horizontal and vertical components (Cartesian components) method and angles are frequently quoted in radians. The general equation of a phasor is:

$$a = A \sin(\omega t + \theta)$$

This will be explained later when we look at graphs. For the moment, recognize that A is the **amplitude** (modulus) and θ is the **phase angle** (argument) of the phasor.

EXAMPLE 5.8

Two electric currents $I_1 = 12 \sin(\omega t + \frac{2\pi}{3})$ and $I_2 = 18 \sin(\omega t - \frac{\pi}{4})$ meet at a node. Sketch a diagram showing the phasors of these currents then find the resultant of the two.

A sketch is always useful to get a clear picture of the situation. It shows which of the four quadrants are involved and provides the opportunity of a rough check by the parallelogram law.

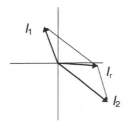

Our sketch indicates that the resultant, which we have called I_r, is likely to have an angle that is slightly negative, or at least very close to zero, and an amplitude that is somewhere between 12 and 18, say around 10.

Resolving horizontal and vertical components:

$$x_1 = r_1 \cos\theta_1 = 12\cos\frac{2\pi}{3} = -6.00 \qquad y_1 = r_1 \sin\theta_1 = 12\sin\frac{2\pi}{3} = 10.39$$

$$x_2 = r_2 \cos\theta_2 = 18\cos-\frac{\pi}{4} = 12.73 \qquad y_2 = r_2 \sin\theta_2 = 18\sin-\frac{\pi}{4} = -12.73$$

$$x_r = -6 + 12.73 = 6.73 \qquad\qquad y_r = 10.39 - 12.73 = -2.34$$

So the x-component of the resultant is positive 6.73 and the y-component is negative 2.34. Before converting to polar form we need another sketch.

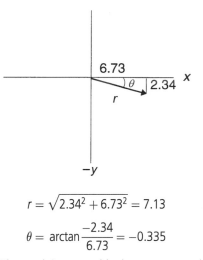

$$r = \sqrt{2.34^2 + 6.73^2} = 7.13$$

$$\theta = \arctan\frac{-2.34}{6.73} = -0.335$$

−0.335 radians is about 19°. The result is reasonably close to our rough check so we can conclude that the resultant current has

$$7.13\text{A} \angle -0.335$$

or alternatively,

$$I_r = 7.13\sin(\omega t - 0.335)$$

Now do some revision. Go back over the worked examples 5.1 to 5.8 in this chapter and see if you can do them on your own.

Assignment IV

This assignment, like the previous one, is an open-book test. It requires you to use trigonometry to solve two engineering problems. One is a structural engineering problem, the other an electrical problem. You should use your own notes, textbooks or any other reference material you think may be appropriate. However, it is important that you work through this alone under normal examination conditions, allowing yourself about $1\frac{1}{2}$ hours.

Assignments are normally graded on the basis of two things: process and outcomes.

For process:

- Plan your work by starting with appropriate sketches, selecting and stating the trigonometric rules, ratios and theorems you think will lead you to a solution.
- Seek out information from sources available to you and handle the trigonometry and given data by algebraic manipulation so that it will lead directly to a solution.
- Evaluate results using rough checks and/or sketches and by comparing with the practical situation to see if they make sense.

For outcomes:

- You should obtain results that are correct, giving them to an appropriate degree of accuracy, including correct units. You should present your results, neatly, clearly and concisely.

Get involved and enjoy it!

1. THE STRUCTURAL PROBLEM

A high voltage power transmission line is being constructed. Because of land restriction it is necessary to change the direction in which the power lines run so a special tower is erected at the point where the lines change direction. The change in direction is from SE to E. The structural engineers know that the horizontal component of the combined loading of the cables is 1.25 MN in the NW direction and 960 kN in the E direction.

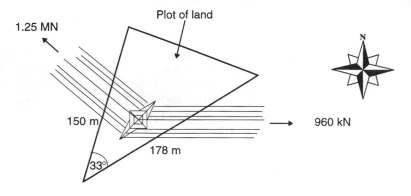

You, as part of the team of engineers involved, have been assigned the task of calculating the minimum tensile support required to overcome the combined loading force of the power cables. Remember that in specifying a force you must give its magnitude and direction.

NB: The degrees of a compass are referenced clockwise from North.

The support tower is to be erected on a triangular plot of land, as shown. Your company has asked you to calculate the area of the land because it will be charged rental on a per-square-metre basis.

2. THE ELECTRICAL PROBLEM

Transmission line systems consist of three sets of power cables designed to carry currents that are separated in phase, often called the *red*, *yellow* and *blue* phases. The three phase currents are then combined in a common *neutral* cable which provides a return path for any out-of-balance current. A simple circuit diagram of such an arrangement is as follows:

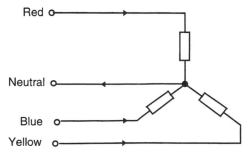

Under ideal conditions the three phase currents will be *balanced* so that they are equal in magnitude and spaced equally apart in phase. However, electricity demand is unpredictable so individual currents can be quite different at any particular point in time.

Measurements taken at a substation at peak demand, indicate the following:

$$\text{red phase} \qquad 1255\text{A} \angle -12°$$
$$\text{yellow phase} \quad 1107\text{A} \angle -131°$$
$$\text{blue phase} \qquad 1304\text{A} \angle 98°$$

Calculate the magnitude and phase of the sum of these currents when they combine in the *neutral* cable.

NB: Start by sketching, roughly to scale, a phasor diagram of the three currents. Constructing two parallelograms will enable you to estimate the amplitude and phase of the resultant current and provide you with a basis for evaluating your final result.

chapter

Graphs

Graphs are an important mathematical tool. Although seldom used to obtain accurate numerical solutions, they are very useful in a number of other ways.

- They provide a quick and clear picture of the relationship between variables.
- Designers can use them as models to apply the *what if...* technique. What if a constant is changed? What if a variable goes beyond a certain limit? Computer generated graphs, in particular, provide an instant picture of a situation that depends on a number of things.
- Graphs can be used to interpret experimental data and help formulate general equations from a set of observations.
- **Extrapolation** provides a means of predicting a situation that depends on values which lie beyond normally available data.
- **Interpolation** allows us to interpret situations that exist between those defined by discrete data.

Graphs of engineering data are plotted using **Cartesian coordinates** or **polar coordinates**. Both of these were dealt with in the chapter on trigonometry where you saw that it is possible to specify the position of a point, the magnitude and direction of a vector or the amplitude and phase of a phasor in two different ways.

Cartesian graphs with their x- and y-axes are already familiar so they need no introduction. Polar graphs are probably new to you so they deserve some attention.

6.1 Polar graphs

Instead of specifying points in terms of their (x, y) components, with polar graphs we specify points in terms of their $r\angle\theta$ components, where:

- r is the **modulus** or distance from the origin
- θ is the **argument** or angular displacement in an anticlockwise direction from the datum.

To plot polar graphs we need polar graph paper, which is constructed with a circular grid and radii from the origin – called the **pole**.

EXAMPLE 6.1

Plot the polar graph of $r = 3 \sin \theta$ for values of θ between 0 and 360° at 30° intervals (see Fig. 6.1).

The first step in plotting any graph is to draw up a table of values, starting with the independent variable. Here, we have values given in degrees so we just make sure the calculator is in degree mode:

θ (°)	0	30	60	90	120	150
$3 \sin \theta$	0	1.5	2.6	3.0	2.6	1.5
θ (°)	180	210	240	270	300	330
$3 \sin \theta$	0	−1.5	−2.6	−3.0	−2.6	−1.5

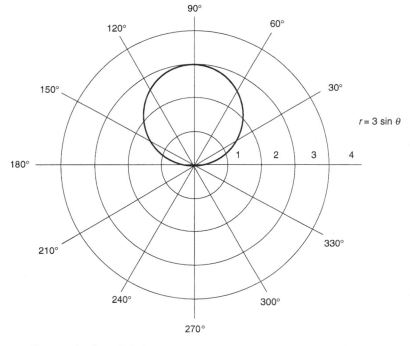

Figure 6.1 The graph of $r = 3 \sin \theta$

The polar graph of a sine function consists of two circles that are superimposed on one another. This is because negative values of r are drawn turned through 180°. The same shape is obtained with a cosine function except the circles are shifted 90° clockwise.

There are one or two special cases: for example, what is the graph of $r = a$? or $\theta = a$? Study them carefully.

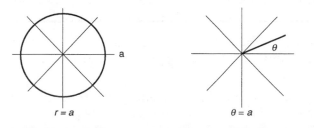

The graphs of $r = a$ and $\theta = a$ where a is just a constant are mathematically trivial. In other words they represent situations that are of little interest. However, they are useful for developing understanding. Can you think of equivalent situations with Cartesian graphs?

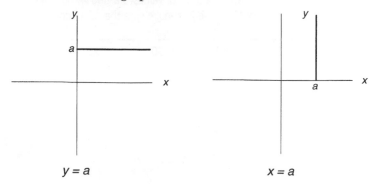

$y = a$ $x = a$

With $y = a$, whatever the value of x, y is fixed at a value a. Similarly with $x = a$ the value of x is held constant. It is useful to compare these with their polar equivalents: when r is fixed at a we get a circle about the pole, when θ is constant we get a line.

An important polar plot is that of the general equation:

$$r = a\theta + b$$

What kind of equation does this remind you of?

$$y = mx + c$$

which is the law of the straight line and gives us a Cartesian graph that is linear. The following exercise will show you how the polar equivalent looks.

PRACTICE EXERCISE 6.1

Plot, on polar graph paper, see Fig. 6.2, a polar graph of $r = 2\theta + 1$ for values of θ that are equivalent to between 0 and 540° at intervals of 45°.

Before you begin you must be clear about one thing. **Degrees cannot be mixed with numbers, so when using angles directly in calculations we must convert them to radians.** Why was it not necessary to change degrees to radians in Example 6.1?

Because the sine of an angle is not in degrees, it is just a number

Solution

Table of values:

θ (rad)	0	$\frac{\pi}{4}$	$\frac{\pi}{2}$	$\frac{3\pi}{4}$	π	$\frac{5\pi}{4}$	$\frac{3\pi}{2}$
$2\theta + 1$	1	2.6	4.1	5.7	7.3	8.9	10.4
θ (rad)	$7\frac{\pi}{4}$	2π	$9\frac{\pi}{4}$	$5\frac{\pi}{2}$	$11\frac{\pi}{4}$	3π	
$2\theta + 1$	12.0	13.6	15.1	16.7	18.3	19.8	

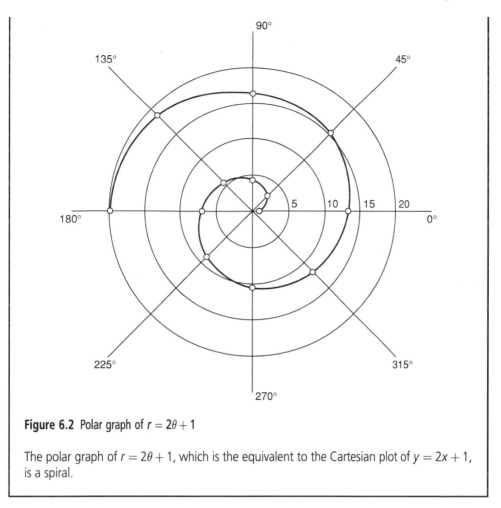

Figure 6.2 Polar graph of $r = 2\theta + 1$

The polar graph of $r = 2\theta + 1$, which is the equivalent to the Cartesian plot of $y = 2x + 1$, is a spiral.

Polar graphs are useful in mechanical, electrical, control and telecommunications engineering. The spiral in Practice Exercise 6.1 is an example of a profile that might be used in the design of a cam or gear teeth. Another example is the plot of a polar diagram of the reception pattern around a radio antenna:

Figure 6.3 shows that the direction of maximum reception is 'easterly'.

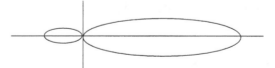

Figure 6.3 Reception pattern around a radio antenna

PRACTICE EXERCISE 6.2

You will need to find some polar graph paper for this Practice Exercise.

(a) Plot the graph of $r = 2 + \cos\theta$.

(b) Plot the graph of $r = -2\theta$ for θ between 0 and 450°. Note that a negative r is drawn turned through 180°.
(c) Plot the graph of $r = 2 + \theta$, for θ between 0 and $-270°$ at intervals of 30°.

Remember that negative angles define a clockwise rotation of r.

6.2 Graphs of functions

We now return to Cartesian graphs and look at the way they represent some important **functions**.

The word *function* is used in mathematics to indicate the existence of a link between two variables. When it is known that one variable is dependent upon another, such as y is often dependent on x, then we can say that *y is a function of x*. This is usually abbreviated as

$$y = f(x)$$

This does not mean that y is $f \times x$. It is a special notation which abbreviates the words:

'y is some function of x'

The statement that 'y is some function of x' is not specific, it simply means that y depends on x in some way. So what are these 'functions' specifically? You have come across some of the most important ones already. These are: linear functions, quadratic functions, exponential and trigonometric functions. When we know the nature of the function we can use functional notation specifically, for example, if:

$$f(x) = 2x$$

and y is a function of x then we can write that $y = 2x$, which you will recognize by now as being a simple linear equation. The use of functional notation can be a useful means of abbreviating long expressions. For example, if we know that:

$$f(x) = 7.5x^2 - 315x + 158$$

we can avoid writing the expression on the RHS by simply referring to it as $f(x)$. For instance, in the table in Practice Exercise 6.1, we could have just written $f(\theta)$ instead of $2\theta + 1$.

Linear functions, $f(x) = mx + c$

The important features of Fig. 6.4 and any graph are always:

- **The gradient.** Previously we called this the *rise/run*. Now we apply a more formal notation, $\delta y/\delta x$. The Greek letter 'delta' (δ) is used to signify 'a small change in'. So δy means a small change in y and δx means a small change in x. With a linear function the gradient is a constant, $\delta y/\delta x = m$.
- **The intercept on the y-axis.** Here we have c. It is the value of y when x is zero. Think of it as a *y-shift* or *offset*.

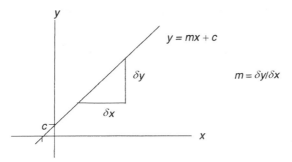

Figure 6.4 Linear functions

- **The intercept on the x-axis.** This is the value of x when y is zero. It is the 'solution of x'.

PRACTICE EXERCISE 6.3

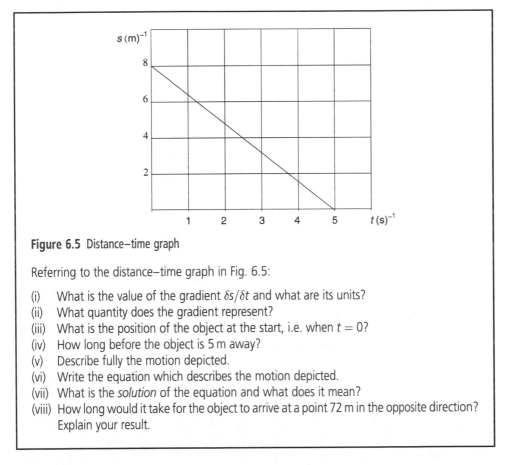

Figure 6.5 Distance–time graph

Referring to the distance–time graph in Fig. 6.5:

(i)   What is the value of the gradient $\delta s/\delta t$ and what are its units?
(ii) What quantity does the gradient represent?
(iii) What is the position of the object at the start, i.e. when $t = 0$?
(iv) How long before the object is 5 m away?
(v)   Describe fully the motion depicted.
(vi) Write the equation which describes the motion depicted.
(vii) What is the *solution* of the equation and what does it mean?
(viii) How long would it take for the object to arrive at a point 72 m in the opposite direction? Explain your result.

Solutions

(i) $\delta s/\delta t = -8/5 = -1.6\,\mathrm{m\,s^{-1}}$.
(ii) The gradient is a change in distance with respect to time, which is speed.

(iii) It is 8 m away from the point of reference (origin, on the graph).
(iv) Approximately 1.9 s.
(v) It is the motion of an approaching object moving at constant speed from a position 8 m away. The speed is constant because the gradient is a constant.
(vi) Comparing with the general linear function, $m = -1.6$, $c = 8$; so: $s = -1.6t + 8$
(vii) The solution is the value of t when $s = 0$. It is 5. In this case it is the time it takes the object to arrive.

(viii) $s = -1.6t + 8$ so $t = \dfrac{8 - s}{1.6} = \dfrac{8 - (-72)}{1.6} = 50\,s$

In 50 s the object would have travelled a total distance of 80 m. Having approached from 8 m away it will have moved a further 72 m in the opposite ($-$ve) direction.

That was a fairly thorough treatment of a simple linear function. It emphasized the important aspects of graphs.

There are many other relationships in engineering which are described by linear functions: stress/strain within the limit of elasticity, voltage/current in a resistance, e.m.f./velocity of a conductor in a uniform magnetic field, and so on.

Let's look back at simultaneous linear equations.

Simultaneous linear equations revisited

In Chapter 4, we examined two methods that are used to solve linear equations in two unknowns. What were these methods?

Substitution and elimination methods

They are both analytical methods for finding a solution that satisfies two equations that link the same two variables. Figure 6.6 shows what a simultaneous solution looks like on a graph:

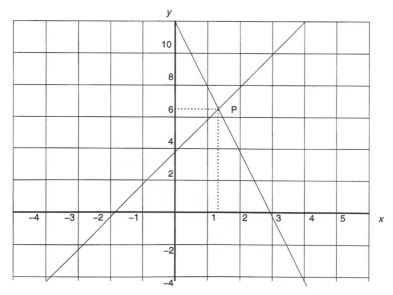

Figure 6.6 Simultaneous solution

PRACTICE EXERCISE 6.4

(i)   Find the equations of the two graphs plotted in Fig. 6.6.
(ii) Find the solution that satisfies the two equations.

Solutions

(i) $y = -4x + 12$
      $y = 2x + 4$
(ii) At point P the values of x and y are common to both lines so these are the values that simultaneously satisfy both equations: $x = 1.3$, $y = 6.6$.
      Check: $-4(1.3) + 12 = 6.8$
      $\qquad\quad 2(1.3) + 4 = 6.6$

The results are only approximately correct because of the resolution error introduced when reading off graphs. The solution which we obtained by one of the analytical methods was $x = 1.33$, $y = 6.67$, correct to three significant figures.

What could be said about linear equations that had the same gradient? Sketch the following **family of graphs** and see what you think.

$$y = 2x - 2, \qquad y = 2x - 1, \qquad y = 2x,$$
$$y = 2x + 1, \qquad y = 2x + 2, \qquad y = 2x + 3$$

They have no solutions. These equations are said to be **inconsistent**.

Quadratic functions, $f(x) = ax^2 + bx + c$

This is the first important group of **non-linear** functions (Fig. 6.7).

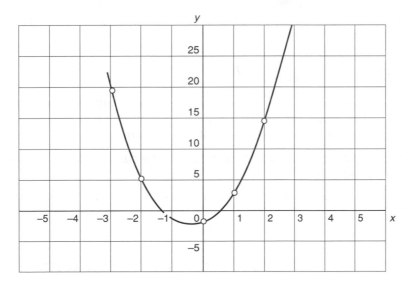

Figure 6.7 A non-linear function

PRACTICE EXERCISE 6.5

Plot, on a sheet of graph paper, the graph of $y = 3x^2 + 2x - 2$ for values of x between -3 and 3. From the graph, determine the approximate solution of x.

Solution

Table of values:

x	-3	-2	-1	0	1	2	3
$y = f(x)$	19	6	-1	-2	3	14	31

Compare your graph with the one plotted in Fig. 6.7.

The solution of an equation is the value of x when $y = 0$. In this case it is the value of x when $3x^2 + 2x - 2 = 0$. This occurs where the graph crosses the x-axis.

I make: $x = -1.2, 0.5$ approximately.

So that deals with graphical solutions of quadratic functions. What about the important features of a quadratic curve? How does it compare with a linear graph? The most important thing to remember is that:

The *gradient* of any non-linear function is not a constant, it is a *variable*.

You should be able to judge this for yourself by looking at the graph you have plotted. To the left of the **vertex** (also called **turning point**) the gradient is steeply negative; this gradient reduces to zero at the vertex and then increases, becoming steeply positive to the right of the vertex. At any point of a quadratic graph, the gradient is uniquely different.

The constants of a parabola

A **parabola** is the name given to the shape of the graph of a quadratic function. Unlike the linear graph it has no constant like m which defines a fixed value of $\delta y / \delta x$ but it does have the constants a, b and c.

a – Increasing values increase the steepness of the curve and $-$ve values invert the parabola (Fig. 6.8)

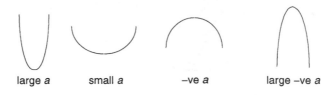

large *a*　　　small *a*　　　$-$ve *a*　　　large $-$ve *a*

Figure 6.8 Various *a* values

b – Shifts the curve to left or right (Fig. 6.9)

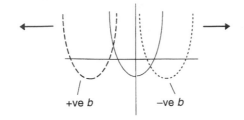

+ve b −ve b

Figure 6.9 Various b values

c – This is our old friend, intercept on the y-axis. Different values produce a family of curves shifted up and down the y-axis (Fig. 6.10)

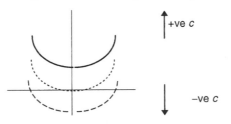

+ve c

−ve c

Figure 6.10 Various c values

The gradient of a parabola

Gradient has a special significance because it represents the rate of change of a function. In Practice Exercise 6.3 you found the gradient of the linear function by taking $\delta s/\delta t$ at any point on the graph. You found its value to be $1.6\,\mathrm{m\,s^{-1}}$. It was the rate of change of distance with respect to time which we call **speed**.

With a parabola (and other non-linear graphs) we can estimate the rate of change by sketching a tangent to the curve and measuring δy and δx at selected points. Take the graph of $y = 2x^2$ (Fig. 6.11):

When $x = 0$, $\delta y/\delta x = 0/\delta x = 0$

When $x = -1$, $\delta y/\delta x = -4/1 = -4$

When $x = 1$, $\delta y/\delta x = 4/1 = 4$

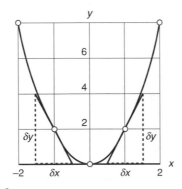

Figure 6.11 Graph of $y = 2x^2$

To find the rate of change of a non-linear function from the gradient of its graph is difficult and produces results that are inaccurate. **Differential calculus** is an analytical method that enables us to find rates of change precisely. We will use it in the next chapter. In the meantime, why the fuss about rates of change? The next exercise should explain.

PRACTICE EXERCISE 6.6

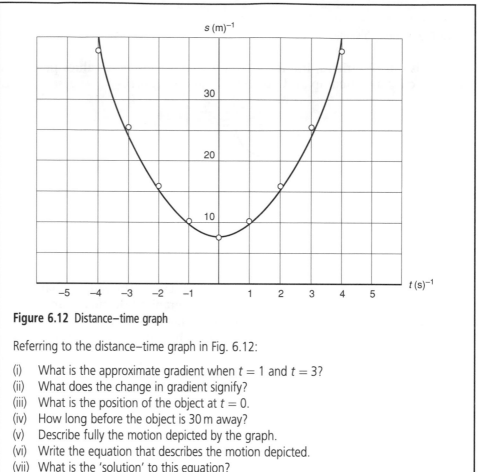

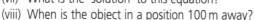
Figure 6.12 Distance–time graph

Referring to the distance–time graph in Fig. 6.12:

(i)   What is the approximate gradient when $t = 1$ and $t = 3$?
(ii)   What does the change in gradient signify?
(iii) What is the position of the object at $t = 0$.
(iv)   How long before the object is 30 m away?
(v)   Describe fully the motion depicted by the graph.
(vi)   Write the equation that describes the motion depicted.
(vii) What is the 'solution' to this equation?
(viii) When is the object in a position 100 m away?

Solution

(i)   At $t = 1$, $\delta s/\delta t = 4$, at $t = 3$, $\delta s/\delta t = 12$.
(ii)   Since the gradient is changing, speed is changing so the object is accelerating.
(iii) At $t = 0$, the object is 8 m away.
(iv)   The object will be 30 m away in approximately 3.4 s.
(v)   The motion is of an object that approaches an observer, passes her at a distance of 8 m and then moves away. The sort of thing you would witness if you watched a passing train or watching a bungee diver falling towards the ground.

(vi) The graph is a parabola so its equation has the general form of a quadratic,

$$s = at^2 + bt + c$$

The intercept on the y-axis is 8 so $c = 8$. The graph is symmetrical about the y-axis so $b = 0$, giving $s = at^2 + 8$. When $t = 1$, $s = 10$; so: $10 = a + 8$ and $a = 2$. So the particular form of this equation is:

$$s = 2t^2 + 8$$

(vii) The equation of this graph has no 'real' solution since the graph does not cross the x-axis. Specifically, there is no value for t when $s = 0$.

(viii) $s = 2t^2 + 8$ so $t = \pm\sqrt{\dfrac{s-8}{2}} = \sqrt{\dfrac{100-8}{2}} = \pm 6.78\,s$

The most important point to emerge from Practice Exercise 6.6 is the fact that the gradient of a graph represents another quantity. Exercise 6.6 and Exercise 6.3 both show that the gradient,

$$\frac{\delta s}{\delta t} = v\,\mathrm{m\,s}^{-1}$$

The rate of change of position with respect to time is **velocity**. The difference between Practice Exercise 6.3 and 6.6 is that in Exercise 6.3 the motion was at a constant velocity because the gradient of the distance–time graph was constant. In Practice Exercise 6.6, the velocity is changing because the gradient is changing so the graph depicts the motion of an object accelerating.

Plotting a graph of the *rates of change* of the distance–time graph in Fig. 6.12 gives us the speed–time graph (Fig. 6.13).

This turns out to be a linear function, i.e. $\delta v / \delta t$, *the rate of change of velocity with respect to time, is constant.*

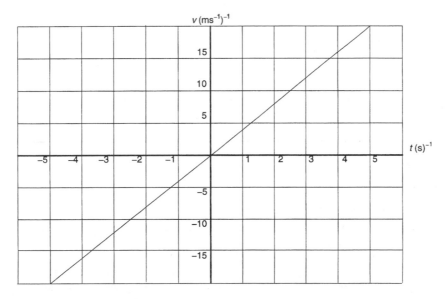

Figure 6.13 Speed–time graph

PRACTICE EXERCISE 6.7

> (i)   What quantity is $\delta v/\delta t$ in Fig. 6.13?
> (ii) Determine its value from the graph above.
> (iii) What else can be said about the motion of this object?

Solution

> (i) $\delta v/\delta t$ is acceleration in units of $\mathrm{m\,s}^{-2}$.
> (ii) From the graph $\delta v/\delta t = 20/5 = 4\,\mathrm{m\,s}^{-2}$.
> (iii) The acceleration of the object is constant.

So quadratic functions are those whose rates of change (the gradients of their graphs) are not constant. The rate of change of a function is important because it represents another quantity. In the case of linear displacement (s) with respect to time (t):

$$\frac{\delta s}{\delta t} = v \quad \text{and} \quad \frac{\delta v}{\delta t} = a$$

Simultaneous linear–quadratic equations

We have seen how to solve simultaneous linear equations graphically. Now look how we solve a pair of simultaneous equations when one is linear and the other quadratic.

PRACTICE EXERCISE 6.8

> Plot on graph paper, against the same axes, the graphs of
>
> $$y = -x^2 + 3x + 4$$
>
> $$y = \tfrac{1}{2}x - 4$$
>
> for values of x between -2 and 5. From the graph determine the solution(s) that simultaneously satisfy both equations. Use back-substitution to check your solutions.

Solutions

> The first step is to construct a table of values:
>
> | x | -2 | -1 | 0 | 1 | 2 | 3 | 4 | 5 |
> |---|---|---|---|---|---|---|---|---|
> | $-x^2 + 3x + 4$ | -6 | 0 | 4 | 6 | 6 | 4 | 0 | -6 |
> | $\tfrac{1}{2}x - 4$ | -5 | | -4 | | | | -2 | |
>
> Note that the second equation is linear so only two values are needed; three to be sure.

There are two points at which the graphs cross, so both offer alternative solutions since at both points the corresponding values of (x, y) satisfy the two equations. I make,

$$x = -1.9, \quad y = -4.9$$
$$x = 4.2, \quad y = -1.8$$

Check: substituting into the linear equation (because it is easier),

$$\tfrac{1}{2}(-1.9) - 4 = -4.95$$
$$\tfrac{1}{2}(4.2) - 4 = -1.9$$

The check indicates that my results are roughly correct. The actual values that I obtained analytically are $x = -1.84$, $y = -4.92$ and $x = 4.34$, $y = -1.83$. So the graphical solution is accurate to within ±0.1, which is $\pm10\%$ of the values obtained. Only by using large scales and very careful plotting can we get better accuracies.

The solution of simultaneous equations is not to be confused with the solution of a quadratic equation that finds the roots of that equation. What are the roots of the quadratic in Practice Exercise 6.8?

$$x = -1, 4$$

The solution of a quadratic equation is the value of x when $y = 0$. So graphically, this is the value of x where the graph crosses the x-axis.

Before we leave polynomials (linear and quadratic functions are polynomial) there is one other function that is of interest.

The 1/x function, $f(x) = ax^{-1}$

This, as you know, can be written as $f(x) = \dfrac{a}{x}$; its curve is the **rectangular hyperbola** (see Fig. 6.14).

The interesting conditions occur when $x = 0$ and when the value of x is such that $y = 0$. Think about them. A third condition, which is marked by dotted

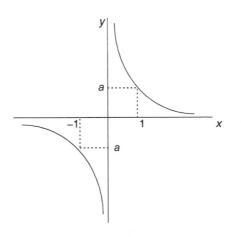

Figure 6.14 $f(x) = a/x$, the rectangular hyperbola

lines, is when $x = 1$; you can see that $y = a$. Negative values of x give a mirror image. An important engineering relationship which gives us a graph like this, is that of frequency and periodic time. What is it?

$$f = \frac{1}{T}$$

Think about the three special conditions. For example, what would you expect the frequency to be when periodic time tends to zero?

Exponential function, $f(x) = A\,e^{kx}$

This is the non-linear function that you first met in algebra so we will not spend too long on this, except to emphasize its most important feature.

> **For any value of x the rate of change of an exponential function is proportional to the value of x.**

To get a clear picture of what this means, work carefully through the next exercise (see Fig. 6.15).

PRACTICE EXERCISE 6.9

On a sheet of graph paper, carefully plot a graph of $y = e^x$ for x between -1 and 3 at intervals of 0.5. Then answer the following questions.

(i)   What is the value of the coefficient A in this equation? How can you tell its value from the graph?
(ii) What is the value of the coefficient k in this equation? How does it affect the curve?
(iii) By drawing tangents to the curve at $x = 0.5$, $x = 2$ and $x = 2.75$ determine the values of $\delta y/\delta x$ at those points. What do these values tell you about the rate of change of an exponential function?

Solutions

x	−1.0	−0.5	0	0.5	1.0	1.5	2.0	2.5	3.0
e^x	0.4	0.6	1.0	1.7	2.7	4.5	7.4	12.2	20.1

(i)   The coefficient of e is 1. Once again, this is the intercept on the y-axis. It is the point where the curve crosses the y-axis.
(ii) The coefficient of x is also 1. A positive value makes this a growth curve. If k is greater than 1 the steepness of the curve is increased.
(iii) At:

$$x = 0.5 \qquad \frac{\delta y}{\delta x} \approx 1.8$$

$$x = 2 \qquad \frac{\delta y}{\delta x} \approx 7.5$$

$$x = 2.75 \qquad \frac{\delta y}{\delta x} \approx 16$$

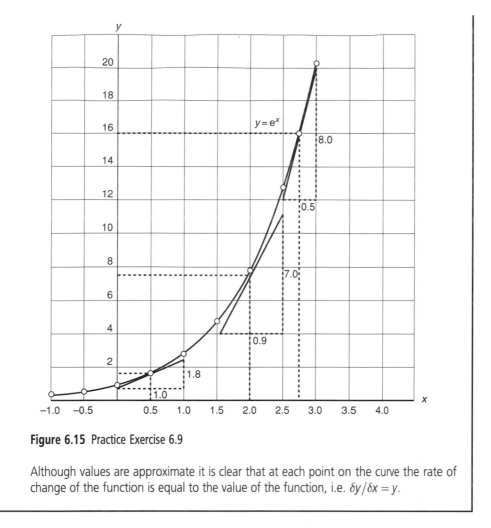

Figure 6.15 Practice Exercise 6.9

Although values are approximate it is clear that at each point on the curve the rate of change of the function is equal to the value of the function, i.e. $\delta y/\delta x = y$.

Your answer to the last part of Practice Exercise 6.9 highlights the most important and unique feature of exponentials. In this example the rate of change is equal to the function but if the coefficients A and k are anything other than 1 then the rate of change is proportional to the function. For example, if k was 2 then the rate of change would be twice the value of the function.

You may remember that there are three types of exponential function. Figure 6.16 summarizes the points of interest.

You need to remember that exponential functions can be to any base, although base 10 or base e are the ones that you are most likely to deal with. Of the two, base e is the more important because of the way its rate of change is linked to the value of the function.

Logarithmic function, $f(x) = A \ln x$

Logarithms are the inverse function of exponentials. Since we used base e to illustrate graphs of exponentials we should stick to base e for logs; in other words, natural logs (ln) (see Fig. 6.17).

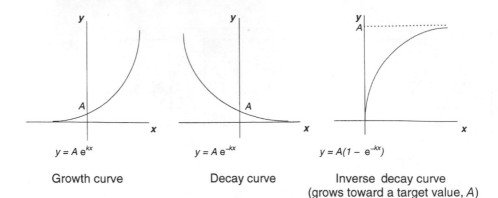

$$y = A\,e^{kx}$$

$$y = A\,e^{-kx}$$

$$y = A(1 - e^{-kx})$$

Growth curve   Decay curve   Inverse decay curve
(grows toward a target value, A)

Figure 6.16 Three types of exponential function

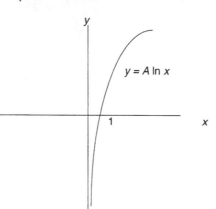

$$y = A \ln x$$

Figure 6.17 Graph of $y = A \ln x$

The distinctive features of a logarithmic graph (to any base) are:

- the log of a number larger than 1 is positive,
- the log of a number less than 1 is negative
- $\log 1 = 0$
- logs of negative numbers do not exist.

Trigonometric function, $f(t) = c + A \sin(\omega t + \theta)$

In Chapter 5, on trigonometry, you saw the expression $A \sin(\omega t + \theta)$ used to represent a phasor. Now, look carefully at the graph of the phasor as it rotates through its full cycle (Fig. 6.18).

If you have not seen one of these diagrams before it may look rather confusing but be patient; we can break it down to things which are familiar. Figure 6.18 models the function, $a = \sin(\omega t)$ in two different ways.

- On the left is the model of a rotating phasor $1\angle 0°$. It is not a polar plot. It simply shows the position of a phasor against a set of Cartesian coordinates as it rotates from 0 to 360°.
- The angular velocity of rotation is $\omega\,\text{rad}\,\text{s}^{-1}$. Multiplying velocity by time gives distance; in this case the distance is **angular displacement**, ωt.

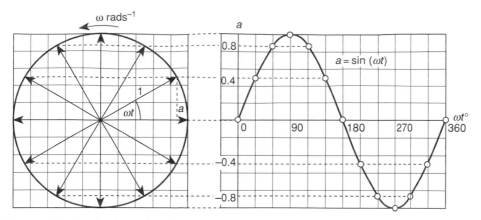

Figure 6.18 Graph of $a = \sin(\omega t)$

- 1, is the length of the hypotenuse of a right-angle triangle and a is the side opposite the angle ωt. Multiplying the sine of the angle by the hypotenuse gives the side opposite. This is where we get the equation $a = 1\sin(\omega t)$.
- The position of the rotating phasor is shown at intervals of 30°: $\omega t = 0°$, $\omega t = 30°$, $\omega t = 60°$ and so on. Degrees are used instead of radians to give a clearer picture. Normally ωt would be in radians.
- On the right is an ordinary Cartesian graph of the function $a = \sin(\omega t)$ whose value is plotted at the same 30° intervals of ωt.
- When ωt is 0°, $a = 0$, when $\omega t = 30°$, $a = 0.5\ldots$ So the value of a on the graph corresponds to the length of the opposite side in the triangle formed by the rotating phasor.
- Joining up the points on the graph produces a graph which we call a **sinusoid**. The graph traces out the shape, or **waveform** of a sine wave.
- The sine wave is very special. It is the most important of all waveforms. It is fundamental to all aspects of electrical and mechanical engineering; indeed, it is fundamental to all things that have **periodicity**.

The constants of a sinusoid

The phasor diagram and graph above represent the equation:

$$a = f(t) = \sin(\omega t)$$

Compare this with the general form:

$$a = c + A\sin(\omega t + \theta)$$

If $a = f(t)$, t is the independent variable and a is the dependent variable. This is why a is often called the **instantaneous value**; it is the value of the sinusoid at some instant in time, t. Now look at the graph in Fig. 6.18 and evaluate the constants c, A, θ and ω.

$$c = 0 \qquad A = 1 \qquad \theta = 0 \qquad \omega \text{ is not known}$$

The first three should present you with no difficulty but ω deserves some explanation. Normally it is the **angular velocity** (sometimes called **angular frequency**) and it depends on frequency.

$$\omega = 2\pi f \text{ rad s}^{-1}$$

In this case we don't know its value because we made the product ωt equal to $0°$, $30°$, $60°$, etc... for convenience.

So what effect do the other three constants, c, A and θ, have on the graph of a sinusoid? The following example should make this clear.

EXAMPLE 6.2

Sketch and label, against the axes provided, graphs of

(i) $r_1 = 10\sin(\omega t)$
(ii) $r_2 = 10\sin(\omega t + \frac{\pi}{3})$
(iii) $r_3 = 10 + 10\sin(\omega t)$

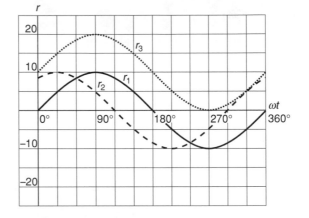

Figure 6.19 Example 6.2

---

c is a y-shift which is equivalent to the intercept on the y-axis (but not the same). It is, in fact, the **average value** of r which has the effect of shifting the curve up or down the y-axis.

A is the **amplitude** of the sinusoid; the **peak value**, positive or negative. In this example it is 10.

θ is the **phase angle**. This is a horizontal offset which has the effect of shifting the curve left (positive phase shift) or right (negative phase shift) along the x-axis. It is equivalent to the intercept on the x-axis.

Other trigonometric functions

$$f(t) = c + A\cos(\omega t + \theta)$$

This is equivalent to a sine function with $\theta = 90°$. In other words, its graph is a sine wave phase advanced by $90°$. Verify this for yourself, look at the sketch (Fig. 6.20) of the graph of:

$$r = 20\cos(\omega t)$$

The effects of the constants with the cosine function are exactly the same as they are with the sine function.

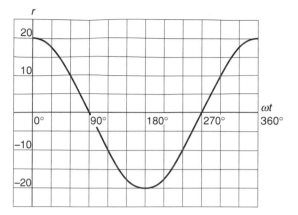

Figure 6.20 $r = 20\cos(\omega t)$

$$f(t) = c + A\tan(\omega t + \theta)$$

Figure 6.21 shows the graph of:

$$r = 10\tan(\omega t)$$

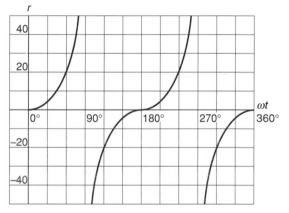

Figure 6.21 $r = 10\tan(\omega t)$

The distinct shape of the tangent curve is best explained by the trig. identity:

$$\tan A \equiv \frac{\sin A}{\cos A}$$

In the first quadrant, as the angle A approaches 90°, $\cos A$ approaches 0 so $\tan A$ tends to infinity. Why should the tangent of an angle slightly above 90° be close to negative infinity?

Because in the 2nd quadrant the cosine is negative

Area under a curve

You should, by now, understand the idea that constants determine the shape of the graph of a function. In addition, the rate of change of a function can be found from the gradient of the graph, $\delta y / \delta x$. With linear functions the rate of

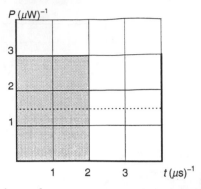

Figure 6.22 Power in a pulsed waveform

change is another constant but with non-linear functions the rate of change is a variable. We will return to rates of change when we start to look at differential calculus. In the meantime we need to take a look at areas under graphs.

Figure 6.22 is a graph of the power in a pulsed waveform of the type we get when we transmit pulses of light in fibre-optic cable. The pulse has an ON time of 2 μs and an OFF time of 2 μs so its periodic time is 4 μs. Suppose we wish to know the average power of the pulses. Multiplying the width of the pulse by its height gives us the area:

$$2 \times 10^{-6} \times 3 \times 10^{-6} = 6 \times 10^{-12}$$

This is, in fact, 6 pJ of energy because we have multiplied power by time. Now, since the length of one cycle is 4 μs, taking the total energy and dividing by the length of the cycle gives:

$$\frac{6 \times 10^{-12}}{4 \times 10^{-6}} = 1.5\,\mu W$$

The result is in watts because we have divided energy by time. 1.5 μW is in fact the average power of the waveform shown. The example shows that for any function:

$$\text{average} = \frac{\text{area under the graph}}{\text{length of base}}$$

When a graph is rectangular, finding the average is quite straightforward; but let's return to the velocity–time graph we considered in Exercise 6.7 (see Fig. 6.23).

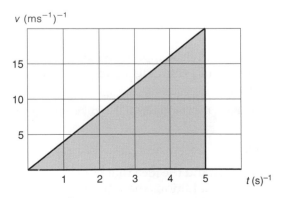

Figure 6.23 Velocity–time graph

PRACTICE EXERCISE 6.10

Find the average velocity of the object whose motion is depicted by the graph in Fig. 6.23, over the interval between $t = 0$ and $t = 5$.

Solution

The area under the graph over the specified interval is just a triangle. So,

$$A = \tfrac{1}{2}bh = \tfrac{1}{2} \times 5 \times 20 = 50\,\text{m}$$

$$V_{av} = \tfrac{50}{5} = 10\,\text{ms}^{-1}$$

So, with any linear graph, we can find the area under the graph by applying geometry or trigonometry. But what happens when the graph is non-linear? Consider the graph of a sinusoidal current whose equation is:

$$i = \sin(100\pi t)$$

Don't be put off; here we have $c = 0$, $\theta = 0$, $A = 1$ and $\omega = 100\pi$, which is the angular velocity of current supplied at 50 Hz (remember that $\omega = 2\pi f$). Look at the plot of the graph (Fig. 6.24) of this current over the interval $t = 0$ to $t = 10\,\text{m s}$.

t (ms)$^{-1}$	0	1	2	3	4	5	6	7	8	9	10
i (A)$^{-1}$	0	0.31	0.59	0.81	0.95	1.0	0.95	0.81	0.59	0.31	0

Note that since the calculations involved the direct use of angles we must work in radians. Plotting the graph of this current gives the curve shown in Fig. 6.24.

The interval chosen covers the time of just a half cycle of the sine wave, which suits our purpose. Now, to find the average current over this half cycle we need the area under the curve. Because the curve is non-linear we cannot use geometry or trigonometry to get an exact value but we can approximate: we choose to apply the **mid-ordinate rule**.

Mid-ordinate rule

The following is the procedure:

Divide the area into equal strips b

Measure the height of the middle of each strip $y_1, y_2, y_3, \ldots$ These are the *mid-ordinates*.

The approximate area is the sum of the areas of the strips.

So, area:

$$A = by_1 + by_2 + by_3 + \cdots$$
$$= b(y_1 + y_2 + y_3 + \cdots)$$

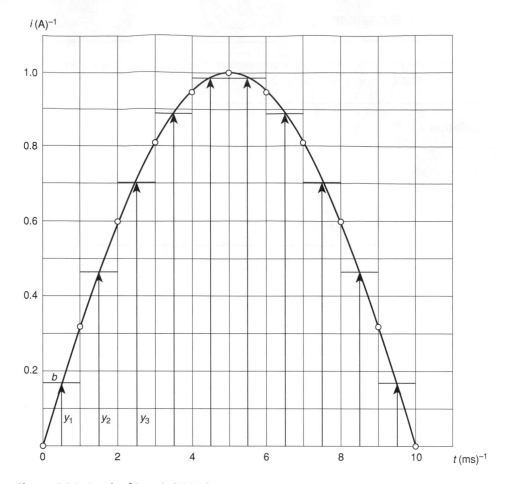

Figure 6.24 Graph of i = sin(100πt)

Stated in words:

   area = width of each strip × the sum of the mid-ordinates

Look back at Fig. 6.24, choosing 0.5, 1.5, 2.5 ... as the mid-ordinates. We can calculate their values using the equation of the curve (this is more accurate than measuring them off the graph); and the width of each strip $b = 1 \times 10^{-3}$.

$$A = 1 \times 10^{-3}(0.1564 + 0.4540 + 0.7071 + 0.8910 + 0.9877$$

$$+ 0.9877 + 0.9810 + 0.7071 + 0.4540 + 0.1564)$$

$$= 1 \times 10^{-3}(6.3924)$$

$$= 6.39 \times 10^{-3}$$

To get the average current we divide by the length of base under the graph.

$$I_{av} = \frac{6.39 \times 10^{-3}}{10 \times 10^{-3}} \approx 0.639 \text{ A}$$

The result is approximate because the mid-ordinate rule is a numerical approximation method. The exact value, using **integral calculus**, gives $2/\pi$ which is 0.637 amperes correct to three significant figures. How could the accuracy of the mid-ordinate rule be increased?

By dividing the area into more strips, the more the better

The trouble is that improving accuracy in this way increases computational effort.

The example illustrated involved a half-cycle of a sinusoidal current and we found its average value to be approximately 0.637 of the peak value. What would be the average value of a full cycle?

The average value of a full cycle of a pure sinusoid is zero

This is because the area under the second half-cycle is *negative* and equal to the area under the first half-cycle. The only time an average value of a sine wave makes any practical sense is when we deal with half-cycles.

The process of finding the area under the graph of a function is called **integration**. The method which you have seen can be applied numerically or graphically (by measuring lengths from a graph). There are two other numerical methods, which are dealt with in Chapter 10: they are the **trapezoidal rule** and **Simpson's rule**. However, you must understand that numerical integration can only provide approximate solutions. The only way of finding the exact area under a curve is to apply **integral calculus**. You must wait until Chapter 8 to see how that works.

6.3   Converting functions to a linear form

The study of graphs, so far, has assumed that we know the nature of the function before a graph is constructed. In other words we have an equation we can use to compute graphical data; if x equals some value, we can use the equation to calculate the corresponding value of y. A set of values assembled in a table can be used to plot the points of a graph that represents the known function.

We will finish this chapter by tackling non-linear relationships whose functions are not known. For example, a practical study or experiment may yield data for a pair of variables which are linked by some unknown function (or, in fact, they may not be linked at all!).

EXAMPLE 6.3

Readings taken in a practical investigation of two quantities x and y produced the following values:

x	5	10	15	20	25
y	6.25	10.00	16.25	25.00	36.25

The investigators suspect that:

$$y \propto x^2$$

in other words the $f(x)$ is possibly a quadratic so that the general form of the equation linking x and y is

$$y = ax^2 + b$$

Your task is to establish whether the relationship between the variables is indeed a quadratic and if so, to determine the values of the constants a and b to find the particular equation which links x and y.

Solution

Let $z = x^2$. This will give a linear relationship in the form of: $y = az + b$.

Draw up a new table of values linking z and y.

x	5	10	15	20	25
z	25	100	225	400	625
y	6.25	10.00	16.25	25.00	36.25

Plot the graph of $y = az + b$

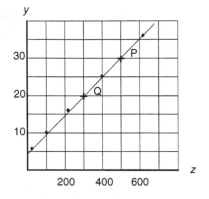

The graph of $y = az + b$ is a straight line so this confirms that the law connecting x and y must be of the form $y = ax^2 + b$.

To find the values of the constants pick two points on the line P and Q. Do not be tempted to use pairs of original data – your graph, if correctly drawn, will tend to iron out experimental error contained in the raw data.

Now we have two equations in two unknowns which can be solved simultaneously:

At P $\quad 30 = a(500) + b$

At Q $\quad 20 = a(300) + b$

Subtracting Q from P gives

$$10 = 200a$$

$$a = 0.05$$

Hence,

$$b = 20 - 300(0.05) = 5$$

Having obtained the values of the constants we can now write the particular equation for x and y:

$$y = 0.05x^2 + 5$$

Example 6.3 shows that it is possible to formulate an equation from a set of data by linearizing the data, constructing a linear graph and using this graph to obtain the values of the constants.

You may have noticed that the value of a could have been obtained by taking the gradient of the graph $\delta y/\delta z$ and the value of c is the intercept on the y-axis. However, if the origin of the graph is not zero this would not work. The method of selecting two points to write simultaneous equations is one that works in all cases, so that is the recommended procedure.

Linearizing an exponential

You have seen that an exponential function is one in which the variable is the exponent (power) of a base and the base is usually e.

Given the general form of an exponential:

$$y = A\,e^{kx}$$

it is possible to convert this to a linear $(y = mx + c)$ form by taking logs:

$$\log y = kx\log e + \log A$$

and if we take logs to the base of e, $\log_e e = 1$ so:

$$\log_e y = kx + \log_e A$$
$$\underset{y}{\diagup}\quad \underset{m}{\diagup}\ \underset{x}{\diagdown}\quad \underset{c}{\diagdown}$$

If the relationship between x and y has been correctly assessed, this equation will produce a linear graph from which it is possible to establish the values of the constants k and A in a manner similar to that used in Example 6.3.

PRACTICE EXERCISE 6.11

A set of measurements of the charging current in a capacitive circuit was taken at intervals of 10 s. It is thought that the relationship between current and time is an exponential of the form:

$$i = I\,e^{kt}$$

Use the following table of results to confirm this relationship and hence determine the values of the constants I and k.

t (seconds)	10	20	30	40	50
i (amperes)	26.2	21.5	17.6	12.4	11.8

Solution

Taking logs:

$$\ln i = kt + \ln I$$

should produce a linear graph if we are dealing with an exponential and if the graph is of the $\log i$ against t:

t	10	20	30	40	50
$\ln i$	3.266	3.068	2.868	2.518	2.468

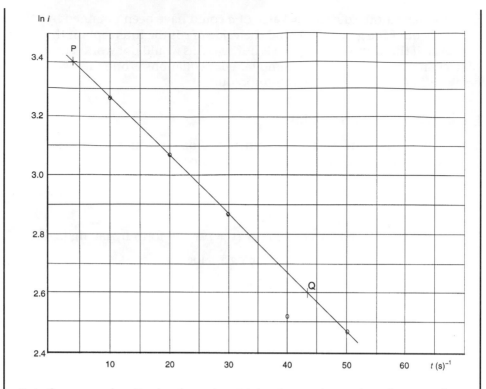

Note the rogue value. Graphs of experimental data frequently contain such rogue values: such values should be shown but ignored when the graph is plotted.

The graph follows the straight line law confirming that the original function is an exponential.

Taking two points on the graph P and Q gives:

$$3.40 = k(3) + \ln I$$
$$2.57 = k(45) + \ln I$$

Subtracting,

$$0.83 = -42k$$
$$k = -20.0 \times 10^{-3}$$

Substituting,

$$3.40 = -0.02(3) + \ln I$$
$$\ln I = 3.40 + 0.06 = 3.46$$
$$I = e^{3.46} = 31.8$$

Correct to two significant figures the constants are $k = -0.020$ and $I = 32$, so the required equation is:

$$i = 32 \, e^{-0.02t}$$

How would you check this equation?

Put in a value from the original set of data and see if
the equation generates the correct corresponding value.

Putting 10 for t in the equation of Exercise 6.11 gives:

$$32\,e^{-0.02(10)} = 26.2$$

Now, revise by tackling the following problems.

PRACTICE EXERCISE 6.12 (REVISION)

(a) Which of the following could be the graph of $y = \dfrac{x(x-4)}{3}$

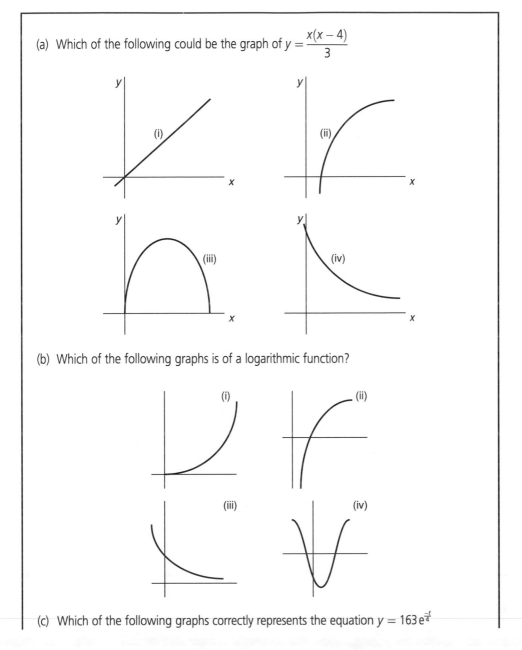

(b) Which of the following graphs is of a logarithmic function?

(c) Which of the following graphs correctly represents the equation $y = 163\,e^{\frac{-t}{4}}$

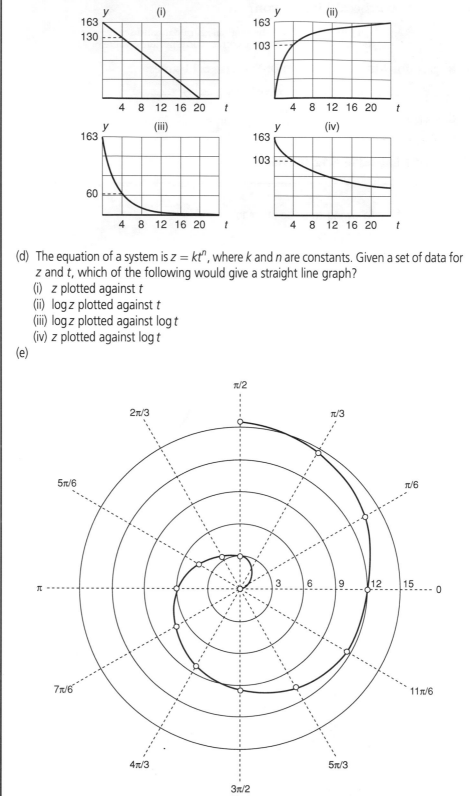

(d) The equation of a system is $z = kt^n$, where k and n are constants. Given a set of data for z and t, which of the following would give a straight line graph?
(i) z plotted against t
(ii) $\log z$ plotted against t
(iii) $\log z$ plotted against $\log t$
(iv) z plotted against $\log t$

(e)

The equation of the polar graph shown is:

(i) $r = 2\theta$

(ii) $r = e^{2\theta}$

(iii) $r = 2\ln\theta$

(iv) $r = 2\log\theta$

Solutions

(a) (iii); (b) (ii); (c) (iii); (d) (iii); (e) (i)

Assignment V

Graphical methods

This assignment requires you to use graphs to model engineering situations and solve engineering problems. In carrying out this assignment, there are certain minimum performance criteria which you should meet. They are listed below.

- Data are represented graphically.
- Functions and graphs are used to solve engineering problems with correct units and indication of error.
- A trigonometric function is represented using a rotating vector.
- Data are reduced to a linear form.
- Areas under graphs of functions are interpreted to solve engineering problems.

If, in the end, your assignment does not meet these criteria you should go back, revise what you have studied in trigonometry and tackle the problem(s) again.

In addition to these performance criteria, remember the four themes which should run through the assignments that you do.

Planning
Formulate the correct algebraic equations and use them to generate tables of values. Start with sketches. Plan your graphs so that they are drawn to a scale and size that focuses on the area of interest. At the same time, the graphs should be of a sensible size and positioned centrally on the page.

Information seeking and handling
Select and use the given information. Work and give answers to an appropriate degree of accuracy. Numerical values should be assigned correct SI units but the scales of graphs must be dimensionless.

Evaluation
Apply checking procedures to ensure your results are correct. You should also consider whether your solution makes sense within the problem situation.

Your evaluation of the solution to the sports field problem should involve some discussion which is supported by sketches. Your solution to this problem should also give an indication of error.

Outcomes
Your product should be correct, accurate and professionally presented.

This assignment will take a little of your time, make the most of it.

1. A SPORTS FIELD

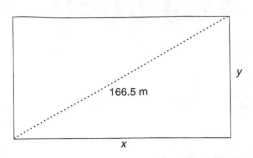

The outer perimeter of a rectangular sports field is required to be 460 m. The diagonal distance between the corners of the field is to be 166.5 m. Now the perimeter is just $2x + 2y = 460$ which simplifies to

$$x + y = 230$$

and the diagonal forms the hypotenuse to the sides x and y so:

$$x^2 + y^2 = 166.5^2$$

Plot carefully and accurately, against the same set of axes, the graphs of these two equations. Use the graphs to find the dimensions of the sports field, x and y.

Your solution should be stated using correct units and an indication of error.

Some notes about graph plotting:
- If you choose to use a software package to generate graphs you should make sure that it will give you results that are satisfactory, bearing in mind the points which follow.
- Start with a table of values showing the equation(s) used.
- Choose scales that are as large as possible but sensible for the size of the graph paper used.
- Plot only those values which are of interest – this often means that the origin of an axis is not zero. Selecting which values to plot is often a matter of trial and error.
- The points of a graph should be very small: dots within circles (the circles are there to draw our attention to the dots) or very small crosses.
- Join up the points with a thin, smooth curve. Never zigzag.
- Make it a best fit curve. It should average out any errors so there should be an equal number of points either side.
- Ignore rogue values. Values which are clearly outside the main trend should be plotted but ignored when drawing the curve.
- Label the graphs, either by name or by the equation which they represent.
- Label the axes. They should be calibrated in numbers so divide the dimension of an axis by its unit. For example, a time axis in units of seconds should be labelled as t/s or $t \ (s)^{-1}$.
- Clearly show the points of interest, with the aid of dotted lines, if necessary.

2. A SINUSOID

The following equation gives voltage v as a function of time t in seconds.

$$v = 3\sin\left(100\pi t + \frac{\pi}{4}\right) \text{ volts}$$

(a) Plot the graph of this function between $t = -2.5\,\text{ms}$ and $t = 17.5\,\text{ms}$ at intervals of 1.25 ms.

(b) Draw and label the phasor of this function showing its angular velocity, amplitude and phase angle at time $t = 0$. Position the phasor so that it relates to the graph and is of the same scale.

(c) Plot, using the same set of axes, the graph of the square of this function; this is the graph of v^2.

(d) Use the mid-ordinate rule to find the approximate area under the v^2 curve and then calculate the mean of v^2.

(e) Hence find the root-mean-square value of the voltage using $V_{\text{rms}} = \sqrt{\overline{v^2}}$.

Note

You can save yourself a considerable amount of time in applying the mid-ordinate rule if you recognize that the area under the v^2 waveform consists of four identical parts.

3. A FACTORY LOAD

The power and the current supplied to a factory load is monitored under various conditions in order to confirm that the load is resistive and obeys the general law:

$$P = I^n R$$

The data are assembled in the following table.

I (A)	25	28	36	41	46	49
P (kW)	7.4	9.5	15.6	20.1	25.5	28.9

Confirm the law by converting it to a suitable form and constructing a graph which gives a linear relationship, and then use the graph to determine the particular law which links P and I.

chapter

7 Differential calculus

You now have a good idea about rates of change and average values of functions. You know how to evaluate them graphically. Let's recap.

- The approximate rate of change of a function can be found by measuring the gradient of the graph of the function.
- The gradient of a graph is a 'small change in y over a small change in x', $\delta y / \delta x$, which was first introduced as 'rise over run'.
- $\delta y / \delta x$ can be positive or negative.
- $\delta y / \delta x$ of linear functions is a constant.
- $\delta y / \delta x$ of non-linear functions is a variable.
- If a function describes the link between two physical quantities, its rate of change is another physical quantity. The rate of change of distance with respect to time is speed. The rate of change of speed with respect to time is acceleration. The rate of change of electric charge over time is current.
- The approximate average value of a function can be found by taking the area under the graph of the function and dividing by the length of base.
- Graphical and numerical approximation methods for finding areas include the mid-ordinate and trapezoidal rules.
- If a function describes the link between two physical quantities, the area under the graph of the function usually represents another physical quantity. The area under a power–time curve represents energy. The area under a speed–time graph represents distance and the area under a current–time graph is electric charge.

Differential calculus provides us with an elegant method for finding rates of change analytically. Instead of measuring lengths off graphs we can write algebraic equations and use them to evaluate a rate of change as precisely as we like.

Integral calculus works in a similar way; it allows us to use algebraic equations to evaluate the integral of a function. The integral is the area under the graph of the function.

The word *calculus* is an abbreviation of *infinitesimal calculus* which means calculation with numbers which are infinitesimally small. Differential calculus is probably the single most important mathematical invention of modern times, and is attributed to two men working independently. In Germany,

Leibnitz first published an account of it in 1684, while in England, Newton published a separate account of it in 1693. Although Newton was second in publishing his work, his notes indicate that he had used calculus in 1669, so it is now generally agreed that he was the first to have invented it.

In this chapter and the next, you will be introduced to differential and integral calculus. Apart from being a vital tool in engineering, calculus is a fascinating topic on its own.

7.1 Differentiating from first principles

Consider the equation, $y = x^2$, whose graph is sketched in Fig. 7.1.

The two points on the graph, P and Q, have an average gradient between them. It is $\delta y / \delta x$. The closer that Q is to P, the smaller the length of the section PQ and the more accurately does the average gradient represent the actual gradient at point P.

Now imagine that Q is brought closer and closer to P so that the length of the section of the curve, PQ, becomes so small that its length tends to zero.

That sets the scene. Now for some algebra. We write down the Cartesian coordinates of point Q as shown.

Coordinates of Q are:

$$(x + \delta x), (y + \delta y)$$

We substitute the coordinates of Q into the equation of the curve. This puts the y-coordinate of Q on the left and the x-coordinate of Q on the right. Then expand the brackets.

$$y = x^2$$
$$y + \delta y = (x + \delta x)^2$$
$$y + \delta y = (x + \delta x)(x + \delta x)$$
$$y + \delta y = x^2 + x\delta x + x\delta x + (\delta x)^2$$
$$y + \delta y = x^2 + 2x\delta x + \delta x^2$$

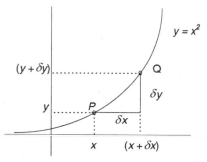

Figure 7.1 Graph of $y = x^2$

Since $y = x^2$, removing y from the left allows us to remove x^2 from the right.

$$\delta y = 2x\delta x + \delta x^2$$

Dividing by δx forms an equation which gives the average rate of change of y with respect to x between the points P and Q.

$$\frac{\delta y}{\delta x} = 2x + \delta x$$

Now, as Q is brought closer to P so that the length of PQ tends to zero, the length of δx also tends to zero. This changes our gradient equation. δx disappears from the right and the term on the left no longer represents average gradient. As 'δx tends in the limit to zero' a **differential equation** is formed.

$$\delta x \xrightarrow{\lim} 0, \qquad \frac{\delta y}{\delta x} = \frac{dy}{dx} = 2x$$

Because we are now looking at a section of the curve which is a point, the differential equation describes a gradient which is precise. We indicate this by writing it as dy/dx instead of $\delta y/\delta x$.

dy/dx is called the **differential coefficient** of y with respect to x. It is also called the **1st derivative** of $f(x)$.

So what is so special about dy/dx? Having found an algebraic expression for dy/dx we have a formula that allows us to calculate the rate of change of a function for any value of x. For the simple quadratic function which we just differentiated, $y = x^2$:

the rate of change of y with respect to x is given by $\dfrac{dy}{dx} = 2x$

Notice, because we are dealing with a non-linear function, dy/dx is not a constant – it depends on x.

EXAMPLE 7.1

Differentiate from first principles $y = 2x$.

Try it on your own before looking at my working. Use Fig. 7.2 as a prompt.

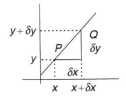

Figure 7.2 Graph of $y = 2x$

$$\text{At } Q \quad y + \delta y = 2(x + \delta x)$$
$$y + \delta y = 2x + 2\delta x$$
$$\delta y = 2\delta x$$
$$\frac{\delta y}{\delta x} = 2$$

$$\delta x \xrightarrow{\text{lim}} 0, \qquad \frac{\delta y}{\delta x} = \frac{dy}{dx} = 2$$

You may feel that this example was a waste of time. After all, we already knew that the rate of change of linear function is the value of m, the coefficient of x. However, we now have the comfort of knowing that differentiation works for linear as well as quadratic functions. Try another example.

EXAMPLE 7.2

Differentiate from first principles $y = x^3$.

$$\text{At } Q \qquad y + \delta y = (x + \delta x)^3$$

$$y + \delta y = (x + \delta x)(x + \delta x)(x + \delta x)$$

$$y + \delta y = (x^2 + 2x\delta x + \delta x^2)(x + \delta x)$$

$$y + \delta y = x^3 + 3x^2\delta x + 3x\delta x^2 + \delta x^3$$

$$\delta y = 3x^2\delta x + 3x\delta x^2 + \delta x^3$$

$$\frac{\delta y}{\delta x} = 3x^2 + 3x\delta x + \delta x^2$$

$$\delta x \xrightarrow{\text{lim}} 0, \qquad \frac{\delta y}{\delta x} = \frac{dy}{dx} = 3x^2$$

So for $y = 2x$, $dy/dx = 2$, for $y = x^2$, $dy/dx = 2x$ and for $y = x^3$, $dy/dx = 3x^2$.

You should be able to see a pattern emerging. The power of $f(x)$ becomes a coefficient of the derived function and the power of the derived function is reduced by 1.

The words are clumsy so we summarize them in a formula: a standard derivative.

7.2 Standard derivatives

Derivative of $y = ax^n$

For a function which has the general form $y = ax^n$:

$$\frac{dy}{dx} = anx^{n-1}$$

So, you might be relieved to know, it is not necessary to differentiate from first principles each time we wish to find the rate of change of a polynomial. There is, in existence, a standard derivative or formula that can be applied to any function that has the general form $f(x) = ax^n$.

PRACTICE EXERCISE 7.1

Use the standard derivative of a polynomial to find the differential coefficient of:

(i) $y = 7x^5$
(ii) $y = 18.2x^{-2}$
(iii) $y = \frac{1}{2}x^2$
(iv) $y = 2/x$
(v) $s = 4.9t^2$
(vi) $r = s$
(vii) $r = 3s$
(viii) $y = 3$

Solutions

(i) $\dfrac{dy}{dx} = 35x^4$

(ii) $\dfrac{dy}{dx} = -36.4x^{-3} = -\dfrac{36.4}{x^3}$

(iii) $\dfrac{dy}{dx} = x$

(iv) $\dfrac{dy}{dx} = -2x^{-2} = -\dfrac{2}{x^2}$

(v) $\dfrac{ds}{dt} = 9.8t$

(vi) $\dfrac{dr}{ds} = s^0 = 1$ (Linear function – gradient is a constant)

(vii) $\dfrac{dr}{ds} = 3s^0 = 3$ (Linear function – gradient is a constant)

(viii) $\dfrac{dy}{dx} = 3(0)x^{-1} = 0$ (Function is a constant – gradient is zero)

Spend a little time thinking about some of the results you obtained in Practice Exercise 7.1. The last one is particularly interesting. If a function is a constant then it does not change so its rate of change must be zero!

Derivative of $y = A\,e^{kx}$

For a function which has the general form $y = A\,e^{kx}$:

$$\frac{dy}{dx} = kA\,e^{kx}$$

You saw in Chapter 6 that the rate of change of an exponential function is proportional to the value of the function. So we can move on.

But before we do so, look how we multiplied by the coefficient of the variable. Differentiating an exponential requires us to multiply by k. The explanation for this will come later. For the moment just accept that in finding the derivative of an exponential we must multiply by the coefficient of the variable.

PRACTICE EXERCISE 7.2

Differentiate:

(i) $y = 2e^{3x}$

(ii) $s = 4e^{-2t}$

(iii) $v = \dfrac{5}{e^{3u}}$

(iv) $m = \dfrac{1}{2e^{-\frac{1}{2}n}}$

(v) $z = 3.71e^{-1.76t}$

Solutions

(i) $\dfrac{dy}{dx} = 6e^{3x}$

(ii) $\dfrac{ds}{dt} = -8e^{-2t}$

(iii) $v = 5e^{-3u}, \dfrac{dv}{du} = -15e^{-3u} = -\dfrac{15}{e^{3u}}$

(iv) $m = \frac{1}{2}e^{\frac{1}{2}n}, \dfrac{dm}{dn} = \frac{1}{4}e^{\frac{1}{2}n}$

(v) $\dfrac{dz}{dt} = -6.53e^{-1.76t} = -\dfrac{6.53}{e^{1.76t}}$

Derivatives of $a = A\sin(\omega t + \theta)$ and $a = A\cos(\omega t + \theta)$

For a function which has the general form $a = A\sin(\omega t + \theta)$:

$$\frac{da}{dt} = \omega A\cos(\omega t + \theta)$$

and for $a = A\cos(\omega t + \theta)$:

$$\frac{da}{dt} = -\omega A\sin(\omega t + \theta)$$

These tell us that the rate of change of a sine function is its cosine function and the rate of change of a cosine function is *minus* its sine function. Now this deserves some explanation and it is best done graphically.

First, note how we have to multiply by the coefficient of the variable again. In this case we multiply the derivative by ω. Remember this, it will come up later.

Now, look carefully at the sine wave whose gradient is indicated by the arrows in Fig. 7.3. Compare it with the cosine wave below it:

- at 0° the rate of change of the sine is maximum positive and the value of the cosine is also maximum positive,
- at 90° the rate of change of the sine is zero and the value of the cosine is also zero,

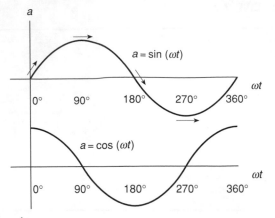

Figure 7.3 Sine and cosine curves

- at 180° the rate of change of the sine is maximum negative and the value of the cosine is also maximum negative,
- at 270° the rate of change of the sine is zero and the value of the cosine is also zero.

Conclusion:

**the *rate of change of a sine* function
is equal to *the value of the cosine* function.**

Now look at the cosine wave and its sine counterpart below it (Fig. 7.4):

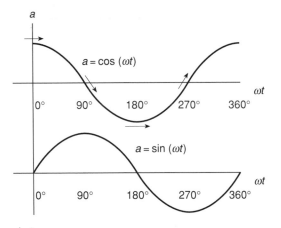

Figure 7.4 Cosine and sine curves

- at 0° the rate of change of the cosine wave is zero and the value of the sine is also zero,
- at 90° the rate of change of the cosine wave is a maximum negative and the value of the sine wave is maximum positive,
- at 180° the rate of change of the cosine wave is zero and so is the value of the sine,
- at 270° the rate of change of the cosine wave is a maximum positive and the value of the sine is a maximum negative.

Conclusion:

> the *rate of change of the cosine* function is
> equal to *minus the value of the sine* function.

PRACTICE EXERCISE 7.3

Differentiate:

(a) $a = 2\sin t$

(b) $i = I\sin\omega t$

(c) $v = 340\sin 100\pi t$

(d) $x = 12\cos 2t$

(e) $y = 3\cos(2\theta + \varphi)$

(f) $i = LI\sin\left(\omega t + \dfrac{\pi}{3}\right)$

(g) $y = A\cos 0.4x$

Solutions

(a) $\dfrac{da}{dt} = 2\cos t$

(b) $\dfrac{di}{dt} = \omega I\cos\omega t$

(c) $\dfrac{dv}{dt} = 3.4\pi \times 10^3 \cos 100\pi t$

(d) $\dfrac{dx}{dt} = -24\sin 2t$

(e) $\dfrac{dy}{d\theta} = -6\sin(2\theta + \varphi)$

(f) $\dfrac{di}{dt} = \omega LI\cos\left(\omega t + \dfrac{\pi}{3}\right)$

(g) $\dfrac{dy}{dx} = -0.4A\sin 0.4x$

Differentiating a sum of terms

So far you have dealt with the differentiation of a function when it consists of a single term. How do we deal with functions that contain a sum of terms such as that shown below?

$$y = ax^2 + bx + c$$

There is no need to worry. The rate of change of a function is simply the sum of the rates of change of its terms. All we have to do is to deal with each term separately.

Differentiating the quadratic term, the linear term and the constant term separately gives:

$$\frac{dy}{dx} = 2ax + b$$

Notice how the constant dropped out because its rate of change is zero and the rate of change of the linear term is just the constant, b.

PRACTICE EXERCISE 7.4

Differentiate:

(a) $y = 5\sin x + \cos x$
(b) $s = \frac{1}{2}t^2 - 6t + 12$
(c) $y = 3x^3 - 7x^2 + 8x - 35$
(d) $y = 2x^2 - 5e^{-3x} + 2\cos 3x$

Solutions

(a) $\dfrac{dy}{dx} = 5\cos x - \sin x$

(b) $\dfrac{ds}{dt} = t - 6$

(c) $\dfrac{dy}{dx} = 9x^2 - 14x + 8$

(d) $\dfrac{dy}{dx} = 4x + 15e^{-3x} - 6\sin 3x$

PRACTICE EXERCISE 7.5

(a) The extension, in metres of an undamped vibrating spring, is given by:

$$x = 0.43\cos(0.2t + 0.12) + 2.5$$

Find an expression for the rate at which the extension of the spring is changing and then calculate the speed of the spring when $t = 0$ and when $t = 3\,\text{s}$.

(b) Differentiate with respect to θ:

$$y = 2\sin 3\theta - 3\cos 3\theta + 5e^{-2\theta} - \frac{12}{\theta^4}$$

(c) Differentiate:

$$s = \sqrt{t} + \frac{3}{\sqrt{t}} - 9.81$$

Remember that $\sqrt{x} = x^{\frac{1}{2}}$.

(d) Differentiate:

$$y = \frac{x^2 + \sqrt{x}}{x^3}$$

(e) Differentiate:

$$q = \frac{10 + p^3}{p}$$

(f) Find the gradient of the curve of:

$$y = 3x^2 + 7x - 3$$

at the points where $x = -3$ and $x = 2$.

(g) Find the values of x for which the gradient of the curve of:

$$y = 3 + 5x - 2x^2$$

is -2.5, 0, and 1.5.

(h) The current in a circuit at a time t is given by: $i = 0.16(1 - e^{-40t})$ amperes. How fast is the current changing when $t = 50\,ms$?

(i) The induced e.m.f. in an inductive circuit is given by the equation:

$$e = L\frac{di}{dt} \text{ volts}$$

Given that the current in the circuit: $i = 32\sin 100\pi t$ A, find the induced e.m.f. in an inductance of $L = 8.2$ henrys at a time $t = 10\,ms$.

Solutions

(a) $-10.3\,mm\,s^{-1}$, $-56\,mm\,s^{-1}$

(b) $\frac{dy}{d\theta} = 6\cos 3\theta + 9\sin 3\theta - 10\,e^{-2\theta} + 48\theta^{-5}$

(c) $\frac{ds}{dt} = \frac{1}{2\sqrt{t}} - \frac{3}{2\sqrt{t^3}}$

(d) $\frac{dy}{dx} = -\frac{1}{x^2} - \frac{5}{2\sqrt{x^7}}$

(e) $\frac{dq}{dp} = 2p - \frac{10}{p^2}$

(f) -11, 19

(g) 1.87, 1.25, 0.875

(h) $866\,mA\,s^{-1}$

(i) $-82.4\,kV$

7.3 Turning points

The two graphs shown in Fig. 7.5 are typical of quadratic and cubic functions. Look at the points on the curves which are marked with arrowed lines.

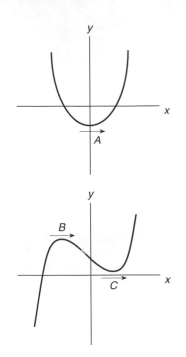

Figure 7.5 Examples of quadratic and cubic functions

A   Here $dy/dx = 0$. The point on the curve is a **turning point**, (also called a **stationary point**) and the value of y is a minimum.

B   Again $dy/dx = 0$ and this is a **turning point** but this time the value is a *local* maximum. Note that the maximum is not an *absolute* maximum because the curve rises to some value very much greater.

C   As with A, $dy/dx = 0$, but this time the turning point represents a local minimum not an absolute minimum.

So turning points are those points at which the rate of change is zero and which represent some maximum or minimum value of a function. In a moment you will see how useful these are but first you need to know how to identify turning points without having to draw graphs. What do we know about each of the turning points shown above?

$$\textbf{At a turning point the first derivative of a function } \frac{dy}{dx} = 0$$

This is always the case, so to find a turning point we simply find the value of x that gives $dy/dx = 0$. Look at the next example, it is quite straightforward.

EXAMPLE 7.3

Find the turning point(s) of: $s = 3t^2 - 12t + 5$

First we want the derivative so we differentiate. At a turning point $ds/dt = 0$ so we equate the derivative with 0.

$$\frac{ds}{dt} = 6t - 12$$

$$6t - 12 = 0$$

$$t = 2$$

So the turning point occurs at $t = 2$. To find if this corresponds to a minimum or maximum value of s the best thing to do is to find the **second derivative**.

Second derivative

This is written as shown. The superscript *does not mean 'squared'*.

$$\frac{d^2y}{dx^2}$$

The second derivative is obtained by differentiating the function a second time so it represents the rate of change of a rate of change. Take Example 7.3 again and differentiate it twice.

$$s = 3t^2 - 12t + 5$$

$$\frac{ds}{dt} = 6t - 12$$

$$\frac{d^2s}{dt^2} = 6$$

Straightforward enough, but what does it mean? Well, since 6 is a positive value it tells us that the turning point is a minimum. In fact, the general rule is:

At a minimum, the second derivative is positive.
At a maximum the second derivative is negative.

Study the graphs in Fig. 7.6 (on page 166) carefully to see why this is so.

Relating the graphs of $f(x)$ and its first and second derivatives should make things clear for you. Remember that the graph of the first derivative is a graph of the rate of change of $f(x)$. Similarly, the graph of the second derivative is a graph of the rate of change of the first derivative.

The next worked example puts the whole thing into a practical context. Spend some time on it.

EXAMPLE 7.4

The motion of a particular object is fully described by:

$$s = t^3 + 2t^2 - 3t + 4$$

where s is distance in metres and t is time in seconds.

(i)   Find the first derivative of $f(t)$.
(ii) Find the second derivative of $f(t)$.
(iii) Find the velocity of the object at time $t = 2$ s and $t = -2$ s.
(iv) Find the acceleration of the object at time $t = 2$ s and $t = -2$ s.

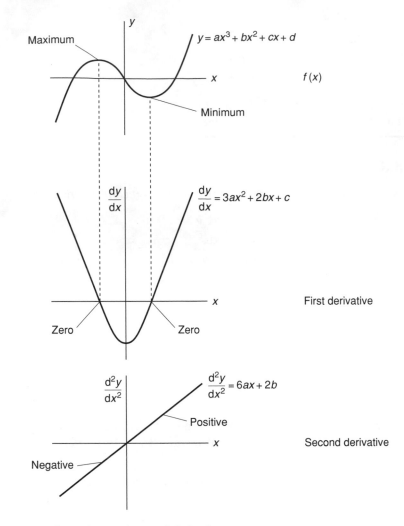

Figure 7.6 Turning points and second derivatives

(v) The distance of the object when $t = 0$.
(vi) The time(s) when the velocity of the object is zero.
(vii) The maximum (local) and minimum (local) distances of the object.
(viii) Sketch the distance–time graph from the information you have gained above.

Solutions

(i) $\dfrac{ds}{dt} = 3t^2 + 4t - 3$

(ii) $\dfrac{d^2s}{dt^2} = 6t + 4$

(iii) When $t = 2$, $v = \dfrac{ds}{dt} = 3(2)^2 + 4(2) - 3 = 17\,\text{m s}^{-1}$

   when $t = -2$, $v = 3(-2)^2 + 4(-2) - 3 = 1\,\text{m s}^{-1}$

(iv) When $t = 2$, $a = 6(2) + 4 = 16\,\text{m s}^{-2}$

   when $t = -2$, $a = 6(-2) + 4 = -8\,\text{m s}^{-2}$

(v) When $t = 0$, $s = 4$ m

(vi) These are at the turning points: when $\dfrac{ds}{dt} = 0$,

$$3t^2 + 4t - 3 = 0$$

$$t = \frac{-4 \pm \sqrt{16 - 4(3)(-3)}}{6}$$

$$= -0.667 \pm 1.202$$

$$= 0.535\,\text{s}, \; -1.869\,\text{s}.$$

(vii) When $t = 0.535$, $\dfrac{d^2s}{dt^2} = 6(0.535) + 4 = 7.21$. This is positive so it corresponds to a minimum.

$s_{min} = (0.535)^3 + 2(0.535)^2 - 3(0.535) + 4 = 3.12$ m and

$s_{max} = (-1.869)^3 + 2(-1.869)^2 - 3(-1.869) + 4 = 10.06$ m

(viii)

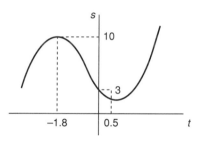

Now finish off with a set of practice exercises. Take your time – some of the problems may challenge you but, with patience, you will gain the satisfaction of having solved them all.

PRACTICE EXERCISE 7.6

(a) Find the 2nd derivative of the following:

(i) $s = 4t^{-\frac{3}{2}}$

(ii) $y = \dfrac{5}{7\sqrt[3]{x}} - \dfrac{3}{2x^2}$

(iii) $x = 37\cos(1.57t + 0.707) + 15.3t^2 - \dfrac{1}{\sqrt{t}} + 23.4t - 74.8$

(iv) $y = \dfrac{3}{4\,e^{-2x}} + \dfrac{1}{2}x^3 + \sqrt{x} + 8 - 3\sin 2x$

(v) $r = as^4 + \dfrac{bs^2 - cs}{ds^3} - A\,e^{-ks} + B + C\cos \omega s$, with respect to s

(b) The vertical displacement of a model in a wave tank is given by $y = 25\sin 0.1t$ mm. The vertical velocity of the model is:

(i) $250\cos 0.1t$

(ii) $-2.5\cos 0.1t$

(iii) $2.5\cos 0.1t$

(iv) $250\cos 0.1t$

(c) An alternating voltage v volts varies with time t seconds according to the equation $v = 100 \sin 2\pi f t$, where f is frequency in hertz. If $t = 12\,\text{ms}$ and $f = 50\,\text{Hz}$ the rate of change of voltage is:

(i) $-25\,416\,\text{V}\,\text{s}^{-1}$
(ii) $-18\,466\,\text{V}\,\text{s}^{-1}$
(iii) $2065.6\,\text{V}\,\text{s}^{-1}$
(iv) $31\,348\,\text{V}\,\text{s}^{-1}$

(d) A small open topped container is made from a piece of sheet steel 60 mm square by cutting and folding as shown.

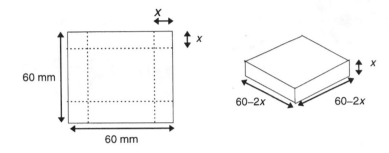

The value of x which gives the maximum volume is:

(i)   5 mm
(ii)   10 mm
(iii) 15 mm
(iv)   20 mm

(e) The total power in an electric circuit depends on current according to $P = EI - I^2 R_s$, where E is the e.m.f., I is the current and R_s is the resistance of the power source.

Show that the condition of maximum power occurs when $R_s = \dfrac{E}{2I}$

Solutions

(a) (i) $\dfrac{d^2s}{dt^2} = \dfrac{15}{\sqrt{t^7}}$; (ii) $\dfrac{d^2y}{dx^2} = \dfrac{20}{63\sqrt[3]{x^7}} - \dfrac{9}{x^4}$

(iii) $\dfrac{d^2x}{dt^2} = -91.2 \cos(1.57t + 0.707) + 30.6 - \dfrac{3}{4}t^{-\frac{5}{2}}$

(iv) $\dfrac{d^2y}{dx^2} = 3\,e^{2x} + 3x + \dfrac{1}{4}x^{-\frac{3}{2}} + 12 \sin 2x$

(v) $\dfrac{d^2r}{ds^2} = 12as^2 + \dfrac{2b}{d}s^{-3} - \dfrac{6c}{d}s^{-4} - k^2A\,e^{-ks} - \omega^2 C \cos \omega s$

(b) $v = 2.5 \cos 0.1t$; (c) $-25\,416\,\text{V}\,\text{s}^{-1}$; (d) $x = 10$

(e) $\dfrac{dP}{dI} = E - 2R_sI$, $\dfrac{d^2P}{dI^2} = -2R_s$, so P_{max} occurs when $E - 2R_sI = 0$,

$\therefore R_s = \dfrac{E}{2I}$ for max power.

Before tackling the next assignment you should go back over this chapter and revise the important points. Give the subject the attention it deserves. Don't skip over anything.

It can be a good idea to go over the same problems more than once. So if you are uncertain, do some of the problems in Practice Exercises 7.5 and 7.6 again.

Assignment VI

Analytical methods

This assignment requires you to use functions to model engineering situations and solve engineering problems. Note the performance criteria that should be met before you go on to the next topic.

- Functions and graphs are used to model engineering situations.
- Rates of change of functions are interpreted to solve engineering problems.
- Differentiation is used to solve engineering problems.

In the three problems that follow most of the planning has been done already, but you should still consider the other three themes that determine the quality of your assignment work.

Information handling

Select and use the given information. Work and give answers to an appropriate degree of accuracy. Numerical values should be assigned correct SI units.

Evaluation

Apply checking procedures to ensure your results are correct. You should also consider whether your solution makes sense within the problem situation.

Outcomes

Aim for a product that is professional in substance and in appearance.

1. AN EQUATION OF MOTION

The motion of an object is given by:

$$s = 3t^3 - 12t^2 + \frac{14}{\sqrt{t}} + 9\,e^{-0.2t}$$

where s is distance in metres and t is time in seconds. Find:

(i) the velocity of the object after 2 seconds.
(ii) the acceleration after 2 seconds.

2. CAPACITOR CURRENT

A circuit containing capacitance will draw a current:

$$i = C\frac{dv}{dt} \text{ amperes}$$

where C is capacitance in farads, t is time in seconds and

$$v = V\sin(\omega t + \varphi) \text{ volts}$$

Find the current drawn by a capacitor at $t = 3\,\text{ms}$ if:

$$C = 220\,\mu\text{F}$$

$$V = 12\,\text{V}$$

$$\omega = 314\,\text{rad}\,\text{s}^{-1}$$

$$\varphi = \pi/2\,\text{rad}$$

3. MAXIMUM AREA

A rectangle of sides x and y is to have a perimeter of 460 m. Show that the area of the rectangle is:

$$A = 230x - x^2$$

then find the dimensions of x and y that will give the maximum possible area.

Calculate the maximum area and sketch a graph of $f(x)$, clearly labelling the maximum point and its gradient.

Integral calculus

8.1 Introduction

According to Isaac Newton's **fundamental theorem of calculus**, integration is the inverse of differentiation. This means that if we integrate a differential coefficient of some function, we obtain the original function as shown with the cubic below.

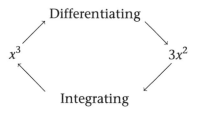

Take a quartic as another example.

$$y = \frac{x^4}{4}$$

Differentiating gives a cubic.

$$\frac{dy}{dx} = x^3$$

Multiplying by dx gives what is called the **differential** of the quartic.

$$dy = x^3\, dx$$

Now, by definition, if we integrate the differential we get back to the original quartic.

$$y = \int (x^3)\, dx = \frac{x^4}{4}$$

The $\int$ symbol is derived from the medieval letter 's' which stands for 'the sum of...'. The significance of this will become clear when you apply integration to find the area below a graph. Now, as with **standard derivatives** it would be useful to have **standard integrals** which would allow us to apply a simple formula when we wish to integrate a particular function. Let's see how we can get a standard integral for a simple algebraic term.

If integration is to be the inverse of differentiation then, for an algebraic function having the general form shown, differentiation gives x^n.

$$y = \frac{x^{n+1}}{n+1}$$

This is because:

$$\frac{dy}{dx} = (n+1)\frac{x^{n+1-1}}{(n+1)} = x^n$$

We then obtain the differential of y with respect to x.

$$dy = x^n\,dx$$

And integration leads us back to the original function.

$$y = \int (x^n)\,dx = \frac{x^{n+1}}{n+1}$$

In other words:

$$\frac{x^{n+1}}{n+1} \quad \text{is the integral of } x^n$$

8.2 Standard integrals

So, we have established the standard integral for an algebraic function.

By applying the same reasoning to the other standard derivatives we can obtain standard derivatives for the trig. and exponential functions as well.

If $y = x^n$, then $\int (x^n)\,dx = \frac{x^{n+1}}{n+1}$

If $y = \cos x$, then $\int (\cos x)\,dx = \sin x$

If $y = \sin x$, then $\int (\sin x)\,dx = -\cos x$

If $y = e^x$, then $\int (e^x)\,dx = e^x$

Think carefully about why the sine integrates into the negative of the cosine. Differentiating a cosine led to a sine and involved a change in sign. Integrating must reverse this process. So integrating a sine must also involve a change in sign. For instance, if we start with a positive sine, integration must involve a change and give us a negative cosine. This can be confusing so try to remember the following simple rule:

'Integration changes the sign of a sine'

Another point worth noting is that the integral of an exponential function is identical to the function if the coefficient of x is 1.

8.3 Constant of integration

Remember that differentiation eliminates a constant term. So if integration is the inverse of differentiation, it must replace the constant term.

Although we may not know the numerical value of that term we can indicate its existence by adding some constant, say C, to the integral. We call this term the **constant of integration**.

Given:

$$y = \frac{x^2}{2} + 5, \quad \frac{dy}{dx} = x$$

and:

$$\int x \, dx = \frac{x^2}{2} + C$$

Given some additional information, it is then possible to evaluate this constant. More about that later.

PRACTICE EXERCISE 8.1

Integrate the following with respect to x:

(a) $y = x^5$
(b) $y = \sqrt{x}$
(c) $y = \dfrac{1}{x^3}$
(d) $y = \cos x$

Solutions

(a) $\displaystyle\int x^5 \, dx = \frac{x^6}{6} + C$

(b) $\displaystyle\int \sqrt{x} \, dx = \int x^{\frac{1}{2}} \, dx = \frac{x^{\frac{3}{2}}}{\frac{3}{2}} + C = \frac{2x^{\frac{3}{2}}}{3} + C$

(c) $\displaystyle\int \frac{1}{x^3} \, dx = \int x^{-3} \, dx = \frac{x^{-2}}{-2} + C = -\frac{1}{2x^2} + C$

(d) $\displaystyle\int (\cos x) \, dx = \sin x + C$

Constant coefficient

If differentiation has no effect on a constant coefficient then integration can have no effect as well. So a constant coefficient may be written outside the integral.

PRACTICE EXERCISE 8.2

Complete the following integrations:

(a) $\int 3x^2 \, dx$

(b) $\int (4 \sin \theta) \, d\theta$

Solutions

(a) $\int 3x^2 \, dx = 3 \int x^2 \, dx = \dfrac{3x^3}{3} + C = x^3 + C$

(b) $\int (4 \sin \theta) \, d\theta = 4 \int (\sin \theta) \, d\theta = -4 \cos \theta + C$

8.4　Integral of a sum of terms

Again, we are guided by the rules of differentiation. The integral of a sum of terms is the sum of the individual integrals.

PRACTICE EXERCISE 8.3

Complete the following:

(a) $\int (x^2 + x) \, dx$

(b) $\int (3x^4 + e^x - 6) \, dx$

(c) $\int (3 \sin t - 5 \cos t) \, dt$

(d) $\int (2x + 5)^2 \, dx$

Solutions

(a) $\int (x^2 + x) \, dx = \int x^2 \, dx + \int x \, dx = \dfrac{x^3}{3} + \dfrac{x^2}{2} + C$

(b) $\int (3x^4 + e^x - 6) \, dx = \dfrac{3x^5}{5} + e^x - 6x + C$

(c) $\int (3\sin t - 5\cos t)\,dt = -3\cos t - 5\sin t + C$

(d) $\int (2x+5)^2\,dx = \int (4x^2 + 20x + 25)\,dx = \dfrac{4x^3}{3} + 10x^2 + 25x + C$

Notice that it is not necessary to rewrite the integral as a sum of separate integrals. You can go straight into the integration just dealing with each term separately.

Now for some more practice.

PRACTICE EXERCISE 8.4

Integrate the following with respect to the given variable.

(a) x^3; (b) t^7; (c) $\sqrt{x}$; (d) x^{-2}; (e) $\dfrac{1}{x^4}$; (f) $\dfrac{1}{x^{-2.5}}$; (g) $\dfrac{1}{\sqrt[3]{z}}$

(h) $6x^3$; (i) $2u^3 + e^u$; (j) $16x^3 - x^2 + 12$; (k) $2t + \frac{1}{2}t^2$

(l) $\dfrac{2}{\sqrt{x}} - \dfrac{3x^2}{4} + 2\cos x$; (m) $3\cos\theta - 5\sin\theta - e^\theta$; (n) $(x+3)(x-5)$

Solutions

(a) $\dfrac{x^4}{4} + C$; (b) $\dfrac{t^8}{8} + C$; (c) $2\sqrt{x^3} + C$; (d) $-\dfrac{1}{x} + C$; (e) $-\dfrac{1}{3x^3} + C$; (f) $\dfrac{x^{3.5}}{3.5} + C$

(g) $\dfrac{3}{2}\sqrt[3]{z^2} + C$; (h) $\dfrac{3}{2}x^4 + C$; (i) $\frac{1}{2}u^4 + e^u + C$; (j) $4x^4 - \frac{1}{3}x^3 + 12x + C$

(k) $t^2 + \frac{1}{6}t^3 + C$; (l) $4\sqrt{x} - \frac{1}{4}x^3 + 2\sin x + C$; (m) $3\sin\theta + 5\cos\theta - e^\theta + C$

(n) $\frac{1}{3}x^3 - x^2 - 15x + C$

8.5 Evaluating the constant of integration

We can find the value of constant C providing we are given corresponding values of x and y. Take a specific example.

Find the equation of a curve whose gradient at point $(2, 3)$ is x^2.

Knowing the gradient allows us to write the differential coefficient of the function.

$$\frac{dy}{dx} = x^2$$

Integrating gives a general equation for y with C the unknown term as normal. If we wish to know the value of C, we must transpose to make this the subject of the equation.

$$y = \int x^2 \, dx = \frac{x^3}{3} + C$$

$$y = \frac{1}{3}x^3 + C$$

$$C = y - \frac{1}{3}x^3$$

Now, we happen to know that when $x = 2$, $y = 3$, so we can substitute these values into the equation for C and evaluate.

$$C = 3 - \frac{1}{3}(2)^3 = \frac{9-8}{3} = \frac{1}{3}$$

$$y = \frac{1}{3}x^3 + \frac{1}{3}$$

The solution is a particular equation in which we have the value of C. Here, we may simplify by taking out the common factor.

$$y = \frac{1}{3}(x^3 + 1)$$

PRACTICE EXERCISE 8.5

(a) The gradient of a curve is given by $3x$. Find the equation of the curve given that it passes through point $(3, 1)$.

(b) If $\frac{dy}{dx} = x^2 - 4$ find an expression for y if it is known that the curve passes through point $(2, 5)$.

(c) If it is known that $y = 1$ when $x = 0$ find the equation of the curve whose gradient is given by $(3 + x)^2$.

(d) If $\frac{dy}{dx} = \sin x$ what is the equation of the curve that passes through $(2, 3)$.

(e) Velocity is the rate of change of displacement. Find the displacement equation if velocity is given by $v = u + at$ and it is known that displacement s is zero when t is zero.

(f) If $\frac{dv}{dt} = e^t + \cos t$ find the value of v when $t = 1$, given that the curve of v passes through point $(\pi, 3)$.

Solutions

(a) $y = \frac{1}{2}(3x^2 - 25)$; (b) $y = \frac{1}{3}(x^3 - 12x + 31)$; (c) $y = 1 + 9x + 3x^2 + \frac{1}{3}x^3$
(d) $y = 2.58 - \cos x$; (e) $s = ut + \frac{1}{2}at^2$; (f) $v = -16.5$

8.6 The definite integral

So far, we have been dealing with what is called an **indefinite integral**. For some function of x, $f(x)$, the indefinite integral is written down in the general form shown.

General form of an indefinite integral:

$$\int f(x)\,dx$$

The **definite integral** takes the second form shown, in which a and b are the limiting values of x. a specifies the **lower** limit and b specifies the **upper** limit of x.

General form of a definite integral:

$$\int_a^b f(x)\,dx$$

The significance of this will be explained in a moment. First, learn how to solve a definite integral by example. It's not difficult.

EXAMPLE 8.1

Evaluate: $\displaystyle\int_2^3 x^2\,dx$

The first step is to integrate in the normal way, leaving the indefinite integral within square brackets labelled with the upper and lower limits.

Next, substitute the upper limit for x in the integral.

Then subtract the integral with the lower limit substituted for x.

The result is a number. This is the definite integral.

Solution

$$\int_2^3 x^2\,dx = \left[\frac{x^3}{3} + C\right]_2^3 = \left(\frac{(3)^3}{3} + C\right) - \left(\frac{(2)^3}{3} + C\right)$$

$$= \left(\frac{27}{3} + C\right) - \left(\frac{8}{3} + C\right)$$

$$= \frac{27 - 8}{3} + C - C$$

$$= \frac{19}{3} \approx 6.3$$

Take note of the steps:

- Always use square brackets to indicate that you have the indefinite integral and you are about to substitute the limits to obtain the definite integral.

- The constant of integration cancels to zero. This always happens so it is customary to omit C altogether when finding a definite integral.
- A definite integral is always a numerical value – not an algebraic expression like the indefinite integral.

PRACTICE EXERCISE 8.6

Evaluate the following:

(a) $\int_1^2 (3x^2 - 2x + 5)\, dx$

(b) $\int_0^{\frac{\pi}{2}} (\sin\theta)\, d\theta$

Solutions

(a) $\int_1^2 (3x^2 - 2x + 5)\, dx = \left[x^3 - x^2 + 5x \right]_1^2$

$= [(2)^3 - (2)^2 + 5(2)] - [(1)^3 - (1)^2 + 5(1)]$

$= 8 - 4 + 10 - 1 + 1 - 5$

$= 9$

(b) $\int_0^{\frac{\pi}{2}} (\sin\theta)\, d\theta = \left[-\cos\theta \right]_0^{\frac{\pi}{2}}$

$= \left(-\cos\frac{\pi}{2} \right) - (-\cos 0)$

$= 0 - (-1)$

$= 1$

You will see the practical importance of definite integrals after you have practised some more with the technique.

PRACTICE EXERCISE 8.7

Evaluate the following correct to 3 s.f.

(a) $\int_3^4 x^2\, dx$; (b) $\int_{-1}^2 (3x - 4)\, dx$; (c) $\int_4^5 (\sqrt{x} + e^x)\, dx$; (d) $\int_0^{\frac{\pi}{2}} \cos x\, dx$; (e) $\int_0^{\pi} (\sin x)\, dx$

Solutions

(a) 12.3; (b) −7.50; (c) 95.9; (d) 1.00; (e) 2.00

8.7 Area under the curve of a function

You have seen that differential calculus is an analytical method for finding the exact gradient of a curve. This is better than the graphical method of trying to find the gradient by measuring the slope of a graph. And you should also know that the gradient of the graph of a function is a rate of change.

You have used graphical and numerical integration (the mid-ordinate rule) to find the area under a curve by dividing the area into strips and then adding the areas of the strips.

You should also appreciate that the area under a curve, which represents the relationship between two quantities, provides us with a third quantity. For example, the area under a power–time graph gives us energy. The area under a velocity–time graph gives us distance travelled. Now you will see how integral calculus provides us with an analytical method for calculating areas precisely.

Given some function of x, say $f(x) = y$ we can formulate an expression for the definite integral of that function. And look at what this means!

$$y = f(x)$$

$$\int_a^b f(x)\,dx = \int_a^b y\,dx$$

In words: 'it is the sum of y times dx, between x is equal to a and x is equal to b.'

Now, graphically, the product, $y\,dx$ is the area of a strip whose width, dx, is *infinitesimally small* (see Fig. 8.1).

The integral, $\int y\,dx$, is the sum of those very small strips, and the definite integral, $\int_a^b y\,dx$, is the sum of those very small strips between prescribed limits of x.

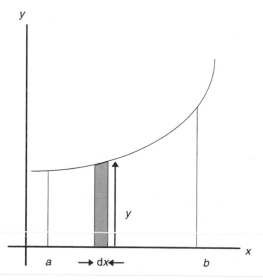

Figure 8.1 Integral calculus

You know from the mid-ordinate rule that the narrower the strips, the greater their number and the closer we get to the true value of the area.

With integral calculus, since the width of each strip:

$$\delta x \xrightarrow{\text{lim}} 0$$

then the number of strips must also tend in the limit to infinity and the area obtained must be the exact, true value (see Fig. 8.1).

Breathtaking stuff!

EXAMPLE 8.2

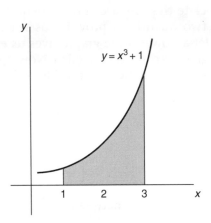

Find the area under the curve shown, between the limits of $x = 1$ and $x = 3$.

Solution

$$\text{Area} = \int_a^b y \, dx$$

$$= \int_1^3 (x^3 + 1) \, dx$$

$$= \left[\frac{x^4}{4} + x \right]_1^3$$

$$= \left(\frac{3^4}{4} + 3 \right) - \left(\frac{1}{4} + 1 \right)$$

$$= 22 \text{ units}^2$$

QUESTION

If the graph in the above example was one of velocity against time, what would be the unit of the definite integral?

ANSWER

The definite integral is area so it would be the product of velocity and time ($\text{m s}^{-1} \times \text{s} = \text{m}$) so the result is distance in metres.

EXAMPLE 8.3

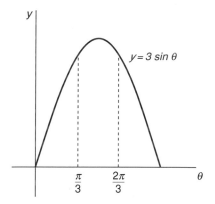

Find the area under the curve shown, between the limits of $\theta = \pi/3$, and $\theta = 2\pi/3$.

Solution

$$\text{Area} = \int_a^b y \, d\theta$$

$$= \int_{\frac{\pi}{3}}^{\frac{2\pi}{3}} (3 \sin \theta) \, d\theta$$

$$= -3 \left[\cos \theta \right]_{\frac{\pi}{3}}^{\frac{2\pi}{3}}$$

$$= -3 \left(\left(\cos \frac{2\pi}{3} \right) - \left(\cos \frac{\pi}{3} \right) \right)$$

$$= 3 \text{ units}^2$$

You should be in a habit of checking results, where possible. With the last example it's quite straightforward. The width of the area is one third of π, slightly over 1. The average height is slightly under 3. The product of roughly 1 and roughly 3 is about 3! Satisfying isn't it?

PRACTICE EXERCISE 8.8

(a) Find the area below the curve of $y = x^2 - e^x$ the x-axis and the boundaries formed by $x = 0$ and $x = 2$.
(b) Find the area under the curve of $y = 5x^2 - 3x + 10$ between the limits of $x = -3$ and $x = -1$.
(c) Find the energy under the power curve $P = 27t^2$ in the range of $t = 0$ and $t = 10$ s.
(d) Find the area under the sinusoid $v = 12 \sin t - 15 \cos t$ over the interval between $t = 0$ and $t = 20$ ms.
(e) Find the distance travelled by an object whose speed is given by
$$v = 3 e^t - 15t + \frac{2}{\sqrt{t}} \text{ over the first 5 seconds of motion.}$$
(f) Find the area under one complete cycle of the sinusoid $y = \sin \theta$.

Solutions

(a) −3.72;   (b) 75.3;   (c) 9 kJ;   (d) −0.298;   (e) 270;   (f) 0

You may have been surprised by some of the results you obtained in Practice Exercise 8.8. For example, does a negative area make sense?

Yes it does, it is the area of a curve lying *below* the x-axis

What about the area below a sine wave being zero?

If equal portions of the curve lie below and above the x-axis
then overall the area is zero

That concludes your introduction to elementary calculus. Up to this point, we have concentrated on the technique rather than its application in engineering. Later, after some revision, you will learn more about the methods and how to apply them to practical problems.

Meanwhile, I hope you appreciate how heavily calculus relies on earlier topics. It is vital to know how to apply the rules of directed numbers, algebra, indices and trigonometry. This is a good time for you to take stock. If you had any difficulty with the calculus, go back and polish up your underpinning knowledge.

Remember, a great deal can be learned simply by doing the same problems again, several times, if necessary.

Further calculus

At this stage you know how to apply four standard derivatives and four standard integrals to simple algebraic, trigonometric and exponential functions. Here is one more: a logarithmic function (see Figs 9.1 and 9.2).

For a simple function like,

$$y = \ln x$$

The standard derivative is,

$$\frac{dy}{dx} = \frac{1}{x}$$

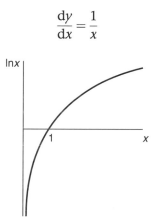

Figure 9.1 Graph of $y = \ln x$

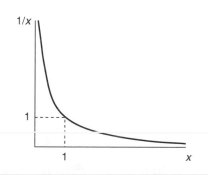

Figure 9.2 Graph of $y = 1/x$

You get this without proof but sketching the logarithmic curve and the rectangular hyperbola over corresponding values of x should be sufficient. If you are not convinced, plot a graph of the natural log of x and measure its gradient at selected points by drawing tangents. You will find that at any point the numerical value you obtain is equal to the reciprocal of x. Now, since integration is the reverse of differentiation it follows that:

If,

$$y = \frac{1}{x}$$

then,

$$\int y \, dx = \ln x + C$$

In case you have forgotten, the natural log (ln) is the same thing as the exponential log ($\log_e$).

Now this last integral comes in very handy. Try using the standard integral of an algebraic to integrate:

$$y = \frac{1}{x}$$

Using,

$$\int ax^n \, dx = \frac{a}{n+1} x^{n+1} + C$$

to find,

$$\int \frac{1}{x} \, dx$$

gives,

$$\int x^{-1} \, dx = \frac{x^0}{0} \dots ?$$

which is mathematically not defined.

It does not work! So we must add a restriction to our list of standard integrals.

If,

$$y = ax^n$$

then,

$$\int y \, dx = \frac{ax^{n+1}}{n+1} + C \quad (n \neq -1)$$

Now let's see how we apply the standard derivatives to slightly more complicated functions.

9.1 General theorems of differentiation

'Function of a function' rule

In the equation shown, we say that y is a function of x. In other words, y depends on x in some way. And then if we say that the function of x is $\sin x$, we specify the way in which y depends on x.

$$y = \sin x$$

$$y = f(x)$$

$$f(x) = \sin x$$

Now look at the equation

$$y = \sin 2x$$

If we wished to evaluate it with a conventional calculator it would be necessary to do so in two steps. First we need to find the value of $2x$ and then take the sine of that. This is an example of a **function of a function**.

For convenience we call the 'inner' function u so that y is a function of u and at the same time, u is a function of x. In this case the function of x is simply $2x$ and the function of u is $\sin u$.

Let,

$$y = \sin u, \quad \text{where } u = 2x$$

so,

$$y = F[u] \quad \text{and} \quad u = f(x)$$

hence

$$y = F[f(x)]$$

When differentiating a function of a function like this one, we deal with each function separately. Let's see how it is done.

If,

$$y = \sin 2x$$

we rewrite the equation making y the function of u and differentiate y with respect to u.

Then we write u as a function of x and differentiate u with respect to x.

Let,

$$y = \sin u, \quad \text{so } \frac{dy}{du} = \cos u$$

with

$$u = 2x, \quad \text{so } \frac{du}{dx} = 2$$

Now, multiplying the two derivatives is equal to multiplying the two differential coefficients.

$$2\cos u = \frac{dy}{du} \times \frac{du}{dx} = \frac{dy}{dx}$$

Look what happens – du cancels to 1 and we have the differential coefficient of y with respect to x:

$$\therefore \quad \frac{dy}{dx} = 2\cos u = 2\cos 2x$$

So here is the rule for differentiating a function of a function. It is sometimes called the chain rule.

If,

$$y = F[f(x)] \quad \frac{dy}{dx} = \frac{dy}{du}\frac{du}{dx}$$

EXAMPLE 9.1

Differentiate by applying the function of a function rule,

(a) $y = \cos 3\theta$
(b) $a = \sin \omega t$

Solutions

(a) $y = \cos u \qquad \dfrac{dy}{du} = -\sin u \qquad u = 3\theta \qquad \dfrac{du}{d\theta} = 3$

$\dfrac{dy}{du}\dfrac{du}{d\theta} = \dfrac{dy}{d\theta} = -\sin u(3) = -3\sin u \qquad \dfrac{dy}{d\theta} = -3\sin 3\theta$

(b) $a = \sin u \qquad \dfrac{da}{du} = \cos u \qquad u = \omega t \qquad \dfrac{du}{dt} = \omega$

$\dfrac{da}{du}\dfrac{du}{dt} = \dfrac{da}{dt} = \omega \cos u = \omega \cos \omega t$

You may recognize something you have seen before. Differentiating a function of a function of this type involves multiplying by the coefficient of the independent variable. There is no need to go through the business of applying the function of a function rule every time. By writing a standard derivative which takes account of the coefficient it is possible to **differentiate by recognition**.

If,

$$y = \sin ax$$

then,

$$\frac{dy}{dx} = a\cos ax$$

and if,

$$y = \cos ax$$

then,

$$\frac{dy}{dx} = -a\sin ax$$

EXAMPLE 9.2

Differentiate:

(a) $y = 2 e^{0.6x}$
(b) $y = 3 \sin 100\pi t$
(c) $y = 2 \ln 5x$
(d) $y = \cos^2 \theta$
(e) $y = (2 + 3x)^3$
(f) $x = 3 e^{\sin 2t}$

Worked Solutions

(a) $y = 2 e^u$,　$\dfrac{dy}{du} = 2 e^u$　$u = 0.6x$,　$\dfrac{du}{dx} = 0.6$　$\dfrac{dy}{dx} = 2 e^u (0.6) = 1.2 e^{0.6x}$

(b) $\dfrac{dy}{dt} = 300\pi \cos 100\pi t$

(c) $y = 2 \ln u$　$\dfrac{dy}{du} = \dfrac{2}{u}$,　$u = 5x$　$\dfrac{du}{dx} = 5$　$\dfrac{dy}{dx} = \dfrac{2}{u}(5) = \dfrac{10}{u} = \dfrac{10}{5x} = \dfrac{2}{x}$

(d) $y = \cos^2 \theta = (\cos \theta)^2$　$y = u^2$　$\dfrac{dy}{du} = 2u$,　$u = \cos \theta$　$\dfrac{du}{d\theta} = -\sin \theta$

$\dfrac{dy}{d\theta} = 2u(-\sin \theta) = -2u \sin \theta = -2 \cos \theta \sin \theta$

(e) $y = u^3$　$\dfrac{dy}{du} = 3u^2$　$u = 2 + 3x$　$\dfrac{du}{dx} = 3$

$\dfrac{dy}{dx} = 3u^2 (3) = 9u^2 = 9(2 + 3x)^2 = (6 + 9x)^2$

(f) $x = 3 e^u$　$\dfrac{dx}{du} = 3 e^u$　$u = \sin 2t$　$\dfrac{du}{dt} = 2 \cos 2t$

$\dfrac{dx}{dt} = 3 e^u (2 \cos 2t) = 6 \cos 2t (e^u) = 6 \cos 2t (e^{\sin 2t})$

Since integration is the reverse of differentiation, when integrating, the coefficient of the variable does not form a product but the opposite, a quotient. So now we are in a position to construct a set of tables of standard derivatives and integrals (see Tables 9.1 and 9.2). Copy them out and *remember them*.

Table 9.1. Standard derivatives

$y = f(x)$	$\dfrac{dy}{dx}$
ax^n	anx^{n-1}
$\sin ax$	$a \cos ax$
$\cos ax$	$-a \sin ax$
e^{kx}	$k e^{kx}$
$\ln ax$	$\dfrac{1}{x}$

Table 9.2. Standard integrals

$y = f(x)$	$\int y \, dx$
ax^n	$\dfrac{ax^{n+1}}{n+1} + C \; (n \neq -1)$
$\sin ax$	$-\dfrac{1}{a}\cos ax + C$
$\cos ax$	$\dfrac{1}{a}\sin ax + C$
e^{kx}	$\dfrac{1}{k}e^x + C$
$\dfrac{1}{x}$	$\ln x + C$

Product rule

If u and v are both functions of x and:

$$y = uv$$

then,

$$\frac{dy}{dx} = u\frac{dv}{dx} + v\frac{du}{dx}$$

EXAMPLE 9.3

Differentiate:

(a) $y = x^3 \ln x$
(b) $y = \cos\theta \sin\theta$

Solution

(a) Let $u = x^3$, so $\dfrac{du}{dx} = 3x^2$ and $v = \ln x$, so $\dfrac{dv}{dx} = \dfrac{1}{x}$

$$\frac{dy}{dx} = u\frac{dv}{dx} + v\frac{du}{dx} = x^3\frac{1}{x} + \ln x(3x^2) = x^2 + 3x^2 \ln x$$

$$\frac{dy}{dx} = x^2(1 + 3\ln x)$$

(b) $u = \cos\theta \quad \dfrac{du}{d\theta} = -\sin\theta \quad v = \sin\theta \quad \dfrac{dv}{d\theta} = \cos\theta$

$$\frac{dy}{d\theta} = u\frac{dv}{d\theta} + v\frac{du}{d\theta} = \cos\theta\cos\theta + \sin\theta(-\sin\theta)$$

$$\frac{dy}{d\theta} = \cos^2\theta - \sin^2\theta$$

EXAMPLE 9.4

Differentiate:

(a) $y = \sqrt{x}\,e^x$
(b) $y = (1 - 3t^2)\sin t$

Solution

(a) $u = x^{\frac{1}{2}}$ $\quad \dfrac{du}{dx} = \dfrac{1}{2}x^{-\frac{1}{2}}$ $\quad v = e^x$ $\quad \dfrac{dv}{dx} = e^x$

$$\frac{dy}{dx} = u\frac{dv}{dx} + v\frac{du}{dx} = x^{\frac{1}{2}}e^x + e^x\left(\frac{1}{2}x^{-\frac{1}{2}}\right)$$

$$\frac{dy}{dx} = e^x\left(\sqrt{x} + \frac{1}{2\sqrt{x}}\right)$$

(b) $u = 1 - 3t^2$ $\quad \dfrac{du}{dt} = -6t$ $\quad v = \sin t$ $\quad \dfrac{dv}{dt} = \cos t$

$$\frac{dy}{dt} = 1 - 3t^2(\cos t) + \sin t(-6t) = (1 - 3t^2)\cos t - 6t\sin t$$

Quotient rule

This states that if u and v are both functions of x and,

$$y = \frac{u}{v}$$

then,

$$\frac{dy}{dx} = \frac{v\dfrac{du}{dx} - u\dfrac{dv}{dx}}{v^2}$$

If you are one of those people who does not like to accept rules without explanation, turn to Appendix IV and see how the Product and Quotient Rules are derived.

EXAMPLE 9.5

Differentiate:

(a) $y = \dfrac{x^2}{1 + x^3}$

(b) $y = \dfrac{e^x}{1 + 2e^x}$

Solution

(a) $u = x^2$, $\quad \dfrac{du}{dx} = 2x$, $\quad v = 1 + x^3$, $\quad \dfrac{dv}{dx} = 3x^2$

$$\frac{dy}{dx} = \frac{v\frac{du}{dx} - u\frac{dv}{dx}}{v^2} = \frac{(1+x^3)(2x) - x^2 3x^2}{(1+x^3)^2}$$

$$\frac{dy}{dx} = \frac{2x + 2x^4 - 3x^4}{(1+x^3)^2} = \frac{2x - x^4}{(1+x^3)^2}$$

(b) $u = e^x$, $\frac{du}{dx} = e^x$, $v = 1 + 2e^x$, $\frac{dv}{dx} = 2e^x$

$$\frac{dy}{dx} = \frac{(1+e^x)e^x - e^x 2e^x}{(1+e^x)^2} = \frac{e^x + e^{2x} - 2e^{2x}}{(1+e^x)^2}$$

$$\frac{dy}{dx} = \frac{e^x}{(1+e^x)^2}$$

EXAMPLE 9.6

Differentiate:

$$y = \tan x$$

Solution

From the trig. identity $\tan\theta \equiv \frac{\sin\theta}{\cos\theta}$

So, $y = \tan x = \frac{\sin x}{\cos x}$

$u = \sin x$, $\frac{du}{dx} = \cos x$, $v = \cos x$, $\frac{dv}{dx} = -\sin x$

$$\frac{dy}{dx} = \frac{v\frac{du}{dx} - u\frac{dv}{dx}}{v^2} = \frac{\cos x \cos x - \sin x(-\sin x)}{\cos^2 x}$$

$$\frac{dy}{dx} = \frac{\cos^2 x + \sin^2 x}{\cos^2 x} = \frac{1}{\cos^2 x} \quad (\text{since, } \cos^2\theta + \sin^2\theta = 1)$$

$$\frac{dy}{dx} = \sec^2 x \quad \left(\text{since, } \sec\theta = \frac{1}{\cos\theta}\right)$$

The previous exercise shows how the rules of differentiation, together with some trigonometric identities, can be used to generate further standard derivatives. We now have a sixth derivative that we can add to our stock.

If,

$$y = \tan x$$

then,

$$\frac{dy}{dx} = \sec^2 x$$

PRACTICE EXERCISE 9.1

Differentiate the following with respect to the appropriate variable.

(a) $y = 7x^2 - \dfrac{3}{x^2}$;   (b) $y = (1 + 3t)^2$;   (c) $y = 2\sqrt{x}(1 - 3x + 4x^3)$

(d) $y = 7\sec\theta$;   (e) $y = (1 - 2x)\ln x$;   (f) $y = \dfrac{1 - 2x}{1 + 2x}$

(g) $s = \dfrac{2 + \ln t}{4 + 2\ln t}$;   (h) $p = \dfrac{1}{(1 + q)(1 - q)}$;   (i) $x = \dfrac{1}{(2 + t)^2}$

Solutions

(a) $14x + \dfrac{6}{x^3}$;   (b) $6(3t + 1)$;   (c) $2\sqrt{x}(12x^2 - 3) + \dfrac{1}{\sqrt{x}}(4x^3 - 3x + 1)$;   (d) $7\sec\theta\tan\theta$

(e) $\dfrac{1}{x}(1 - 2x) - 2\ln x$;   (f) $-\dfrac{4}{(1 + 2x)^2}$;   (g) 0;   (h) $\dfrac{2q}{(1 - q^2)^2}$;   (i) $-\dfrac{2}{(2 + t)^3}$

The following problems will require the use of all the rules of differentiation which you have seen so far.

PRACTICE EXERCISE 9.2

(a) $y = (x + 3x^2)^5$;   (b) $y = (1 - 2x + x^2)^{-3}$;   (c) $y = (3x - 1)\ln 2x$

(d) $y = \dfrac{x^2}{e^{3x}}$;   (e) $y = \dfrac{3x + 5}{\ln(3x^2)}$;   (f) $y = e^{10t}\sin\left(3t + \dfrac{\pi}{3}\right)$

(g) The relationship between anode current I and the supply voltage V of a certain thyristor may be taken as $I = 6 + 3V + \frac{1}{2}V^2$. Find the rate of change of the anode current when the supply voltage is 2. Find, also, the value of V that makes $dI/dV = 1$.

(h) A body moving through a liquid has its position from a fixed point given by $s = 12(1 - e^{-t})$ m, where t is time in seconds. Determine an expression for both the velocity and acceleration of the body. Calculate the time at which the velocity of the body is $6\,\text{m s}^{-1}$.

(i) Differentiate the function $s = t(2t - 3)$ and find the gradient of the curve of this function at $t = 2$ seconds.

(j) A coil has a self-inductance of 4H and a resistance of 200 Ω. A d.c. supply of 100 V is applied to the coil. The instantaneous current iA is given by $i = \dfrac{E}{R}(1 - e^{-\frac{Rt}{L}})$ where t is time in seconds. Deduce an expression for the rate of change of current and find the value of the current 10 ms after the supply is switched on.

Solutions

(a) $\dfrac{dy}{dx} = 5(x + 3x^2)^4(1 + 6x)$;   (b) $\dfrac{dy}{dx} = \dfrac{6 - 6x}{(1 - 2x + x^2)^4}$;   (c) $\dfrac{dy}{dx} = 3 - \dfrac{1}{x} + 3\ln 2x$

(d) $\dfrac{dy}{dx} = \dfrac{2x - 3x^2}{e^{3x}}$; (e) $\dfrac{dy}{dx} = \dfrac{3}{\ln(3x^2)} - \dfrac{6x + 10}{x[\ln(3x^2)]^2}$

(f) $\dfrac{dy}{dt} = e^{10t}\left[3\cos\left(3t + \dfrac{\pi}{3}\right) + 10\sin\left(3t + \dfrac{\pi}{3}\right)\right]$

(g) $5AV^{-1}$, $-2\,V$; (h) $v = 12\,e^{-t}$, $a = -12\,e^{-t}$, $693\,\mathrm{ms}$; (i) 5; (j) $\dfrac{E}{L}\,e^{-\frac{Rt}{L}}$, $15.2\,\mathrm{A\,s^{-1}}$

Now, let's revisit an important application of differentiation – turning points.

EXAMPLE 9.7

A cylindrical tank with an open top is to be made to hold 300 cubic metres of liquid. Find the dimensions of the tank so that its surface area is a minimum.

Solution

First sketch the problem situation and write down the facts. Assign symbols. Let the radius be r and the height of the tank, h.

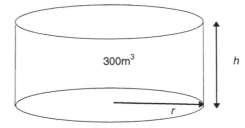

Formulate an equation of the function whose minimum is required. The total surface area is the sum of the area of the bottom and the area of the cylinder wall.

$$\text{total surface area} = \text{area of circle} + \text{surface area of cylinder}$$

$$A = \pi r^2 + 2\pi rh$$

The area equation yields two unknown terms, h and r.

Knowing the volume allows us to write an expression for h. Substituting this into the area equation gives us area as a function of radius alone.

$$V = \pi r^2 h = 300$$

$$h = \frac{300}{\pi r^2} \quad \text{and} \quad A = \pi r^2 + 2\pi r\left(\frac{300}{\pi r^2}\right)$$

$$A = \pi r^2 + 600r^{-1} \qquad \frac{dA}{dr} = 2\pi r - 600r^{-2}$$

We know that at turning points the rate of change of A with respect to r is equal to zero. This allows us to find the value of r which corresponds to a minimum or maximum area.

At turning points, $\dfrac{dA}{dr} = 0$

so,

$$2\pi r - 600r^{-2} = 0$$

$$2\pi r^3 - 600 = 0$$

$$2\pi r^3 = 600$$

$$r = \sqrt[3]{\frac{600}{2\pi}} = \left(\frac{300}{\pi}\right)^{\frac{1}{3}} = 4.57\,\text{m}$$

The second derivative is positive

$$\frac{d^2A}{dr^2} = 2\pi + 1200r^{-3}$$

so $r = 4.57$ corresponds to a minimum area.

We also want the height of the cylinder. We have a ready made equation for this.

$$h = \frac{300}{\pi r^2} = \frac{300}{\pi(4.57)^2} = 4.57\,\text{m}$$

Finally, we check the dimensions against given facts. Here the volume should be $300\,\text{m}^3$. Check,

$$V = \pi r^2 h = \pi(4.57)^2 \times 4.57 = 300$$

The solution is correct to 3 s.f. So required dimensions are $r = 4.57\,\text{m}$ and $h = 4.57\,\text{m}$.

This sort of problem always leaves me wondering why food is nearly always packaged in cans that are tall and narrow. The maths tells us that if the minimum amount of material is to be used in making a baked bean can, the can should be designed to have a diameter equal to its height!

EXAMPLE 9.8

A lever having a weight of 10 N per metre of length is pivoted as shown in Fig. 9.3. Find the length of the lever so that the equilibrium force F is a minimum; hence find the value of the force.

To get started, let x be the length of the lever. Then you must remember that moments are products of force and distance from the pivot.

The distributed force of the weight of the lever acts at the mid-point of the lever, i.e. a distance $x/2$ from the pivot.

Equating downward and upward moments allows you to formulate an equation in which force is a function of distance.

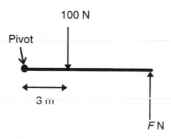

Figure 9.3 Pivoted lever

To find a turning point, differentiate once. To test for a minimum differentiate a second time. Then evaluate your solution by back-substitution.

Solution

Let x be the length of the lever so that,

$$\text{Downward moments} = 100(3) + 10x\left(\frac{x}{2}\right)$$

$$= 300 + 5x^2$$

Upward moments $= Fx$

In equilibrium, $Fx = 300 + 5x^2$

$$F = 300x^{-1} + 5x$$

$$\frac{dF}{dx} = -300x^{-2} + 5$$

At turning points, $5 - 300x^{-2} = 0$

$$5 = 300x^{-2}, \qquad x^{-2} = \frac{5}{300}$$

$$x^2 = \frac{300}{5}, \qquad x = \pm\sqrt{\frac{300}{5}} = 7.75\,\text{m (discounting } -7.75\,\text{m)}$$

$$\frac{d^2F}{dx^2} = 600x^{-3}$$

is positive so the turning point is a minimum.

$$F = \frac{300}{x} + 5x = \frac{300}{7.75} + 5(7.75) = 77.5\,\text{N}$$

Check,

$$Fx = 300 + 5x^2 \approx 600$$

EXAMPLE 9.9

The total hourly cost in £sterling of operating a printing press is calculated according to the formula:

$$C_h = 75 + 0.001N^2,$$

where N is the number of posters printed per hour. Formulate a suitable equation that gives the total cost of printing 2 million posters then determine the value of N which gives the minimum total cost.

Solution

Total hours to produce 2×10^6 posters $= \dfrac{2 \times 10^6}{N}$

Total cost, $C_T = \dfrac{2 \times 10^6}{N}(75 + 0.001N^2) = \dfrac{150 \times 10^6}{N} + 2000N$

$$C_T = 150 \times 10^6 \, N^{-1} + 2000N \qquad \frac{dC_T}{dN} = -150 \times 10^6 N^{-2} + 2000$$

At turning point, $2000 - 150 \times 10^6 \, N^{-2} = 0$,

$$2000N^2 - 150 \times 10^6 = 0$$

$$N = \pm\sqrt{\frac{150 \times 10^6}{2000}} = \pm 274$$

-274 is not rational so $N = 274$

$$\frac{d^2C_T}{dN^2} = 300 \times 10^6 \, N^{-3},$$

+ve value confirms a minimum

9.2   Applied integration

You have seen how it is possible, by intregration, to find the precise area under the curve of a known function. Let's run through it again.

The approximate area under the curve, shown in Fig. 9.4, can be found by taking the sum of the areas of a number of rectangular strips over the region of interest. The region of interest is specified by the lower and upper limits, $x = a$ and $x = b$ respectively. If the area of each strip has a width δx and a height y then we can say that:

$$\text{Area} = \sum_{x=a}^{x=b} y\delta x \text{ approximately}$$

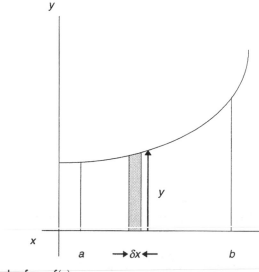

Figure 9.4 Graph of $y = f(x)$

The symbol Σ is the upper case of the Greek letter 'sigma' which represents 'the sum of'. Now, to find the area precisely we need to make δx very small which leads to the formula for integration:

$$\text{As } \lim_{\delta x \to 0}, \quad \text{Area} \approx \sum_{x=a}^{x=b} y\delta x = \int_a^b y\,dx$$

Mean value

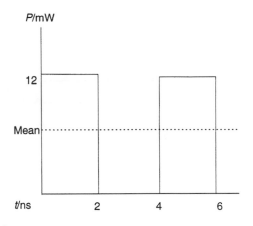

Figure 9.5 Power waveform

Let's return to something we dealt with in Chapter 6. The graph of Fig. 9.5 is a power waveform representing digital pulses of light transmitted along fibre-optic cable.

One cycle is composed of a '1' and a '0'. In the time that a 1 is being transmitted, power is at a peak value of 12 mW, in the time that a 0 is being transmitted power is zero.

Assuming that the transmission consists of a stream of alternating 1s and 0s then it is quite simple to calculate the mean power because the mean power of one cycle is the mean power of all cycles.

We take the area under one pulse. Now, this is power multiplied by time so we have found the energy in the pulse in joules.

Then we divide by the 'width' of a complete cycle. Dividing joules by seconds gives us watts again.

$$\text{Area} = 12 \times 10^{-3} \times 2 \times 10^{-9}$$

$$= 24\,\text{pJ}$$

$$\text{Mean} = \frac{24 \times 10^{-12}}{4 \times 10^{-9}} = 6\,\text{mW}$$

Now, we have a general rule for finding the mean value graphically. Do you remember it?

$$\text{Mean} = \frac{\text{area under curve}}{\text{length of base}}$$

Length of base represents the 'region of interest'.

Now finding the mean of a function which is a square wave is easy. You may have known intuitively, at the start, that the mean power of the light pulses is 6 mW.

How do we deal with other functions? Well, we can find the area under the curve by finding the sum of the areas of a number of strips and then divide by the difference of the limits of the area.

More formally:

$$\text{Mean} = \frac{1}{b-a} \sum_{x=a}^{x=b} y \delta x$$

Take a linear function.

EXAMPLE 9.10

Find the mean value of $y = 2x$ between $x = 0$ and $x = 4$.

To find the precise mean we do not use graphical or numerical methods unless we have to. We integrate analytically wherever possible.

$$\text{Mean} = \frac{1}{b-a} \sum_{x=a}^{x=b} y \delta x$$

$$= \frac{1}{b-a} \int_a^b y \, dx$$

$$= \frac{1}{4-0} \int_0^4 2x \, dx = \frac{1}{4} \left[x^2 \right]_0^4$$

$$= \frac{16}{4} = 4$$

Figure 9.6 is a sketch of the curve of the function in Example 9.10.

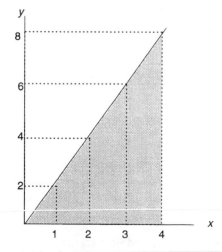

Figure 9.6 Graph of $y = 2x$, as in Example 9.10

Notice that, since this is a linear function, finding its mean involved finding the area of a triangle with a height of 8 and a base of 4. Normally we would do this by using:

$$\text{Area} = \tfrac{1}{2}bh$$

$$= \tfrac{1}{2}(4)(8) = 16$$

and then to get the mean height of the triangle we divide by the length of the base:

$$\text{Mean} = \frac{\text{Area}}{\text{Base}} = \frac{16}{4} = 4$$

However, this short cut would not give us the precise mean of a non-linear function.

EXAMPLE 9.11

The voltage across a charging capacitor is described by the function:

$$v = 12(1 - e^{-0.1t}) \text{ volts}$$

Find the mean voltage across the capacitor over the interval of time between 0 and 50 seconds.

First, expand the brackets to avoid having to use the product rule.

Integrating with respect to t involves a constant and an exponential so we use:

$$\frac{ax^{n+1}}{n+1} \quad \text{and} \quad \frac{1}{k}a\,e^{kx}$$

Then find the area by evaluating the definite integral (remember that we can omit the constant of integration) and finally divide by the difference of the limits of integration.

$$v = 12 - 12\,e^{-0.1t}$$

$$\bar{v} = \frac{1}{50 - 0} \int_0^{50} (12 - 12\,e^{-0.1t})\,dt$$

$$= \frac{1}{50} \left[12t + 120\,e^{-0.1t} \right]_0^{50}$$

$$= \frac{1}{50} [(12(50) + 120\,e^{-0.1(50)}) - (120\,e^0)]$$

$$= \frac{1}{50} [(600 + 0.809) - (120)]$$

$$= 9.62\,\text{V}$$

A sketch of the curve of the function (see Fig. 9.7) shows the capacitor charging toward a steady-state value of 12 V and confirms that our mean value makes sense.

EXAMPLE 9.12

Prove that the mean of a sine wave is equal to zero.

Take the general equation of a sinusoid and integrate over a full cycle, i.e. over the interval between $\theta = 0$ and $\theta = 2\pi$.

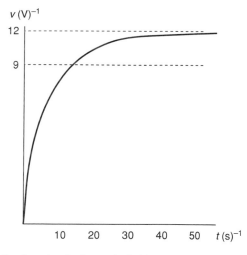

Figure 9.7 Graph of the function in Example 9.11

Solution

$$a = A \sin \theta$$

$$\bar{a} = \frac{1}{2\pi - 0} \int_0^{2\pi} (A \sin \theta) \, d\theta$$

$$= \frac{1}{2\pi} \left[-A \cos \theta \right]_0^{2\pi}$$

$$= \frac{1}{2\pi} (-A \cos 2\pi) - (-A \cos 0)$$

$$= \frac{1}{2\pi} (-1 + 1) = 0$$

Sketching a curve of a sine wave will show that this makes sense. Since the *positive area* is equal to the *negative area*, the average value must be zero.

EXAMPLE 9.13

Find the mean velocity of an object whose position is described by:

$$s = 4t^3 + 3t^2 - 5t + 10$$

over the interval $t = -10\,\text{s}$ and $t = 10\,\text{s}$.

Solution

$$v = \frac{ds}{st} = 12t^2 + 6t - 5$$

$$\bar{v} = \frac{1}{10 - (-10)} \int_{-10}^{10} (12t^2 + 6t - 5) \, dt$$

$$= \frac{1}{20} \left[4t^3 + 3t^2 - 5t \right]_{-10}^{10}$$

$$= \frac{1}{20} [(4(10)^3 + 3(10)^2 - 5(10)) - (4(-10)^3 + 3(-10)^2 - 5(-10))]$$

$$= \tfrac{1}{20}[(4000 + 300 - 50) - (-4000 + 300 + 50)]$$

$$= \tfrac{1}{20}(4250) - (-3650)$$

$$= \tfrac{1}{20}(7900)$$

$$= 395 \,\mathrm{m\,s}^{-1}$$

It is worth examining what we have done here.

By integrating the velocity equation we found the position equation.

In evaluating the definite integral we obtained the position of the object at 10 s as being 4250 m and at −10 s as −3650 m. The difference of these positions is 7900 m, the distance travelled. Dividing distance travelled by time taken, 20 s gave us average velocity over that time interval.

R.M.S. value

Since the average value of an alternating wave is zero, using the average value to calculate the power of a voltage or current is of no use – it has no meaning. We use, instead, the r.m.s. value. The r.m.s. value gives us the a.c. equivalent of d.c. So, for example, power calculations with 240 V r.m.s. give us the equivalent power of 240 V d.c.

To find the r.m.s. value we must:

take the square root of the mean of the squared values

The mathematical notations for this are shown below.

$$\text{r.m.s.} \approx \sqrt{\frac{1}{b-a}\sum_{x=a}^{x=b} y^2 \delta x}$$

or,

$$\text{r.m.s.} = \sqrt{\frac{1}{b-a}\int_a^b y^2 \, dx}$$

EXAMPLE 9.14

Calculate the r.m.s. value of the function:

$$y = x - x^2$$

over the interval $x = -1$ and $x = 1$.

Solution

$$\text{r.m.s.} = \sqrt{\frac{1}{b-a}\int_a^b y^2 \, dx} = \sqrt{\frac{1}{1-(-1)}\int_{-1}^1 (x - x^2)^2 \, dx}$$

$$= \sqrt{\frac{1}{2}\int_{-1}^1 (x - x^2)(x - x^2) \, dx} = \sqrt{\frac{1}{2}\int_{-1}^1 (x^2 - 2x^3 + x^4) \, dx}$$

$$= \sqrt{\frac{1}{2}\left[\frac{x^3}{3} - \frac{x^4}{2} + \frac{x^5}{5}\right]_{-1}^{1}}$$

$$= \sqrt{\frac{1}{2}\left(\frac{1}{3} - \frac{1}{2} + \frac{1}{5}\right) - \left(-\frac{1}{3} - \frac{1}{2} - \frac{1}{5}\right)}$$

$$= \sqrt{\frac{1}{2}\left[\left(\frac{10 - 15 + 6}{30}\right) - \left(\frac{-10 - 15 - 6}{30}\right)\right]}$$

$$= \sqrt{\frac{1}{2}\left(\frac{1}{30} + \frac{31}{30}\right)} = \sqrt{\frac{32}{60}} = 0.730$$

To solve the next problem we need to use an important trigonometric identity for $\sin^2 x$.

$$\sin^2 x \equiv \tfrac{1}{2}(1 - \cos 2x)$$

The identity will be explained in Chapter 14.

EXAMPLE 9.15

Calculate the r.m.s. value of the voltage:

$$v = 10 \sin \theta$$

over the interval of one cycle.

Solution

$$\text{r.m.s.} = \sqrt{\frac{1}{2\pi - 0}\int_0^{2\pi} (10 \sin \theta)^2 \, d\theta}$$

$$= \sqrt{\frac{1}{2\pi}\int_0^{2\pi} (100 \sin^2 \theta) \, d\theta}$$

$$= \sqrt{\frac{100}{2\pi}\int_0^{2\pi} \frac{1}{2}(1 - \cos 2\theta) \, d\theta} = \sqrt{\frac{100}{4\pi}\int_0^{2\pi} (1 - \cos 2\theta) \, d\theta}$$

$$= \sqrt{\frac{25}{\pi}\int_0^{2\pi} (1 - \cos 2\theta) \, d\theta} = \sqrt{\frac{25}{\pi}\left[\theta - \frac{1}{2}\sin 2\theta\right]_0^{2\pi}}$$

$$= \sqrt{\frac{25}{\pi}\left[\left(2\pi - \frac{1}{2}\sin 4\pi\right) - \left(0 - \frac{1}{2}\sin 0\right)\right]}$$

$$= \sqrt{\frac{25}{\pi}(2\pi)} = \sqrt{50} = 7.07 \, \text{V}$$

Two things: we do not have a standard integral to deal with $\sin^2 \theta$ so it was necessary to substitute the appropriate identity before we could perform the integration.

Secondly, the result is 0.707 of the peak value. You have come across this value before.

EXAMPLE 9.16

Show that the r.m.s. value of any sinusoid is $\dfrac{1}{\sqrt{2}}$ times its peak value.

Solution

$a = A\sin\theta$

$$\text{r.m.s.} = \sqrt{\frac{1}{2\pi - 0}\int_0^{2\pi}(A\sin\theta)^2\,d\theta}$$

$$= \sqrt{\frac{A^2}{2\pi}\int_0^{2\pi}(\sin^2\theta)\,d\theta} = \sqrt{\frac{A^2}{2\pi}\int_0^{2\pi}\frac{1}{2}(1 - 2\cos\theta)\,d\theta}$$

$$= \sqrt{\frac{A^2}{4\pi}\Big[\theta - 2\sin\theta\Big]_0^{2\pi}}$$

$$= \sqrt{\frac{A^2}{4\pi}[(2\pi - 2\sin 2\pi) - (0 - 2\sin 0)]}$$

$$= \sqrt{\frac{A^2}{4\pi}(2\pi)} = \sqrt{\frac{A^2}{2}} = \frac{A}{\sqrt{2}} = A\frac{1}{\sqrt{2}}$$

This example shows where we get our conversion factor of 0.707. Note that it *only applies to sinusoids*.

EXAMPLE 9.17

Find the average and the r.m.s. values of the waveform shown in Fig. 9.8.

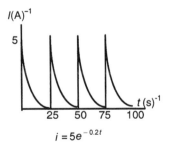

$i = 5e^{-0.2t}$

Figure 9.8 Graph of $i = 5e^{-0.2t}$

Solution

$$\text{average} = \frac{1}{25 - 0}\int_0^{25}5e^{-0.2t}\,dt = \frac{5}{25}\int_0^{25}e^{-0.2t}\,dt$$

$$= \frac{1}{5}\left[-\frac{1}{0.2}e^{-0.2t}\right]_0^{25}$$

$$= \tfrac{1}{5}[(-5e^{-5}) - (-5e^{0})]$$

$$= \tfrac{1}{5}[(-33.7 \times 10^{-3}) + 5] = 993\,\text{mA}$$

$$\text{r.m.s.} = \sqrt{\frac{1}{25-0}\int_0^{25}(5e^{-0.2t})^2\,dt} = \sqrt{\frac{25}{25}\int_0^{25}(e^{-0.4t})\,dt}$$

$$= \sqrt{\int_0^{25}(e^{-0.4t})\,dt} = \sqrt{\left[-\frac{1}{0.4}e^{-0.4t}\right]_0^{25}}$$

$$= \sqrt{(-2.5e^{-10})-(-2.5e^0)} = \sqrt{(-113\times10^{-6})+2.5}$$

$$= 1.58\,\text{A}$$

Note, the function is periodic so we only need to concern ourselves with one cycle.

First moments of area

CENTROIDS

Centroids were introduced in Chapter 3 where we used them to apply Pappus' theorem to find volumes. Let's take another look.

A plane area having negligible uniform thickness is called a **lamina**. The geometric centre of a lamina is called its **centroid**. You may think of the centroid of an area as its centre of gravity – the point at which the force of gravity is focused. The centroid is, in fact, the focal point of any force acting on the area and it is also the point about which the area balances in equilibrium. It is useful to know the position of a centroid when calculating surface areas, volumes, moments of forces and moments of inertia.

With a regular area such as the rectangle shown in Fig. 9.9, the centroid G lies at the point where the lines of symmetry AB and CD intersect.

Note that a line of symmetry is a line that bisects an area.

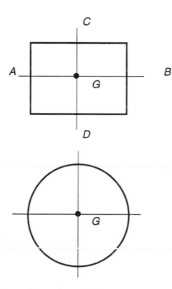

Figure 9.9 Centroids of a rectangle and a circle

The centroid of a circular area is found just as easily. It is the point where the lines of symmetry (the diameters) intersect.

With irregular areas it may be more difficult to find the centroid. However, if we are dealing with an area whose boundary can be described by some known mathematical function we can apply calculus and look for what is known as the **first moment of area**.

FIRST MOMENT OF AREA

This is defined as follows.

> **About a given axis the first moment of area is the product of the area and the perpendicular distance of its centroid from that axis**

The mathematical notation for this is:

1st moment of area about the y-axis $= \bar{x}A$

1st moment of area about the x-axis $= \bar{y}A$

Consider the area, between the limits of a and b, under the curve shown in Fig. 9.10.

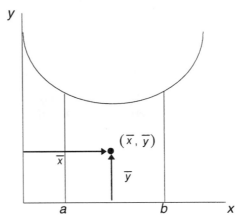

Figure 9.10 Area under a curve

$\bar{x}$ is the distance of the centroid from the y-axis and $\bar{y}$ is its distance from the x-axis so the position of the centroid has coordinates $(\bar{x}, \bar{y})$.

We could divide the area under the curve into thin strips (Fig. 9.11) and consider each strip as a rectangle with an error that tends to zero as the width of each strip tends to zero.

Turning to Fig. 9.11, we also know that the position of the centroid of each rectangular strip is half way along each dimension so we can write down its coordinates.

$$(x + \tfrac{1}{2}\delta x, \tfrac{1}{2}y)$$

Now if δx is very small the position coordinates of the centroid approximate to:

$$(x, \tfrac{1}{2}y)$$

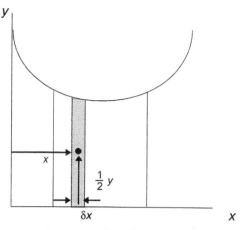

Figure 9.11 Dividing the area under a curve into thin strips

and since the approximate area of each strip:

$$A = y\delta x$$

Then the moment about the y-axis is:

$$xy\delta x$$

This is the moment of area of a rectangular strip, so for the plane area we must integrate to get the sum of these moments.

Now, remember that the 1st moment of area is the product:

$$\bar{x}A$$

So dividing by area gives the distance of the centroid from the y-axis:

$$\frac{\bar{x}A}{A} = \bar{x}$$

And since area:

$$A = \int_a^b y\,dx$$

we obtain a formula which allows us to find a centroid by integration.

$$\text{1st moment of area} = \sum_{x=a}^{x=b} xy\delta x$$

$$\text{when } \lim_{\delta x \to 0}$$

$$\text{1st moment of area} = \int_a^b xy\,dx$$

Distance of a centroid from the y-axis:

$$\bar{x} = \frac{\displaystyle\int_a^b xy\,dx}{\displaystyle\int_a^b y\,dx}$$

So finding the 1st moment of an area about the y-axis enables us to find the x coordinate of the centroid of an area. We simply divide by the area.

What about the y coordinate of the centroid? Look at Fig. 9.11 again.

The first moment of the area of the rectangular strip *about the x-axis* is the product of the distance of the centroid from the x-axis and the area of the strip:

$$\tfrac{1}{2}yy\delta x = \tfrac{1}{2}y^2\delta x$$

Integrating the moments of each elemental strip between the limits of x gives us the 1st moment of area about the x-axis.

$$\text{1st moment of area} = \sum_{x=a}^{x=b} \tfrac{1}{2}y^2\delta x$$

when $\lim\limits_{\delta x \to 0}$

$$\text{1st moment of area} = \int_a^b \tfrac{1}{2}y^2 \, dx$$

Then dividing by area, as before, gives us the formula to find the y coordinate of the centroid by integration.

$$\bar{y} = \frac{\tfrac{1}{2}\int_a^b y^2 \, dx}{\int_a^b y \, dx}$$

EXAMPLE 9.18

Find the x coordinate of the centroid of the area under the curve of

$$y = 4x - 2x^2$$

between $x = 0$ and $x = 2$.

This is a quadratic with roots at $x = 0$ and $x = 2$ and the turning point is a maximum, $y = 2$ at $x = 1$. So we can sketch a curve of the function and get a better idea of what we are doing, see Fig. 9.12.

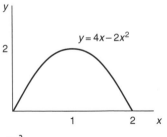

Figure 9.12 Graph of $y = 4x - 2x^2$

Solution

1st moment of area about the y-axis:

$$\int_a^b xy \, dx = \int_0^2 x(4x - 2x^2) \, dx$$

$$= \int_0^2 (4x^2 - 2x^3) \, dx = \left[\frac{4}{3}x^3 - \frac{1}{2}x^4 \right]_0^2$$

$$= \left(\frac{4}{3}(2)^3 - \frac{1}{2}(2)^4\right) - \left(\frac{4}{3}(0)^3 - \frac{1}{2}(0)^4\right)$$

$$= 2.67$$

Area:

$$\int_a^b y \, dx = \int_0^2 (4x - 2x^2) \, dx$$

$$= \left[2x^2 - \frac{2}{3}x^3\right]_0^2 = 2(2)^2 - \frac{2}{3}(2)^3 = 2.67$$

$$\bar{x} = \frac{\int_a^b xy \, dx}{\int_a^b y \, dx} = \frac{2.67}{2.67} = 1$$

EXAMPLE 9.19

Find the y coordinate of the centroid of the area under the curve

$$y = 4x - 2x^2$$

between $x = 0$ and $x = 2$.

Solution

1st moment of area about the x-axis:

$$\frac{1}{2}\int_a^b y^2 \, dx = \frac{1}{2}\int_0^2 (4x - 2x^2)^2 \, dx$$

$$= \frac{1}{2}\int_0^2 (16x^2 - 16x^3 + 4x^4) \, dx = \frac{1}{2}\left[\frac{16}{3}x^3 - 4x^4 + \frac{4}{5}x^5\right]_0^2$$

$$= \frac{1}{2}\left[\left(\frac{16}{3}(2)^3 - 4(2)^4 + \frac{4}{5}(2)^5\right) - \left(\frac{16}{3}(0)^3 - 4(0)^4 + \frac{4}{5}(0)^5\right)\right]$$

$$= 2.13$$

$$\text{Area} = 2.67$$

$$\bar{y} = \frac{\int_a^b y^2 \, dx}{\int_a^b y \, dx} = \frac{2.13}{2.67} = 0.80$$

An evaluation of the solutions obtained in the last two examples indicates that the centroid of the area in question has coordinates (1, 0.8).

Look at the sketch of the curve in Fig. 9.13. Do you feel the result makes sense?

The x-component of the centroid lies along the vertical axis of symmetry of the curve so that is in order.

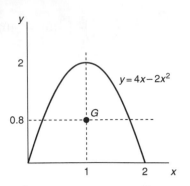

Figure 9.13 Graph of $y = 4x - 2x^2$ with centroid shown

The y-component of the centroid is a little below half the maximum value. Since the area is wider in the lower regions, this seems in order too.

PRACTICE EXERCISE 9.3

Find the coordinates of the centroid of the area under the curve $y = 3x^2$ between $x = 1$ and $x = 4$.

Solution

(3.04, 14.61)

Volumes of revolution

You know how to calculate the area under the curve of some known function.

If the area is now rotated through 2π radians then a solid is formed.

Consider the function whose curve is sketched in Fig. 9.14.

Rotating the strip of width δx through 360° describes a circular disc whose thickness is δx and surface area is πy^2.

Now, multiplying the area of the disc by thickness gives the volume of the disc:

$$\pi y^2 \delta x$$

Integrating a large number of such discs over the region from a to b gives the volume of the solid described by the rotation of the function:

$$V \approx \sum_{x=a}^{x=b} \pi y^2 \delta x$$

when $\lim_{\delta x \to 0}$

$$V = \int_a^b \pi y^2 \, dx$$

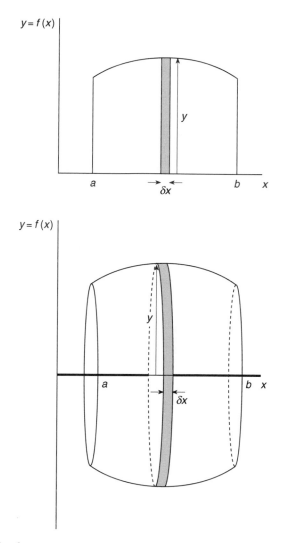

Figure 9.14 Sketch of a curve

EXAMPLE 9.20

Determine the volume generated by rotating the parabola,

$$y^2 = 10x$$

about the x-axis between the points $x = 0$ and $x = 3.5$.

Solution

$$V = \int_a^b \pi y^2 \, dx = \pi \int_0^{3.5} 10x \, dx$$

$$= \pi \left[5x^2 \right]_0^{3.5} = \pi \left(5(3.5)^2 \right) - \left(5(0)^2 \right)$$

$$= 192 \text{ units}^3$$

EXAMPLE 9.21

A parabolic dish is to be manufactured for a communications antenna. The outer dimensions are radius: 10 m, depth: 1 m, hence the curve of the antenna dish is given by:

$$y^2 = 100x$$

as shown in Fig 9.15.

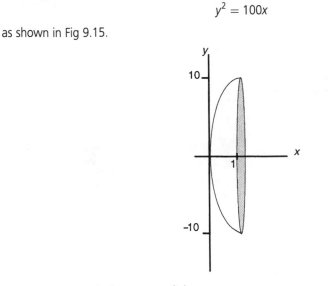

Figure 9.15 Parabolic antenna dish

Calculate the volume of material required to manufacture a dish having a uniform thickness of 10 mm such that the curve of the inner surface has a maximum depth of 0.990 m.

The volume of material is the difference between the outer volume and the inner volume.

To find the outer volume we must integrate between the limits of 0 and 1 m.

The inner volume is found by the same integral with the upper limit changed to 0.99 m.

Solution

$$V = \int_a^b \pi y^2 \, dx$$

Outer volume,

$$\pi \int_0^1 100x \, dx = \pi \left[50x^2 \right]_0^1 = 50\pi \, \text{m}^3$$

$$\text{Inner volume} = \pi \left[50x^2 \right]_0^{0.99} = 153.95 \, \text{m}^3$$

$$\text{Volume of material} = 50\pi - 153.95 = 3.13 \, \text{m}^3$$

A suitable rough check might be to consider the dish to be a circular disc, having a radius of 10 m and thickness of 10 mm. The area of the disc multiplied by the thickness is:

$$\pi(10)^2(0.01) = 3.14$$

Close enough!

As with most problems involving areas and volumes, it is always wise to start by sketching a curve of the area/volume in question. Bear that in mind when you tackle the following problems.

PRACTICE EXERCISE 9.4

(a) Determine the volume generated by the parabola $y^2 = 10x$ when rotating about the x-axis between the points $x = 1$ and $x = 3$.

(b) A function is described by $y = \sqrt{x+1}$. If the curve of this function is rotated once about the horizontal axis find the volume generated between $x = 4$ and $x = 9$.

(c) Simplify the expression for y where $y = \dfrac{x+1}{x}$ and rotate the curve of the function through 2π radians about the x-axis and determine the volume generated between $x = 1.5$ and $x = 2.5$.

Solutions

(a) 40π units^3;   (b) 37.5π units^3;   (c) 7.19 units^3

The following summary outlines the essentials of what has been studied in the last three chapters. You may find it a useful prompt as you work through your next assignment.

9.3   Summary of useful information

$y = f(x)$	$\dfrac{dy}{dx}$
ax^n	anx^{n-1}
$\sin ax$	$a \cos ax$
$\cos ax$	$-a \sin ax$
$\tan ax$	$a \sec^2 ax$
e^{ax}	$a\, e^{ax}$
$\ln ax$	$\dfrac{1}{x}$
uv	$u\dfrac{dv}{dx} + v\dfrac{du}{dx}$
$\dfrac{u}{v}$	$\dfrac{v\dfrac{du}{dx} - u\dfrac{dv}{dx}}{v^2}$
$f(u)$	$\dfrac{dy}{du} \times \dfrac{du}{dx}$

$$\text{If } s = f(t), \ v = \frac{ds}{dt} \text{ and } a = \frac{d^2s}{dt^2}$$

$$\sin^2 x = \tfrac{1}{2}(1 - \cos 2x)$$

$y = f(x)$	$\int y \, dx$
ax^n	$\dfrac{ax^{n+1}}{n+1} + C \ (n \neq -1)$
$\sin ax$	$-\dfrac{1}{a}\cos ax + C$
$\cos ax$	$\dfrac{1}{a}\sin ax + C$
$\sec^2 ax$	$\dfrac{1}{a}\tan ax + C$
e^{ax}	$\dfrac{1}{a}e^{ax} + C$
$\dfrac{1}{x}$	$\ln x + C$

$$\lim_{\delta x \to 0} \sum y\delta x = \int y \, dx$$

$$\text{Area} = \sum_{x=a}^{x=b} y\delta x$$

$$\text{Average} = \frac{1}{b-a}\sum_{x=a}^{x=b} y\delta x$$

$$\text{r.m.s.} = \sqrt{\frac{1}{b-a}\sum_{x=a}^{x=b} y^2\delta x}$$

$$\text{1st moment of area: } \bar{x}A = \sum_{x=a}^{x=b} xy\delta x$$

$$\text{Volume of revolution} = \sum_{x=a}^{x=b} \pi y^2\delta x$$

Assignment VII

Use of calculus

This is a 'desk-top' assignment that requires you to use differential and integral calculus to solve engineering problems. Your work should meet the following performance criteria.

- Representation of engineering situations using functions.
- Use of rules of differentiation to differentiate combinations of functions.
- Integration of standard functions using indefinite and definite integration.
- Modelling engineering situations using calculus.
- Solution of engineering problems using calculus.

If your work does not meet these criteria you should go back over the calculus chapters, revise and then tackle the relevant questions again.

The recommended time of this assignment is 2 hours. You should work alone, under normal exam conditions.

The outcome of this assignment should be eight correct solutions that demonstrate a proper use of the language of engineering mathematics and bring together your scientific and engineering knowledge with observations. Your results should be given to an appropriate degree of accuracy and state the correct SI units where required.

Do your best!

QUESTION 1

The distance of a mass moving in a straight line is modelled by:

$$s = 4t^2 + 3t + 2$$

where s is distance in metres and t is time in seconds. Determine the force acting on a mass of 2000 kg. Sketch time related curves of the position, speed and acceleration of the mass with respect to time.

QUESTION 2

The self-induced e.m.f. in an inductor can be found by using:

$$e = L\frac{di}{dt} \text{ volt}$$

where L is the inductance in henrys. Calculate the induced e.m.f. when an inductor of 270 mH passes a current:

$$i = 12\sin\left(314t + \frac{\pi}{6}\right)\text{ampere}$$

at the instant when time $t = 10$ ms.

QUESTION 3

Neglecting gravitational force, the position of a missile is described by:

$$s = 42t\ln(1 + t)$$

where s is the displacement in metres of the missile and t is time in seconds. Calculate the acceleration of the missile 1 s and 10 s after it has been fired.

QUESTION 4

The total hourly cost in pounds of operating an assembly line for gearboxes is calculated according to the formula:

$$C_h = 200 + 0.1N^3$$

where N is the hourly production rate, i.e. the number of gearboxes assembled per hour. Formulate a suitable equation which gives the total cost of assembling 5000 gearboxes, then determine the value of N which gives the minimum total cost of assembling that number of gearboxes.

QUESTION 5

The speed of an object is described by:

$$v = 40 - 6t + 10\,e^{-0.1t}$$

find the distance of the object at $t = 5$ if it is known that at $t = 10$, $s = 102$.

QUESTION 6

The general equation of a sinusoid is:

$$a = A\sin\phi$$

where a is the instantaneous value and A is the peak value. Find a general equation for the r.m.s. value of a sinusoid.

QUESTION 7

Find the first moment of area about the y-axis of the area under

$$y = 2x^2 - 3x$$

between $x = 0$, $x = 3$ metres.

QUESTION 8

A parabolic dish is to be manufactured for a communications antenna. The outer dimensions are radius: 2 m, depth: 0.5 m, hence the curve of the antenna is given by:

$$y^2 = 8x$$

as shown.

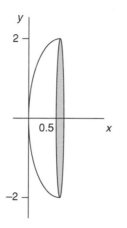

Calculate the volume of material required to manufacture a dish having a uniform thickness of 2 mm such that the curve of the inner surface has a maximum depth of 0.498 m.

End

chapter

10 Numerical methods

Numerical integration

There are many functions that cannot be integrated analytically, either because we do not know (or remember!) the standard integrals or simply because an integral does not exist for a particular function. In such a case we must resort to numerical methods of integration.

You may remember that there are rules for finding the approximate area of an irregular plane. Since these rules provide us with a way of finding an area then they are, in fact, rules for finding a definite integral – approximately at least. Let's deal with them individually.

Mid-ordinate rule

This rule is familiar to you. You used it in Chapter 6 to find the average values of sinusoids.

$$\textbf{Area} = \textbf{width of each strip} \times \textbf{the sum of the mid-ordinates}$$

$$A = by_1 + by_2 + by_3 + \cdots$$
$$= b(y_1 + y_2 + y_3 + \cdots)$$

EXAMPLE 10.1

Use the mid-ordinate rule to find the average value of $y = 2x^2 + 4$ over the interval $x = 0$ and $x = 4$. Then apply integration to find the average value of the function over the same interval and compare the results.

Divide the area into four strips of width $b = 1$. This sets the strip boundaries at $x = 0$, $x = 1$, $x = 2$, $x = 3$, $x = 4$ and the mid-ordinates at $x = 0.5$, $x = 1.5$, $x = 2.5$, $x = 3.5$.

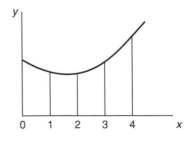

Calculate the values of the ordinates.

$$y = 2x^2 + 4$$

$$y(0.5) = 2(0.5)^2 + 4 = 4.5$$

$$y(1.5) = 2(1.5)^2 + 4 = 8.5$$

$$y(2.5) = 2(2.5)^2 + 4 = 16.5$$

$$y(3.5) = 2(3.5)^2 + 4 = 28.5$$

The average value is found by dividing the area under the curve by the base.

$$Av = \frac{1}{b-a} \quad \text{Area} \approx \frac{1}{4}b(y_1 + y_2 + y_3 + y_4)$$

$$Av \approx \tfrac{1}{4}(4.5 + 8.5 + 16.5 + 28.5) = 14.5$$

The mid-ordinate rule with $b = 1$ gives the average value as 14.5 correct to three s.f.

Applying analytical integration involves taking the definite integral to find the area and also dividing by the length of base.

$$Av = \frac{1}{b-a}\int_0^4 (2x^2 + 4)\,dx = \frac{1}{4}\left[\frac{2}{3}x^3 + 4x\right]_0^4$$

$$= \frac{1}{4}\left(\frac{2}{3}(4)^3 + 4(4)\right) = 14.7 \text{ units}^2$$

We get 14.7 correct to 3 s.f.

Had we divided the area into 40 strips, each having a width of 0.1, then probably correct to 3 s.f. the two results would be identical.

You might be thinking that the analytical method is much easier, particularly if the numerical method requires the calculation of 40 ordinates! Well, you must remember that for this example, the chosen function is one for which we have a standard integral that is easily applied. What would you do with something like $y = \ln x$?

Another point to make is that the power of modern computers is such that we can numerically integrate as accurately as we wish. The 'brute force' computing power of modern machines is setting a new trend in mathematics; a shift from analytical to numerical methods.

Trapezoidal rule

When we apply the mid-ordinate rule we divide an area into strips which approximate a rectangle. With the trapezoidal rule we divide the area into strips which approximate a trapezium (Fig. 10.1).

The trapezoidal rule can be a more accurate method depending upon the shape of the irregular area. It is applied in a similar manner to the mid-ordinate rule:

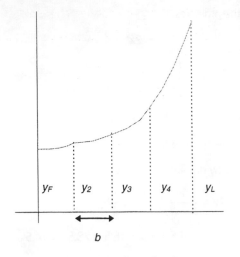

Figure 10.1 Dividing the area into trapezium-shaped strips

- Divide the area into equal strips b
- Treat each area as a trapezium whose area is given by:

$$A_1 = b \times \tfrac{1}{2}(y_1 + y_2)$$

$$A_2 = b \times \tfrac{1}{2}(y_2 + y_3)$$

etc.

- The approximate total area is the sum of the areas of the trapezia.

Total area:

$$A \approx [\tfrac{1}{2}b(y_F + y_2) + \tfrac{1}{2}b(y_2 + y_3) + \tfrac{1}{2}b(y_3 + y_4) + \tfrac{1}{2}b(y_4 + y_L)]$$

$$= b(\tfrac{1}{2}y_F + \tfrac{1}{2}y_2 + \tfrac{1}{2}y_2 + \tfrac{1}{2}y_3 + \tfrac{1}{2}y_3 + \tfrac{1}{2}y_4 + \tfrac{1}{2}y_4 + \tfrac{1}{2}y_L)$$

$$= b(\tfrac{1}{2}y_F + y_2 + y_3 + y_4 + \tfrac{1}{2}y_L)$$

$$= b[\tfrac{1}{2}(y_F + y_L) + y_2 + y_3 + y_4]$$

So in general, by the trapezoidal rule:

$$A \approx b[\tfrac{1}{2}(y_F + y_L) + y_2 + \cdots]$$

**where y_F and y_L are the first and last
and y_2, etc. are the remaining ordinates**

EXAMPLE 10.2

Apply the trapezoidal rule to integrate $f(x) = \ln x$ over the interval $x = 0.5$ and 1.0 to an accuracy of 3 s.f. Start by sketching a curve of the function. Divide the area into five equal strips.

You should remember that the log of 1 is zero and the logs of numbers less than 1 are negative. The region of interest is between x is 0.5 and 1.

Remember, there is no such thing as the log of a negative number.

When doing repeated calculations with the same function it is useful to use functional notation. So if $x = 1$ then $f(x)$ becomes $f(1)$ and so on.

Make the first ordinate $f(0.5)$, the last ordinate $f(1.0)$ and the remaining ordinates: $f(0.6)$, $f(0.7)$, $f(0.8)$ and $f(0.9)$. The width of each strip, $b = 0.1$.

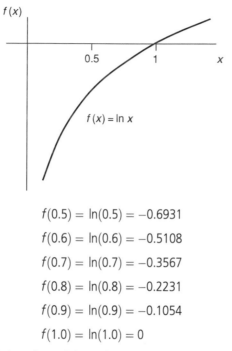

$$f(0.5) = \ln(0.5) = -0.6931$$
$$f(0.6) = \ln(0.6) = -0.5108$$
$$f(0.7) = \ln(0.7) = -0.3567$$
$$f(0.8) = \ln(0.8) = -0.2231$$
$$f(0.9) = \ln(0.9) = -0.1054$$
$$f(1.0) = \ln(1.0) = 0$$

To apply the rule we need the values of the ordinates to put into the formula.

$$\int_{0.5}^{1.0} (\ln x)\, dx \approx b[\tfrac{1}{2}(y_F + y_L) + y_2 + y_3 + y_4 + y_5]$$

$$= 0.1[\tfrac{1}{2}(-0.6931) - 0.5108 - 0.3567 - 0.2231 - 0.1054]$$

$$= -0.154 \, \text{units}^2$$

Here, we work to an accuracy of 4 s.f. to obtain a result expressed to 3 s.f. as required.

Carry out a rough check. Does a negative area make sense?

Yes, the area of interest is below the x-axis. What about the modulus of the result? The area is an approximate triangle of height 0.7 and width 0.5, so taking $\tfrac{1}{2}(bh)$, giving 0.18, which is of the correct order of magnitude.

Simpson's rule

This is the most refined of the three rules. Its derivation is interesting.

Since the mid-ordinate rule approximates each strip to a rectangle the ordinate value used is just a constant.

The trapezoidal rule takes the ordinate of a slope of constant gradient so it assumes a linear function.

Simpson's rule goes one step further and assumes the top of the strip to be a quadratic function.

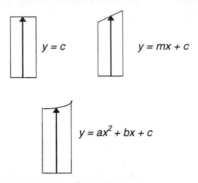

Now, let's integrate the general quadratic function.

We take the upper limit, $x = h$, and the lower limit, $x = -h$, so we are integrating the area shown in Fig. 10.2.

$$\int_{-h}^{h} (ax^2 + bx + c)\, dx = \left[\frac{a}{3}x^3 + \frac{b}{2}x^2 + cx\right]_{-h}^{h}$$

$$= \left(\frac{ah^3}{3} + \frac{bh^2}{2} + ch\right) - \left(\frac{a(-h)^3}{3} + \frac{b(-h)^2}{2} - ch\right)$$

$$= \left(\frac{ah^3}{3} + \frac{bh^2}{2} + ch\right) - \left(\frac{-ah^3}{3} + \frac{bh^2}{2} - ch\right)$$

$$= \frac{2ah^3}{3} + 2ch$$

$$\therefore \quad \text{Area} = \frac{h}{3}(2ah^2 + 6c)$$

Now we look for the values of the three ordinates at $x = -h$, $x = 0$ and $x = h$ which we can call y_0, y_1 and y_n respectively.

$$\text{If, } y = ax^2 + bx + c$$

$$y_0 = ah^2 - bh + c$$

$$y_1 = c$$

$$y_n = ah^2 + bh + c$$

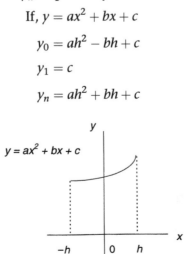

Figure 10.2 Simpson's rule

Adding the ordinates y_0 and y_n eliminates the bh term.

$$y_0 + y_n = 2ah^2 + 2c$$

Subtracting $2c$ gives an expression for $2ah^2$.

$$y_0 + y_n - 2c = 2ah^2$$

We can now substitute this into our area equation.

$$\text{Area} = \frac{h}{3}(y_0 + y_n - 2c + 6c)$$

$$= \frac{h}{3}(y_0 + y_n + 4c)$$

Since $c = y_1$, we have an area equation in terms of three ordinates and h which is the width of two strips separating the ordinates.

$$\therefore \quad \text{Area} = \frac{h}{3}(y_0 + y_n + 4y_1)$$

If we divided each of the two strips into two other strips of equal width the area equation would expand.

$$\text{Area} = \frac{h}{3}[y_1 + y_n + 4(y_2 + y_4) + 2(y_3)]$$

Further subdivisions of the area into an even number of strips yields the general formula for Simpson's rule:

$$\textbf{Area} \approx \frac{h}{3}[y_1 + y_n + 4(y_2 + y_4 + y_6 + \cdots) + 2(y_3 + y_5 + \cdots)]$$

Which is not as frightening as it first looks when we state it in words:

$$\textbf{Area} \approx \frac{1}{3} \textbf{ strip width} \begin{bmatrix} \textbf{sum of 1st and last ordinates} \\ \textbf{+4(sum of even ordinates)} \\ \textbf{+2(sum of remaining odd ordinates)} \end{bmatrix}$$

In which the word *odd* refers to the subscript of y and the first ordinate *must* be y_1.

It is important to remember that Simpson's rule only works with an area divided into an *even* number of strips and hence uses an *odd* number of ordinates.

EXAMPLE 10.3

Use Simpson's rule to find the area under the first quarter of a circle whose radius $r = 1$ (Fig. 10.3).

Integrate by dividing the area into 10 equal strips.

Ten equal strips set the ordinates at $x = 0$, $x = 0.1$, $x = 0.2$, etc. with a strip width of 0.1.

We start by calculating the value of each of the 11 ordinates.

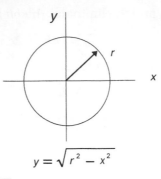

$$y = \sqrt{r^2 - x^2}$$

Figure 10.3 Graph of $y = \sqrt{r^2 - x^2}$

Solution

$$y_1 = \sqrt{1 - 0} = 1.000 \quad y_2 = \sqrt{1 - 0.1^2} = 0.9950 \quad y_3 = \sqrt{1 - 0.2^2} = 0.9798$$

$$y_4 = \sqrt{1 - 0.3^2} = 0.9539 \quad y_5 = \sqrt{1 - 0.4^2} = 0.9165 \quad y_6 = \sqrt{1 - 0.5^2} = 0.8660$$

$$y_7 = \sqrt{1 - 0.6^2} = 0.8000 \quad y_8 = \sqrt{1 - 0.7^2} = 0.7141 \quad y_9 = \sqrt{1 - 0.8^2} = 0.6000$$

$$y_{10} = \sqrt{1 - 0.9^2} = 0.4359 \quad y_{11} = 0$$

$$\text{Area} \approx \frac{0.1}{3}[(1 + 0 + 4(0.995 + 0.9539 + 0.866 + 0.7141 + 0.4359))$$

$$+ 2(0.9798 + 0.9165 + 0.8 + 0.6)]$$

$$= 0.7817$$

We can check this result because the area of a full circle of unit radius is equal to π. Our value for a quarter of a circle gives the area of a full circle as 3.13 correct to 3 s.f. Comparing this to π, we have a relative error of:

$$\frac{3.13 - \pi}{\pi} \times 100 = -0.369\%$$

I leave you to decide whether this is acceptable or not.

PRACTICE EXERCISE 10.1

(a) Use the mid-ordinate rule to find the mean value of $y = x \ln x$ between $x = 4$ and $x = 6$ correct to 3 s.f.

(b) Use the trapezoidal rule to find the mean value of $y = 3 \cos^2 x$ between $x = 0$ and $x = \pi/2$ correct to 3 s.f.

(c) Use Simpson's rule to find the mean speed in the first 10 seconds of $v = \dfrac{10 - t}{30 + t}$ correct to 3 s.f.

(d) Apply all three rules to calculate the r.m.s. value of a current that is described by: $i = 0.075 \sin 100\pi t$ by integrating eight strips over one complete cycle. Compare the three methods and evaluate their accuracies.

Solutions

(a) 8.08; (b) 1.50; (c) 0.147 m s^{-1}; (d) 53.0 mA

10.2 The Newton–Raphson method

This is a numerical method that allows us to find roots of equations. Remember that the roots of a quadratic can be found by formula so quadratic equations present no problem. But what about equations of a higher degree? The roots of a polynomial of a degree that is greater than two are not so easy to find and this is where the Newton–Raphson approximation comes in. The procedure is:

If x_1 is the approximate root of $f(x) = 0$, then a closer approximation to the root, x_2, is given by:

$$x_2 = x_1 - \frac{f(x_1)}{f'(x_1)}$$

Figure 10.4 illustrates the way the method works. It's quite attractive.

The method is sometimes called the **tangent approximation** because it relies on the equation of the straight line which is the tangent to the curve of $f(x) = 0$.

The tangent to the curve at the point x_1 forms a right-angled triangle of sides a, b and c so we have the tangent of θ.

$$\tan \theta = \frac{b}{c} \quad \text{so} \quad c = \frac{b}{\tan \theta}$$

The length of the side b is just the value of y when $x = x_1$ so it is $f(x_1)$.

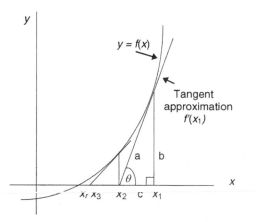

Figure 10.4 The Newton–Raphson method

Also, $\tan\theta$ is the gradient of the curve at $x = x_1$ so it is the first derivative, or $f'(x_1)$.

$$b = f(x_1) \quad \text{and} \quad \tan\theta = f'(x_1)$$

$$\therefore \quad c = \frac{f(x_1)}{f'(x_1)}$$

The value of x_2 is the difference of x_1 and the length c.

$$x_2 = x_1 - c$$

$$\therefore \quad x_2 = x_1 - \frac{f(x_1)}{f'(x_1)}$$

Now, it should be clear that x_2 is a closer approximation to the root x_r.

So if we start off with a first approximation to the root, x_1, we can find a second approximation to the root, x_2. The second approximation then allows us to find a third approximation, x_3. Each approximation leads us closer to the true value of the root, x_r.

The process of repeating a method is called **iteration** so the Newton–Raphson method is an **iterative method of successive approximation**.

The greater the number of iterations, the closer the result converges to the true value. But note, like all numerical methods, the result we obtain is never the exact value. The best we can aim for with numerical methods is a result that is correct to some required number of significant figures.

EXAMPLE 10.4

Find, correct to four significant figures, the root of the equation $x^4 - 11x - 5 = 0$ in the region, $x = 2$ to $x = 3$.

Be neat. Take a methodical approach. The first step is to differentiate and find $f'(x)$.

$$f(x) = x^4 - 11x - 5 = 0$$

$$f'(x) = 4x^3 - 11 = 0$$

Then decide on a value for the first approximation, x_1. Here, we know we must look in the region between $x = 2$ and 3. So we start with those values. x_r must lie closer to 2 than to 3.

First approximation:

$$f(2) = (2)^4 - 11(2) - 5 = -11$$

$$f(3) = (3)^4 - 11(3) - 5 = 43$$

Look at Fig. 10.5.

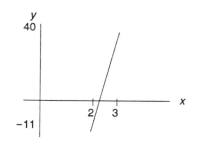

Figure 10.5 Example 10.4

So let's choose $x_1 = 2.2$.

Second approximation:

$$f(2.2) = (2.2)^4 - 11(2.2) - 5 = -5.7744$$

$$f'(2.2) = 4(2.2)^3 - 11 = 31.592$$

The rest is straightforward. We apply the method to evaluate x_2.

$$x_2 = x_1 - \frac{f(x_1)}{f'(x_1)} = 2.2 - \frac{-5.7744}{31.592}$$

$$= 2.3828$$

Then we iterate and find the third approximation which is 2.3590.

Third approximation:

$$f(2.3828) = (2.3828)^4 - 11(2.3828) - 5 = 1.0259$$

$$f'(2.3828) = 4(2.3828)^3 - 11 = 43.1156$$

$$x_3 = 2.3828 - \frac{1.0259}{43.1156} = 2.3590$$

Notice that we must work to a degree of accuracy greater than the required four significant figures.

Fourth approximation:

$$f(2.359) = (2.359)^4 - 11(2.359) - 5 = 0.0189$$

$$f'(2.359) = 4(2.359)^3 - 11 = 41.5102$$

$$x_4 = 2.359 - \frac{0.0189}{41.5102} = 2.3585$$

$$x_r = 2.359$$

The fourth iteration gives us $x_4 = 2.3585$, which, correct to four significant figures is equal to x_3. This tells us that it is time to stop. Any further iterations are unnecessary because they will give us the root to an accuracy which is greater than required.

Probably the most difficult part in applying the method is to find the first approximation(s). In the last example it was known that the required root lay in region between 2 and 3. A simple sketch of the curve showed that it was closer to 2. But what happens when we do not have any information to start with? What could you do to get an idea about how many roots there are and their rough value(s)?

Sketch a curve of the function

This need not be very accurate at all. You just want some idea of where to look. Plotting half a dozen key points is usually sufficient.

EXAMPLE 10.5

Find the root(s) of:

$$y = x^3 + 13.5x^2 + 40x + 16.67$$

correct to three decimal places.

To get some idea of the first approximation of the roots we sketch the curve starting with $x = 0$. Then take $x = 1$ and $x = -1$. We are looking for the points where y changes between positive and negative. So we look for values of x which give a value of y which is close to zero.

$$y(0) = 16.67$$

$$y(1) = 71.17$$

When x is slightly positive y is rising sharply positive so we concentrate on negative values of x.

$$y(-1) = -10.83$$

$$y(-3) = -8.83$$

$$y(-5) = 29.17$$

$$y(-8) = 48.67$$

$$y(10) = -33.33$$

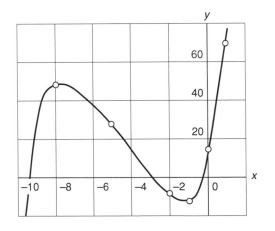

The equation is a cubic so the maximum number of roots is three. Having found them to be approximately -0.5, -3.5 and -9.5 we have the first approximations.

$$y = x^3 + 13.5x^2 + 40x + 16.67$$

$$\frac{dy}{dx} = 3x^2 + 27x + 40$$

We then apply the Newton–Raphson method to each of the three roots.

First root:

$$x_1 = -0.5, \quad x_2 = x_1 - \frac{f(x_1)}{f'(x_1)}$$

$$= -0.5 - \frac{-0.125 + 3.375 - 20 + 16.67}{0.75 - 13.5 + 40} = -0.497$$

$$x_3 = -0.497 - \frac{-0.1228 + 3.3355 - 19.8826 + 16.67}{0.7412 - 13.4207 + 40}$$

$$= -0.497$$

The first root is found in two iterations.

Second root:

$$x_1 = -3.5, \quad x_2 = -3.547, \quad x_3 = -3.546, \quad x_4 = -3.546$$

The second root requires three iterations before we are sure that we have an accuracy of 3 d.p.s.

Third root:

$$x_1 = -9.5, \quad x_2 = -9.457, \quad x_3 = -9.457$$

The third root is found in two iterations.

Solution:

$$x = -0.497, \ -3.546, \ -9.457$$

EXAMPLE 10.6

Find the root(s) of:

$$f(x) = x - 2\sin x$$

The strategy for curve sketching to find the approximate roots can be a bit of an art.

With this problem we have root(s) when:

$$x - 2\sin x = 0$$

so at the roots,

$$x = 2\sin x$$

This provides us with a short cut because we know the roots lie at the point where the graphs of $y = x$ and $y = 2\sin x$ intersect.

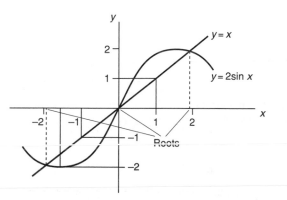

We know that with the linear function, $y = x$ when $x = 1$, $y = 1$, etc.

Also, with $y = 2\sin x$, when $x = 0$, $y = 0$ and when $x = \pi/2$, $y = 2$, etc.

The graphs intersect at $x = 0$, $x = 1.8$ and $x = -1.8$, so these are our first approximations of the roots except that $x = 0$ must be exact.

Solution

$$f(x) = x - 2\sin x$$
$$f'(x) = 1 - 2\cos x$$

First root:

$$x_1 = 1.8, \quad x_2 = x_1 - \frac{f(x_1)}{f'(x_1)} = 1.8 - \frac{1.8 - 2\sin 1.8}{1 - 2\cos 1.8} = 1.9016$$

$$x_3 = 1.9016 - \frac{1.9016 - 2\sin 1.9016}{1 - 2\cos 1.9016} = 1.8955$$

$$x_4 = 1.8955 - \frac{1.8955 - 2\sin 1.8955}{1 - 2\cos 1.8955} = 1.8955$$

Second root: $x = 0$

Third root: $x = -1.8955$

Solution: $x = 1.896, 0, -1.896$

Notice that there was no need to compute the third root. The graph of the function is symmetrical about the origin, so the third root is just a negative value of the first. And if you are not sure about $x = 0$, try back substitution. Put zero into $x - 2\sin x = 0$.

You may have realized that solving polynomials relies heavily on your ability to sketch curves of functions. You have seen a couple of strategies for doing this. Start with the obvious values, $x = 0$, $x = 1$ and $x = -1$. What happens when x is very large and positive? What happens when x is very large and negative? If you are able to identify a sum of two functions, equate them. The intercepts of the graphs of the two functions will point you towards the roots.

The rest is practice. The better you become, the less points you need to plot to arrive at approximate solutions. We will come back to curve sketching in Chapter 19.

PRACTICE EXERCISE 10.2

(a) Find the roots of the equation $x^2 = 4\cos x$ correct to three significant figures.
(b) The motion of a particle under the influence of an electrostatic field is described by:
$y = x^3 + 3x^2 + 5x - 28$. When $x = 2$, y is approximately zero. Find, correct to 2 d.p., the value of x when $y = 0$.
(c) Find the roots of $x\,e^{-x} - 0.1 = 0$ correct to 6 d.p.

Solutions

(a) ±1.20;   (b) 1.93;   (c) 0.111 833, 3.577 152

Assignment VIII

In this assignment you will use numerical techniques to solve engineering problems. The performance criteria that must be met for its satisfactory completion are:

- the engineering situation must be represented in an appropriate form
- graphical methods are used to estimate the solution to an engineering problem
- functions must be integrated using numerical integration
- numerical methods are used to determine the roots of an algebraic function
- numerical solutions are calculated to a given degree of precision
- numerical techniques are used in modelling engineering situations

The assignment is about control engineering. It is in two parts. Both parts require numerical methods to be used.

Plan your solutions. Present your data and formulate algebraic solutions. Curve sketching is an important step in planning solutions.

Information handling is an important aspect of all tasks. Generate and manipulate data that lead to the correct solution. Work and give answers to an appropriate degree of accuracy.

Evaluate your solutions by applying checking procedures to ensure your results are correct. You should also consider whether your solution makes sense within the problem situation.

The outcome of this assignment, as with all engineering assignments, should be a set of solutions that bring together your knowledge of the subject and communicate your ideas in a language that is appropriate to the profession. Numerical results should be stated to an appropriate degree of accuracy and include correct SI units.

Be neat,
pay attention to detail
and
get satisfaction from what you are doing.

PART 1

A control element is found to have an output that is modelled by the equation:

$$v = 0.1\,e^{-\frac{\pi}{6}t}\cos\frac{\pi}{4}t\,\mathrm{m\,s^{-1}}$$

when subjected to a step input. In order to find the position of the output variable at some time (t) we must integrate the function, but this requires integration by parts. You have not studied this technique yet, but you can find a good approximation of the position by numerical integration. Find the displacement of the output variable 1 second after the application of the step input by *three different methods*.

Compare your results and ensure that at least one carefully sketched graph of the function is included with your presentation.

PART 2

As part of a mathematical modelling exercise to evaluate the performance of a process control element it is necessary to find the roots of the cubic equation:

$$f(x) = x^3 - 8x^2 + 16x - 4$$

(a) Sketch a graph of the function and determine a first approximation of all the roots of the equation.

(b) Apply the Newton–Raphson method to estimate the value of the roots to four significant figures.

chapter

11 Series

11.1 Mathematical sequence

A **sequence** is a set of mathematical terms stated in a definite order. Each term of the sequence is formed according to a fixed rule. Such a rule is called a **recurrence relation** so, for example, the recurrence relation:

$$u_{r+1} = 3u_r$$

tells us that the next term of a sequence is 3 times the present term and generates the following sequence.

With $u_r = 2$

$$u_{r+1} = 3u_r$$

gives:

$$2, 6, 18, 54, 162, 486, 1458 \ldots$$

EXAMPLE 11.1

Write the recurrence relation for the sequence:

$$1, 5, 9, 13, 17, 21 \ldots$$

Solution

$$u_{r+1} = u_r + 4$$

Sequences can be finite or infinite.

11.2 Mathematical series

A **series** is formed by the **sum of the terms of a sequence**. Converting the sequence of the last example gives the following series.

$$1 + 5 + 9 + 13 + 17 + 21 + \cdots$$

An **arithmetic series** is formed by *adding* a constant value to each previous term as with $1 + 5 + 9 + \cdots$. It has the general form shown, where a is called the **initial term** and d is the **common difference**.

$$a + (a + d) + (a + 2d) + (a + 3d) + \cdots$$

A **geometric series** is formed by *multiplying* each previous term by a constant value. In the general form of a geometric series a is, once again, the **initial term** and r is called the **common ratio**.

$$a + ar + ar^2 + ar^3 + ar^4 + \cdots$$

Before going any further with series we need to return to binomials and explore what is known as the **binomial theorem**. The theorem forms the basis of an important type of series.

11.3 Binomials

A binomial is simply a polynomial in two terms. Some examples are:

$$a + b, \quad 1 + x, \quad 5y - 2, \quad 3x^2 + 7, \quad 7a^3 + 3a^2$$

Clearing the brackets of a binomial gives a binomial **expansion**.

EXAMPLE 11.2

For practice, expand the following:

$$(a + b)^0$$
$$(a + b)^1$$
$$(a + b)^2$$
$$(a + b)^3$$
$$(a + b)^4$$

Solution

$$(a + b)^0 = 1$$
$$(a + b)^1 = a + b$$
$$(a + b)^2 = (a + b)(a + b) = a^2 + 2ab + b^2$$
$$(a + b)^3 = (a + b)(a + b)(a + b)$$
$$= (a^2 + 2ab + b^2)(a + b)$$
$$= a^3 + a^2b + 2a^2b + 2ab^2 + ab^2 + b^3$$
$$= a^3 + 3a^2b + 3ab^2 + b^3$$

$$(a+b)^4 = (a^3 + 3a^2b + 3ab^2 + b^3)(a+b)$$
$$= a^4 + a^3b + 3a^3b + 3a^2b^2 + 3a^2b^2 + 3ab^3 + ab^3 + b^4$$
$$= a^4 + 4a^3b + 6a^2b^2 + 4ab^3 + b^4$$

The task starts to become tedious as the order of the binomial increases. So we look for short cuts. The coefficients in the expansions seem to follow a pattern. The pattern is revealed in something called **Pascal's triangle**.

$$(a+b)^0 \qquad \qquad 1$$
$$(a+b)^1 \qquad \qquad 1 \quad 1$$
$$(a+b)^2 \qquad \qquad 1 \quad 2 \quad 1$$
$$(a+b)^3 \qquad \qquad 1 \quad 3 \quad 3 \quad 1$$
$$(a+b)^4 \qquad \qquad 1 \quad 4 \quad 6 \quad 4 \quad 1$$
$$(a+b)^5 \qquad \quad 1 \quad 5 \quad 10 \quad 10 \quad 5 \quad 1$$
$$(a+b)^6 \quad \quad 1 \quad 6 \quad 15 \quad 20 \quad 15 \quad 6 \quad 1$$
$$(a+b)^7 \quad 1 \quad 7 \quad 21 \quad 35 \quad 35 \quad 21 \quad 7 \quad 1$$

Pascal's triangle

The triangle shows that:

- The number of terms in each expansion is 1 greater than the power. So for a binomial of order n there are $n + 1$ terms.
- The first and last coefficients are always 1.
- The coefficients form a symmetrical pattern.
- Each coefficient is formed by adding the two coefficients immediately above.

EXAMPLE 11.3

Use Pascal's triangle to obtain the expansion of $(a+b)^8$.

Solution

$$(a+b)^8 = a^8 + 8a^7b + 28a^6b^2 + 56a^5b^3 + 70a^4b^4$$
$$+ 56a^3b^5 + 28a^2b^6 + 8ab^7 + b^8$$

So Pascal's triangle offers an easier means of obtaining a binomial expansion than multiplying out brackets. However, it is still not the most convenient method.

11.4 Binomial theorem

This is sometimes just called the **binomial expansion**. It states that:

$$(a+b)^n = a^n + na^{n-1}b + \frac{n(n-1)}{2!}a^{n-2}b^2 + \frac{n(n-1)(n-2)}{3!}a^{n-3}b^3 + \cdots + b^n$$

The exclamation marks are a convenient notation which mean *factorial*. In general, if n is a positive integer:

$$n! = n(n-1)(n-2)(n-3) \times \cdots \times 1$$

EXAMPLE 11.4

Evaluate:

$$4!, 6! \text{ and } 10!$$

Solution

$$4! = 4 \times 3 \times 2 \times 1 = 24$$

$$6! = 6 \times 5 \times 4 \times 3 \times 2 \times 1 = 720$$

$$10! = 10 \times 9 \times 8 \times 7 \times 6! = 3\,628\,800$$

So it's quite easy. Now look at the binomial theorem again. It allows you to expand any binomial of any order n. You could think of the terms as:

- the first is the n-term (power of a)
- next is the 1-term (there is just 1 n, factorial is 1, the power of a has -1 and the power of b is 1)
- the third is the 2-term (there are 2 ns, factorial is 2, power of a has -2 and power of b is 2)
- the fourth is the 3-term and so on.

EXAMPLE 11.5

Use the binomial theorem to expand $(2+x)^7$.

Solution

Here, $a = 2$, $b = x$ and $n = 7$, so:

$$(2+x)^7 = 2^7 + 7(2)^6 x + \frac{7(6)}{2!}(2)^5 x^2 + \frac{7(6)(5)}{3!}(2)^4 x^3$$

$$+ \frac{7(6)(5)(4)}{4!}(2)^3 x^4 + \frac{7(6)(5)(4)(3)}{5!}(2)^2 x^5$$

$$+ \frac{7(6)(5)(4)(3)(2)}{6!}(2)x^6 + x^7$$

$$= 128 + 448x + 672x^2 + 560x^3 + 280x^4 + 84x^5 + 14x^6 + x^7$$

Notice that: the exponent of a plus the exponent of b always equals n.

EXAMPLE 11.6

Find the terms in the expansion of $(2 - 3x)^4$.

Solution

$$(2-3x)^4 = 2^4 + 4(2)^3(-3x) + \frac{4(3)}{2!}(2)^2(-3x)^2 + \frac{4(3)(2)}{3!}(2)(-3x)^3 + (-3x)^4$$

$$= 16 - 96x + 216x^2 - 216x^3 + 81x^4$$

11.5 The binomial series

Now, there is another way of looking at the binomial theorem. Instead of treating it as a means of finding a binomial expansion we can treat it as a series which represents a function as a polynomial, so if $f(x) = (a+x)^n$, then:

$$(a+x)^n \equiv a^n + na^{n-1}x + \frac{n(n-1)}{2!}a^{n-2}x^2 + \frac{n(n-1)(n-2)}{3!}a^{n-3}x^3 + \cdots + x^n$$

Consider this carefully. It is an **identity** *not an equality*. Remember that an identity holds true for any value of the variable. In this case the RHS is equal to the LHS for any value of x.

So what is the point of an identity? An identity allows us to represent a function in a different form. Series allows us to write: trigonometric, exponential, logarithmic as well as binomial functions as polynomials. One advantage of a polynomial is that, under certain circumstances, it allows us to differentiate over and over again.

Now let's return to the binomial series and look at a special case in which the constant $a = 1$. Write $(1+x)^n$ as a series.

$$(1+x)^n = 1 + nx + \frac{n(n-1)}{2!}x^2 + \frac{n(n-1)(n-2)}{3!}x^3 + \cdots + x^n$$

This is the general case. Now look at some cases where n has particular values. Copy out the examples which follow and study them.

EXAMPLE 11.7

Find the binomial series of:

$$(1+x)^4$$

Solution

$$(1+x)^4 \equiv 1 + 4x + \frac{4(3)}{2!}x^2 + \frac{4(3)(2)}{3!}x^3 + \frac{4(3)(2)(1)}{4!}x^4 + \frac{4(3)(2)(1)(0)}{5!}x^5$$

$$\equiv 1 + 4x + 6x^2 + 4x^3 + x^4$$

The series is finite. It comes to an end when the coefficient of x becomes zero.

EXAMPLE 11.8

Now find the series expansion of:

$$\frac{1}{(1+x)^4}$$

Don't bother going beyond the 5th term.

Solution

$$\frac{1}{(1+x)^4} \equiv (1+x)^{-4}$$

$$(1+x)^{-4} \equiv 1 + (-4)x + \frac{(-4)(-5)}{2!}x^2 + \frac{(-4)(-5)(-6)}{3!}x^3 + \frac{(-4)(-5)(-6)(-7)}{4!}x^4 + \cdots$$

$$\equiv 1 - 4x + 10x^2 - 20x^3 + 35x^4 + \cdots$$

EXAMPLE 11.9

Find the series expansion of:

$$\sqrt{1+x}$$

to five terms.

Solution

$$\sqrt{1+x} \equiv (1+x)^{\frac{1}{2}}$$

$$(1+x)^{\frac{1}{2}} \equiv 1 + \frac{1}{2}x + \frac{\frac{1}{2}(\frac{1}{2}-1)}{2!}x^2 + \frac{\frac{1}{2}(\frac{1}{2}-1)(\frac{1}{2}-2)}{3!}x^3 + \frac{\frac{1}{2}(\frac{1}{2}-1)(\frac{1}{2}-2)(\frac{1}{2}-3)}{4!}x^4 + \cdots$$

$$\equiv 1 + \frac{x}{2} + \frac{-(\frac{1}{4})}{2!}x^2 + \frac{(\frac{3}{8})}{3!}x^3 + \frac{-(\frac{15}{16})}{4!}x^4 + \cdots$$

$$\equiv 1 + \frac{x}{2} - \frac{x^2}{2!(4)} + \frac{3x^3}{3!(8)} - \frac{15x^4}{4!(16)} + \cdots$$

$$\equiv 1 + \frac{x}{2} - \frac{x^2}{8} + \frac{x^3}{16} - \frac{5x^4}{128} + \cdots$$

From the last three examples, you should be able to accept, without proof, that when:

- n is a positive integer, the series is finite,
- n is a negative integer, the series is infinite,
- n is a fraction, the series is infinite

11.6 Convergence

To illustrate what this means and when it occurs, we will take the last three examples one step further.

EXAMPLE 11.10

Use the binomial series to evaluate:

$$(1+x)^4$$

for $x = 2$, $x = 0.5$, $x = -0.5$.

Solution

With $x = 2$

$$(1+x)^4 \equiv 1 + 4x + 6x^2 + 4x^3 + x^4$$
$$= 1 + 4(2) + 6(2)^2 + 4(2)^3 + 2^4$$
$$= 1 + 8 + 24 + 32 + 16 = 81$$

With $x = 0.5$

$$(1+x)^4 \equiv 1 + 4x + 6x^2 + 4x^3 + x^4$$
$$= 1 + 4(0.5) + 6(0.5)^2 + 4(0.5)^3 + 0.5^4$$
$$= 1 + 2 + 1.5 + 0.5 + 0.0625 = 5.0625$$

With $x = -0.5$

$$(1+x)^4 \equiv 1 + 4x + 6x^2 + 4x^3 + x^4$$
$$= 1 + 4(-0.5) + 6(-0.5)^2 + 4(-0.5)^3 + (-0.5)^4$$
$$= 1 - 2 + 1.5 - 0.5 + 0.0625 = 0.0625$$

You may think that this was a waste of time because we could get the same results directly off a calculator. However, a point needs to be made so be patient.

EXAMPLE 11.11

Apply the series expansion to evaluate:

$$\frac{1}{(1+x)^4}$$

for $x = 2$, $x = 0.5$, $x = -0.5$

Solution

With $x = 2$

$$\frac{1}{(1+x)^4} \equiv 1 - 4x + 10x^2 - 20x^3 + 35x^4 + \cdots$$

$$= 1 - 4(2) + 10(2)^2 - 20(2)^3 + 35(2)^4 + \cdots$$
$$= 1 - 8 + 40 - 160 + 560 + \cdots \quad \text{(Terms diverging)}$$

With $x = 0.5$

$$\frac{1}{(1+x)^4} = 1 - 4(0.5) + 10(0.5)^2 - 20(0.5)^3 + 35(0.5)^4 + \cdots$$

$$= 1 - 2 + 2.5 - 2.5 + 2.1875 + \cdots \quad \text{(starting to converge)}$$

With $x = -0.5$

$$\frac{1}{(1+x)^4} = 1 - 4(-0.5) + 10(-0.5)^2 - 20(-0.5)^3 + 35(-0.5)^4 + \cdots$$

$$= 1 + 2 + 2.5 + 2.5 + 2.1875 + \cdots \quad \text{(starting to converge)}$$

So in example 11.10, with n a positive integer we can evaluate for any value of x. But in Example 11.11, with n a negative integer the terms of the expansion grow wildly (diverge) when x is greater than 1. However, they start to settle towards a smaller value (converge) when x is less then 1.

EXAMPLE 11.12

Evaluate the series of:

$$\sqrt{1+x}$$

for $x = 2$, $x = 0.5$, $x = -0.5$.

Solution

$$\sqrt{1+x} \equiv 1 + \frac{x}{2} - \frac{x^2}{8} + \frac{x^3}{16} - \frac{5x^4}{128} + \cdots$$

With $x = 2$,

$$\sqrt{1+x} \equiv 1 + \frac{2}{2} - \frac{(2)^2}{8} + \frac{(2)^3}{16} - \frac{5(2)^4}{128} + \cdots$$

$$= 1 + 1 - 0.5 + 0.5 - 0.625 + \cdots \quad \text{(starting to diverge)}$$

With $x = 0.5$,

$$\sqrt{1+x} \equiv 1 + \frac{0.5}{2} - \frac{(0.5)^2}{8} + \frac{(0.5)^3}{16} - \frac{5(0.5)^4}{128} + \cdots$$

$$= 1 + 0.25 - 0.03125 + 7.8125 \times 10^{-3} - 2.4414 \times 10^{-3} + \cdots \quad \text{(converging quickly)}$$

With $x = -0.5$,

$$\sqrt{1+x} \equiv 1 + \frac{(-0.5)}{2} - \frac{(-0.5)^2}{8} + \frac{(-0.5)^3}{16} - \frac{5(-0.5)^4}{128} + \cdots$$

$$= 1 - 0.25 - 0.03125 - 7.8125 \times 10^{-3} - 2.4414 \times 10^{-3} + \cdots \quad \text{(converging quickly)}$$

The rapid convergence of the series when x is ± 0.5 allows us to obtain approximate values.

When $x = 0.5$ the sum of the terms in the series is 1.22 correct to 3 s.f.

When $x = -0.5$, the sum of the terms is 0.71 correct to 2 s.f.

Compare these values to those obtained with your calculator and you will find that they are exact within the stated degree of accuracy.

Now we can make some important conclusions about the binomial series.

- If n is a positive integer the binomial series is finite and possible to evaluate.
- If n is negative or fractional the series is infinite.
- An infinite series may be convergent or divergent.
- An infinite series is convergent when its terms become progressively smaller so it is possible to evaluate to some required degree of accuracy.
- With an infinite binomial series, convergence occurs when: $|x| < 1$.
- When $|x| \geq 1$ then an infinite binomial series is divergent so a numerical solution is not possible.

PRACTICE EXERCISE 11.1

Expand the following to four terms, using the binomial series:

(a) $(1 + 2x)^9$; (b) $(1 + x)^{-1}$; (c) $(1 + 2x)^{-3}$

(d) $\sqrt[3]{1 + x^2}$; (e) $\dfrac{1}{\sqrt{1 - 3x}}$; (f) $\left(1 - \dfrac{2x}{3}\right)^{\frac{3}{4}}$

Solutions

(a) $1 + 18x + 144x^2 + 672x^3 + \cdots$; (b) $1 - x + x^2 - x^3 + \cdots$

(c) $1 - 6x + 12x^2 - 80x^{-3} + \cdots$; (d) $1 + \dfrac{1}{3}x^2 - \dfrac{1}{9}x^4 + \dfrac{5}{8}x^6 + \cdots$

(e) $1 + \dfrac{3}{2}x + \dfrac{27}{8}x^2 + \dfrac{135}{16}x^3 + \cdots$; (f) $1 - \dfrac{x}{2} + \dfrac{x^2}{24} - \dfrac{5x^3}{432} + \cdots$

11.7 Binomial approximation

A useful application of the binomial series is the **binomial approximation**.

This states that the first two terms of the binomial series provide a close approximation to the value of $(1 + x)^n$ providing x is small.

$$(1 + x)^n \approx 1 + nx$$

Remember that if $|x|$ is less than 1, a series is convergent so each successive term takes the value of the series closer to the true value. However, if x is *very small* compared with 1 then the series converges quickly so that the first two terms can provide a quick approximation. Consider the example that follows.

EXAMPLE 11.13

Find the binomial approximations of:

(a) $(1 + 0.01)^4$
(b) $(1 + 0.05)^4$
(c) $(1 + 0.10)^4$
(d) $(1 + 0.50)^4$

and compare them with calculator values obtained correct to 3 s.f.

Solution

$(1 + x)^n \approx 1 + nx$

(a) $(1 + 0.01)^4 \approx 1 + 4(0.01) = 1.04$

$$\text{calc. value} = 1.04$$

(b) $(1 + 0.05)^4 \approx 1 + 4(0.05) = 1.20$

$$\text{calc. value} = 1.22$$

(c) $(1 + 0.1)^4 \approx 1 + 4(0.1) = 1.40$

$$\text{calc. value} = 1.46$$

(d) $(1 + 0.5)^4 \approx 1 + 4(0.5) = 3.00$

$$\text{calc. value} = 5.06$$

When $x \ll 1$ the approximation is correct to 3 s.f. But how do we define *much smaller than*? The approximation starts to go awry when the difference between x and 1 is one order of magnitude. So if the difference is two orders of magnitude or more we can obtain an approximation which is acceptable in most engineering situations.

Let's look at some practical examples.

EXAMPLE 11.14

Find the relative percentage error in the calculation of the area of a circle if the relative percentage error in the radius measurement is 1%.

Let A be the area of the circle and δA the error in the area. The area calculation will involve squaring the radius plus its error which is $r/100$.

$$A = \pi r^2$$

$$A + \delta A = \pi \left(r + \frac{r}{100} \right)^2 = \pi r^2 \left(1 + \frac{1}{100} \right)^2$$

We write the binomial approximation of the terms in brackets and multiply by A.

$$A + \delta A \approx A\left(1 + \frac{2}{100}\right)$$

$$A + \delta A \approx A + \frac{2A}{100}$$

$$\delta A \approx 2\% \text{ of } A$$

Subtracting A shows that the error in the area is approximately 2%.

EXAMPLE 11.15

The relative percentage error in the measurement of the radius of a sphere is 3.5%. What effect does this have on the relative percentage error in the calculated value of the volume of the sphere.

Solution

$$V = \frac{4}{3}\pi r^3$$

$$V + \delta V = \frac{4}{3}\pi\left(r + \frac{3.5r}{100}\right)^3 \approx \frac{4}{3}\pi r^3\left(1 + (3)\frac{3.5}{100}\right)$$

$$V + \delta V \approx V\left(1 + \frac{10.5}{100}\right) \approx V + \frac{10.5}{100}V$$

$$\delta V \approx 10.5\%$$

EXAMPLE 11.16

The Q-factor of an electric circuit at resonance is calculated using:

$$Q = \frac{1}{R}\sqrt{\frac{L}{C}}$$

Find the approximate error in Q if:

R is 4% high, L is 3% high, C is 6% low.

Solution

$$Q + \delta Q = \frac{1}{R + 0.04R}\sqrt{\frac{L + 0.03L}{C - 0.06C}}$$

$$= \frac{1}{R(1 + 0.04)}\frac{\sqrt{L}\sqrt{1 + 0.03}}{\sqrt{C}\sqrt{1 - 0.06}}$$

$$= \frac{1}{R}\sqrt{\frac{L}{C}}(1 + 0.04)^{-1}(1 + 0.03)^{0.5}(1 - 0.06)^{-0.5}$$

$$\approx Q(1 - 0.04)(1 + 0.015)(1 + 0.03)$$

$$\approx Q(1 - 0.025)(1 + 0.03) \quad \text{neglecting } 0.0006$$

$$\approx Q(1 + 0.005) \quad \text{neglecting } 0.00075$$

$$Q + \delta Q \approx Q + 0.005Q$$

$$\delta Q \approx 0.005Q$$

$$\delta Q \approx 0.5\%$$

Note that in the last example we ignored the values 0.0006 and 0.00075 because they are negligible when compared with the other terms of the expression. After all, if we are approximating in the first place there is no point in fussing with figures that are insignificant.

Example 11.16 also shows that the accumulated error in a calculation is smaller than its component errors because the component errors are favourably combined. However, you should appreciate that the opposite can happen. Component errors can easily combine to amplify the error in a calculation.

11.8 Other series

We now look at Maclaurin's theorem.

Suppose that we wish to express some function as a series. In other words we have $f(x)$ for which we want a polynomial identity, as shown, where a, b, c, $d \ldots$ are constants.

$$f(x) = a + bx + cx^2 + dx^3 + ex^4 + \cdots$$

How do we find the values of the constants?

Well, when x is zero the function is equal to a.

$$f(0) = a \quad \text{so} \quad a = f(0)$$

Now if we differentiate once, a disappears. So when x is zero the first derivative is equal to b.

$$f'(x) = b + 2cx + 3dx^2 + 4ex^3 + \cdots$$

$$f'(0) = b \quad \text{so} \quad b = f'(0)$$

Taking the second derivative eliminates b and allows us to find c.

$$f''(x) = 2c + 3!dx + (4)(3)ex^2 + \cdots$$

$$f''(0) = 2c \quad \text{so} \quad c = \frac{f''(0)}{2}$$

The third derivative leads to a value for d and so on.

$$f'''(x) = 3!d + 4!ex + \cdots$$

$$f'''(0) = 3!d \quad \text{so} \quad d = \frac{f'''(0)}{3!}$$

$$f^{iv}(x) = 4!e + \cdots$$

$$f^{iv}(0) = 4!e \quad \text{so} \quad e = \frac{f^{iv}(0)}{4!}$$

The result we get is **Maclaurin's theorem** which is a general series. It can be used to find series identities for most of the common functions, including the binomial series.

$$f(x) = f(0) + f'(0)x + \frac{f''(0)}{2!}x^2 + \frac{f'''(0)}{3!}x^3 + \frac{f^{iv}(0)}{4!}x^4 + \cdots$$

EXAMPLE 11.17

Use Maclaurin's theorem to find the series expansion of e^x.

Providing you can differentiate as many times as you need, applying Maclaurin's theorem to find a series identity is very straightforward.

Solution

$$f(x) = e^x \quad \text{and} \quad f(0) = e^0 = 1$$
$$f'(x) = e^x \quad \text{and} \quad f'(0) = e^0 = 1$$
$$f''(x) = e^x \quad \text{and} \quad f''(0) = e^0 = 1$$

so

$$e^x = 1 + x + \frac{x^2}{2!} + \frac{x^3}{3!} + \frac{x^4}{4!} + \frac{x^5}{5!} + \cdots$$

Having established the series identity for e^x it is a matter of simple arithmetic to evaluate for any value of x.

EXAMPLE 11.18

Use the first seven terms of the exponential series to find the value of e to 4 d.p.

Solution

$$e = e^1 = 1 + 1 + \frac{1}{2!} + \frac{1}{3!} + \frac{1}{4!} + \frac{1}{5!} + \frac{1}{6!} + \cdots$$
$$= 1 + 1 + 0.5 + 0.16667$$
$$+ 0.04167 + 0.00833 + 0.00139$$
$$= 2.7181$$

EXAMPLE 11.19

Show, by differentiating the exponential series, that the derivative of e^x is equal to e^x.

$$e^x = 1 + x + \frac{x^2}{2!} + \frac{x^3}{3!} + \frac{x^4}{4!} + \frac{x^5}{5!} + \cdots$$
$$\frac{d}{dx}(e^x) = 1 + \frac{2x}{2!} + \frac{3x^2}{3!} + \frac{4x^3}{4!} + \frac{5x^4}{5!} + \cdots$$
$$= 1 + x + \frac{x^2}{2!} + \frac{x^3}{3!} + \frac{x^4}{4!} + \cdots$$

this is equal to the first five terms of the series for e^x so:

$$\frac{d}{dx}(e^x) = e^x$$

The last two examples provide some insight into the way that series can be used. Calculators use series (rather than stored tables) to compute values for e^x. In fact calculators are programmed to evaluate many functions by series: sin, cos, tan and log.

Example 11.19 also provides a neat proof of the unique property of e.

PRACTICE EXERCISE 11.2

(a) Apply Maclaurin's theorem to find the power series for $\sin x$.

(b) Apply Maclaurin's theorem to find the power series for $\cos x$.

(c) Use the series identity of $\sin x$ to evaluate $\sin 30°$ correct to three s.f. Hint: you will need to work in radians.

Solutions

(a) $\sin x = x - \frac{x^3}{3!} + \frac{x^5}{5!} - \frac{x^7}{7!} + \cdots;$ (b) $1 - \frac{x^2}{2!} + \frac{x^4}{4!} - \frac{x^6}{6!};$ (c) 0.500

Assignment IX

In this assignment you will apply series to engineering problems. To meet the performance criteria of this assignment you must:

- identify the parameters of a series
- expand the series and deduce an expression for the nth term
- deduce an expression for the sum of n terms
- solve engineering problems using series
- model engineering situations using series

There are four problems to solve in this assignment.

Plan your solutions by presenting relevant data and formulating algebraic equations before going for a numerical solution.

Generate and manipulate data that leads to the correct solution. Work and give answers to an appropriate degree of accuracy.

Evaluate your solutions by applying checking procedures to ensure your results are correct. You should also consider whether your solution makes sense within the problem situation.

The outcome of this assignment, as with all engineering assignments, should be a set of solutions which bring together your knowledge of the subject and communicate your ideas in the language of engineering mathematics. Numerical results should be stated to an appropriate degree of accuracy and include correct SI units.

Read the questions carefully.

PROBLEM 1

Use the binomial theorem to find the first six terms of the expansion of:

$$(2 + 3x)^{\frac{3}{4}}$$

Simplify the expansion showing each step.

PROBLEM 2

The second moment of area about the line xx is given by the equation:

$$I_{xx} = \frac{bd^3}{12}$$

(a) If d is 4% too large and b is 2.5% too small, what will be the approximate effect on I_{xx}?

(b) If, at a separate instance, d is 2% too large, what change would be required in b to maintain I_{xx} at its original value?

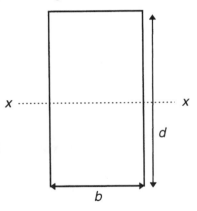

PROBLEM 3

The cut-off frequency of a low pass electronic filter is found from:

$$f_c = \frac{1}{2\pi CR} \text{ Hz}$$

where C is capacitance in farads and R is resistance in ohms. Estimate the maximum relative error in the calculation of the cut-off frequency if the manufacturing tolerance of the capacitor is $\pm 5\%$ of its nominal value and the manufacturing tolerance of the resistor is $\pm 2\%$ of its nominal value.

PROBLEM 4

Use the first five terms of the exponential series to evaluate $e^{\frac{5}{2}}$. Compare this value with the value obtained from your calculator by stating the percentage error.

chapter

12 Revision I

Assignment X

This is a summative assessment in which you will apply calculus, numerical methods and series to engineering problems. To meet the performance criteria for assessment you must:

- use rules of differentiation to differentiate combinations of functions
- integrate standard functions using indefinite and definite integrals
- solve engineering problems using calculus
- use graphical methods to estimate the solution of engineering problems
- use numerical integration to integrate functions
- use numerical methods to determine roots of algebraic functions
- calculate numerical solutions to a given degree of precision
- expand series and deduce an expression for the nth term
- deduce an expression for the sum of n terms
- solve engineering problems using series

The recommended time for this assignment is 2 hours. You should work alone, under normal exam conditions.

Good luck!

QUESTION 1

If, $f(x) = 4x^3 - 2\sin x - e^x + \ln x + 36$, then:

(a) $f'(x) = x^4 + 2\cos x - e^x + x(\ln x + 1) + 36x + C$

(b) $f'(x) = 12x^2 + 2\cos x - e^x + \dfrac{1}{x}$

(c) $f'(x) = 12x^2 - 2\cos x - e^x + \dfrac{1}{x}$

(d) $f'(x) = 12x^2 + 2\cos x - e^x + x$

QUESTION 2

If, $y = (2x^2 - 4)^2$, the function of a function rule gives:

(a) $\dfrac{dy}{dx} = 16x^2$

(b) $\dfrac{dy}{dx} = 16x^3 - 32x$

(c) $\dfrac{dy}{dx} = 8x^3 - 16x$

(d) $\dfrac{dy}{dx} = 16x - 32$

QUESTION 3

Applying the product rule to $s = 4t\ln(t)$, gives:

(a) $\dfrac{ds}{dt} = 4 + 4\ln t$

(b) $\dfrac{ds}{dt} = 4t + \dfrac{4}{t}$

(c) $\dfrac{ds}{dt} = 4 - 4\ln t$

(d) $\dfrac{ds}{dt} = 4 + 4e^t$

QUESTION 4

The gradient of the curve of $y = \dfrac{\sin x}{\cos x}$ is given by:

(a) $\dfrac{1}{\sec^2 x}$

(b) $\tan x$

(c) $\cos^2 x + \sin^2 x$

(d) $\dfrac{\cos^2 x + \sin^2 x}{\cos^2 x}$

QUESTION 5

The distance of an object whose position is given by $s = 2t^2 + 8t$ is:

(a) maximum when t is -4
(b) maximum when t is -2
(c) minimum when t is -4
(d) minimum when t is -2

QUESTION 6

$$\int (3x^2 + 2x - e^x + 17)\,dx =$$

(a) $x^3 + x^2 - e^x + 17x + C$
(b) $x^3 + x^2 - e^x$
(c) $x^3 + x^2 - e^x + C$
(d) $6x + 2 - e^x + C$

QUESTION 7

$$\int_{\frac{\pi}{2}}^{\frac{3\pi}{2}} (2\sin x - 3\cos x)\,dx =$$

(a) 0
(b) 6
(c) 2
(d) 1

QUESTION 8

The r.m.s. value of $\sin 2\theta$, over a full cycle, is:

(a) $\sqrt{\dfrac{1}{\pi} \int_0^{\pi} (\sin^2 2\theta)\,d\theta}$

(b) $\sqrt{\dfrac{1}{2\pi} \int_0^{2\pi} (\sin 2\theta)\,d\theta}$

(c) $\sqrt{\dfrac{1}{2\pi} \int_0^{2\pi} (\sin^2 2\theta)\,d\theta}$

(d) $\sqrt{\dfrac{1}{2\pi} \int_0^{2\pi} (\cos^2 2\theta)\,d\theta}$

QUESTION 9

If velocity $v = 2 + 3t$ then acceleration is:

(a) $2t + \frac{3}{2}t^2$
(b) 3
(c) $2t + \frac{3}{2}t^2 + C$
(d) $3t$

QUESTION 10

For a surface described by $y = 3x$ the distance of the centroid from the y-axis of the area between $x = 0$ and $x = 3$ is:

(a) $\bar{x} = 2$
(b) $\bar{x} = 3$
(c) $\bar{x} = 1$
(d) $\bar{x} = 1.5$

QUESTION 11

Select the one statement which is true.

(a) The mid-ordinate rule approximates a function to be linear.
(b) The trapezoidal rule approximates a function to be a constant.
(c) Simpson's rule approximates a function to be quadratic.
(d) Numerical integration is a method of finding an *indefinite* integral.

QUESTION 12

If Simpson's rule is applied to find the area under the curve of $y = 2 \ln x$ between $x = 1$ and $x = 5$, using four strips, as shown in Fig. 12.1, then, correct to 3 s.f., the area is:

(a) 24.2
(b) 8.16
(c) 7.97
(d) 8.08

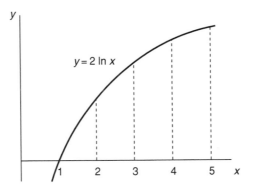

Figure 12.1 Finding the area under the curve using four strips (see Question 12)

QUESTION 13

Equation $y = x^2 - x - 3$ has:

(a) no roots
(b) distinct roots
(c) repeated roots
(d) no exact roots

QUESTION 14

$y = x^3 + x^2 + x + 20$ has only one real root. A first approximation shows that it is close to:

(a) -3
(b) 3
(c) -1
(d) -1.5

QUESTION 15

A first approximation indicates that a root of $x^3 - 9x + 1 = 0$ lies close to 3. To 4 d.p., the Newton–Raphson method gives the second approximation as:

(a) 3.0580
(b) 3.0556
(c) 2.9444
(d) 2.9420

QUESTION 16

The expansion of $(2 + x)^4$ is:

(a) $16 + 32x + 24x^2 + 8x^3 + x^4$
(b) $2 + 4x + 12x^2 + 8x^3 + x^4$
(c) $16 + 32x + 24x^2 + 16x^3 + 6x^4$
(d) $16 + 32x + 48x^2 + 48x^3 + x^4$

QUESTION 17

The binomial series of $(1 + 3x)^9$ is:

(a) indeterminate
(b) divergent
(c) convergent
(d) finite

QUESTION 18

The first four terms of the series of $\sqrt{1 + 2x}$ are:

(a) $1 + 4x - 8x^2 + 16x^3 + \cdots$
(b) $1 + x - \dfrac{x^2}{8} + \dfrac{x^3}{16} + \cdots$
(c) $1 - x + \dfrac{x^2}{8} - \dfrac{x^3}{16} + \cdots$
(d) $1 + 4x + 4x^2 + x^3$

QUESTION 19

With $x = 0.2$ the first four terms of the series of $(1+x)^{-2}$ evaluate to:

(a) 0.688
(b) 0.694
(c) 1.552
(d) 0.448

QUESTION 20

If the relative error in a current measurement (I) is 1.5%, the binomial approximation of the relative error in the resistance calculation $R = \dfrac{P}{I^2}$ is:

(a) -2%
(b) 2%
(c) -3%
(d) 3%

End

chapter

13 Differential equations

Differential equations are equations in which one or more terms are a differential coefficient. Faraday's second law of electrodynamics, which makes the simple statement that a self-induced e.m.f. is proportional to the rate of change of flux, leads to the important equation:

$$E = N\frac{d\Phi}{dt} \text{ volts}$$

This is a differential equation because it contains the differential coefficient of Φ with respect to t.

Differential equations are widely used in engineering. They are the very essence of mathematical modelling. Differential equations can be formed from a known scientific law or from empirical observations that involve a rate of change: temperature over time, friction over speed, current over time and so on.

In this chapter you will learn to solve such equations and to appreciate their importance.

The solution of differential equations requires a working knowledge of basic calculus so let's take a quick look at our standard derivatives and integrals, which were summarized in the tables at the end of Chapter 9. For convenience, here they are again (Table 13.1 and 13.2). They should be quite familiar to you by now!

Table 13.1. Standard differentials

$y = f(x)$	$\dfrac{dy}{dx}$
ax^n	anx^{n-1}
$\sin ax$	$a\cos ax$
$\cos ax$	$-a\sin ax$
$\tan ax$	$a\sec^2 ax$
e^{kx}	ke^{kx}
$\ln ax$	$\dfrac{1}{x}$

Table 13.2. Standard integrals

$y = f(x)$	$\int y\,dx$
ax^n	$\dfrac{ax^{n+1}}{n+1} + C\ (n \neq -1)$
$\sin ax$	$-\dfrac{1}{a}\cos ax + C$
$\cos ax$	$\dfrac{1}{a}\sin ax + C$
$\sec^2 ax$	$\dfrac{1}{a}\tan ax + C$
e^{kx}	$\dfrac{1}{k}e^{kx} + C$
$\dfrac{1}{x}$	$\ln x + C$

Now do the revision problems below. Don't be disappointed if your solutions look different from those given. The algebraic expression that represents a derivative or integral usually requires some tidying-up. This is often a matter of personal preference. However, you should, in general, aim to have an expression that is tidy and compact. Look for common factors. Factorization leads to expressions that contain fewer terms.

PRACTICE EXERCISE 13.1 (Revision)

Differentiate the following with respect to the appropriate variable.

(a) $y = 3x^2 - \dfrac{2}{x^3}$; (b) $y = 2\sec\theta$; (c) $s = \dfrac{3 + \ln t}{2 - 4\ln t}$

(d) $y = (x + 2x^3)^4$; (e) $y = e^{4t}\sin\left(2t + \dfrac{3\pi}{2}\right)$

Solutions

(a) $\dfrac{dy}{dx} = 6\left(x + \dfrac{1}{x^4}\right)$; (b) $\dfrac{dy}{d\theta} = 2\sec\theta\tan\theta$; (c) $\dfrac{ds}{dt} = \dfrac{14}{t(2 - 4\ln t)^2}$

(d) $\dfrac{dy}{dx} = 4(x + 2x^3)^3(1 + 6x^2)$; (e) $\dfrac{dy}{dt} = 2e^{4t}\left[\cos\left(2t + \dfrac{3\pi}{2}\right) + 2\sin\left(2t + \dfrac{3\pi}{2}\right)\right]$

Now the opposite, integration. It is important for you to be familiar enough with calculus so that you do not confuse integration with differentiation. Be patient and work through the next set of revision problems.

PRACTICE EXERCISE 13.2 (Revision)

Find the following integrals.

(a) $\int \dfrac{3}{x^2}\,dx$; (b) $\int (3x-4)^2\,dx$; (c) $\int (3\cos t - 5\sin t)\,dt$

(d) $\int_2^3 \left(\dfrac{x^3 - x^2}{x}\right)\,dx$; (e) $\int_1^3 \left(\dfrac{2}{x} - e^{-5x} + \dfrac{\sqrt{x}}{3} - 3\right)\,dx$

Solutions

(a) $C - \dfrac{3}{x}$; (b) $3x^3 - 12x^2 + 16x + C$; (c) $3\sin t + 5\cos t + C$; (d) $\dfrac{23}{6}$; (e) -2.87

Now you should be warmed up enough to tackle differential equations with ease.

You may have noticed that none of the integration problems involved evaluating the constant of integration. Well, there will be plenty of that shortly. First, a couple of definitions.

13.1 Differential equations classified

We know that differential equations are those equations that contain one or more differential coefficients. They are classified by their **order** and by their **degree**.

THE ORDER OF A DIFFERENTIAL EQUATION

This is the value of the highest derivative which appears in the equation. The example below is a **second-order differential equation**.

$$\frac{d^2y}{dx^2} - 2\frac{dy}{dx} = 2x^2$$

THE DEGREE OF A DIFFERENTIAL EQUATION

This is the highest power of the highest order term of the equation. The example below is a **third-order differential equation of the second degree**.

$$3\left(\frac{d^3y}{dx^3}\right)^2 + 4\left(\frac{dy}{dx}\right)^3 - 2x = 0$$

Notice from this example that the class of equation depends only on the highest order term and its power. Lower order terms do not play a part in the classification and may not even be part of the equation at all.

PRACTICE EXERCISE 13.3

Classify the following equations.

(a) $y = 2x^3 - \dfrac{d^2y}{dx^2} + \left(\dfrac{dy}{dx}\right)^2$

(b) $\dfrac{dQ}{dt} = kQ$

(c) $\left(\dfrac{d^3y}{dx^3}\right)^2 = \dfrac{d^4y}{dx^4}$

Answers

(a) second order, first degree
(b) first order, first degree
(c) fourth order, first degree

For the moment we will concern ourselves with equations of the first order and first degree.

In the following example, we take Faraday's law again and show how a differential equation should be dealt with – from its beginnings in the modelling process to its final solution. It's not difficult.

EXAMPLE 13.1

'The self-induced e.m.f. in a conductor is proportional to the rate
of change of flux which links with that conductor'

It is likely that, before he announced his hypothesis, Michael Faraday must have carried out many experiments under different conditions, using different sized magnets and different lengths of conductors. He must have seen that the faster he moved the magnets, the greater the induced voltage.

The birth of this differential equation was a scientific law based on experiment.

An engineer will convert a hypothesis into something tighter. She can write the proportionality in algebraic terms.

$$e \propto \frac{d\Phi}{dt}$$

The proportionality becomes an equality when we put in a constant. Normally this constant is not known so just a letter, say k, will do. With Faraday's law we now have the benefit of hindsight and we know the constant of proportionality is N, the number of turns of the conductor.

$$e = N\frac{d\Phi}{dt}$$

The differential equation is a mathematical model that links the differentials of flux and time. To solve it we need to find the link between flux and time. We start by making the differential coefficient the subject.

$$\frac{d\Phi}{dt} = \frac{e}{N}$$

We re-arrange further by separating the differentials. Effectively this is a multiplication by dt.

$$d\Phi = \frac{e}{N}\,dt$$

Now it is possible to integrate the equation. To maintain equality we must integrate both sides. Here, we must integrate the LHS with respect to Φ and the RHS with respect to t. We indicate this by writing integration signs.

$$\int d\Phi = \frac{e}{N}\int dt$$

Notice that on the RHS the terms e and N are constant coefficients of the variable t, so for convenience we put them outside the integral. Now, on the left we have *one* dΦ and on the right we have *one* dt. We indicate this by writing the ones.

$$\int 1\,d\Phi = \frac{e}{N}\int 1\,dt$$

Now integrate. On the LHS we integrate 1 with respect to Φ and on the RHS we integrate 1 with respect to t.

$$\Phi + A = \frac{e}{N}t + B$$

Where have the terms A and B come from? Well, each time we integrate an indefinite integral we must put in the constant of integration which you are used to seeing as C. However, in solving differential equations it is customary to start with A and continue with $B, C, D, \ldots$ and so on. This is because integration sometimes yields many constants.

$$\Phi = \frac{e}{N}t + B - A$$

Having said that, look what happens in this case. Subtracting A gives us the term $(B - A)$ which is just some other constant of unknown value which we can write as C. We could have done this in the first place but it's good to know the whole story.

$$\Phi = \frac{e}{N}t + C$$

Now we have what is called the **general solution** of the differential equation. It is general because it contains the constant of integration C.

General solution:

$$\Phi = \frac{e}{N}t + C$$

This is a linear of the form $y = mx + C$. Here our gradient is given by the ratio of e over N, and C is the intercept on the y-axis. If C is not known we know the slope of the graph but not its position. The general solution provides us with a family of graphs that is infinite in number (Fig. 13.1). To tie it down to a single graph we want the **particular solution** that contains the numerical value of C. Do you remember what is required if we wish to evaluate the constant of integration?

We require corresponding values of the variables y and x. Then we can transpose and evaluate C. The corresponding values of the variables are called **boundary conditions**. In this case we want the boundary conditions of Φ and t. A special kind of boundary condition is an **initial condition**.

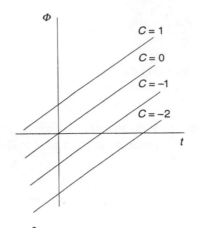

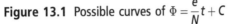

Figure 13.1 Possible curves of $\Phi = \dfrac{e}{N}t + C$

This is the pair of values you get when the independent variable is 0. It is often the easiest kind of boundary condition to find. In this case we might have found that at the start of an experiment, at time $t = 0$, the flux $\Phi = 0$. So we insert those values and look for what is called the **particular solution** of our differential equation.

$$C = \Phi - \frac{et}{N}$$

$$C = 0 - \frac{e(0)}{N} = 0$$

Particular solution:

$$\Phi = \frac{et}{N}$$

So with the initial condition $(0,0)$ we find the value of the constant is also 0.

Let's stay with the same example a little longer because the solution is of some interest.

We have an equation which links flux with time. Not particularly useful as it stands. Magnetic flux is difficult to measure and to measure time may not be practical either. Also, e.m.f. is likely to be the term of greatest interest to us. We may wish to know the output voltage of a simple generator for example. So we apply some simple transposition and substitution to obtain a particular solution in another form.

Multiplying by N and dividing by t makes e the subject.

$$e = \frac{\Phi N}{t}$$

As engineers, we know that flux is the product of flux density and area.

$$\Phi = BA$$

$$e = \frac{BAN}{t}$$

Time in general is given by the ratio of distance moved, say x, over average speed, v.

$$t = \frac{x}{v}$$

$$e = \frac{BAN}{\dfrac{x}{v}} = \frac{BANv}{x}$$

Now, A is the area swept by a conductor of length l moving a distance x, perpendicular to a magnetic flux. So area divided by distance gives the length of the conductor.

$$e = Blv$$

We end up with an equation which should be familiar to you. It is the basic equation which we use to calculate the e.m.f. produced by a generator. It looks completely different to our original particular solution but it is still the particular solution of the differential equation which models Faraday's law.

In this example, we have gone further than you might expect to at this level of study. But it was useful to demonstrate all the steps for solving differential equations in engineering and also the fact that the particular solution may take a number of forms.

Let's summarize the procedure from beginning to end.

- Formulate a proportionality from a scientific law or from experimental data.
- Formulate a differential equation by including a known or unknown constant.
- Rearrange the equation into a form that is suitable for integration.
- Integrate adding a constant of integration – just one will do. This is the general solution.
- Given a boundary condition, evaluate the constant of integration.
- Write the particular solution in a form that you feel is most appropriate to your needs.

13.2 Differential equations of the form $dy/dx = f(x)$

This is the type we solved in the last example. It is the most straightforward because we solve it by direct integration. Go through the examples below.

EXAMPLE 13.2

To solve.

(a) $\dfrac{dy}{dx} = 2x^3$

$dy = 2x^3\,dx$, so $\displaystyle\int dy = 2\int x^3\,dx$

$y = \tfrac{1}{2}x^4 + A$, General solution (G.S.)

This is simple. Multiply by dx. Integrate, on the LHS, 1 with respect to y, on the RHS, x^3 with respect to x. It's a good idea to get into the habit of using A as the first constant of integration. With no boundary conditions only a general solution is possible.

(b) $\dfrac{dy}{dx} = 3x - \sin x$

$dy = (3x - \sin x)\,dx$

$\displaystyle\int dy = \int (3x - \sin x)\,dx$

$y = \tfrac{3}{2}x^2 + \cos x + A$ G.S.

This is also straightforward. Notice how we put brackets around the RHS when multiplying by dx. Then we integrate the whole expression with respect to x. Again, only a general solution is possible.

(c) $\dfrac{dy}{dx} = x^3 - e^{4x}$

with $y = 2$ when $x = 1$

$dy = (x^3 - e^{4x})\,dx$

$\displaystyle\int dy = \int (x^3 - e^{4x})\,dx$

$y = \tfrac{1}{4}x^4 - \tfrac{1}{4}e^{4x} + A$ G.S.

$2 = \tfrac{1}{4}(1)^4 - \tfrac{1}{4}e^{4(1)} + A$

$A = 2 - \tfrac{1}{4} + \tfrac{1}{4}e^4 = 15.4$

$y = \tfrac{1}{4}(x^4 - e^{4x}) + 15.4$ Particular solution (P.S.)

This is the same procedure as above to find the G.S. But now we have boundary conditions so we can insert them into the G.S., transpose and solve for A. Notice we didn't bother to tidy up the G.S. because we know we are aiming for a P.S. With the P.S. we take the opportunity to factorize which makes the expression a little tidier.

(d) $t - 2\dfrac{d\theta}{dt} = \dfrac{1}{3\,e^{2t}} - 5$

with $\theta = \dfrac{\pi}{3}$ when $t = 0$

There is some algebra required before we can solve. We start by subtracting t. Then we divide by -2 to leave the differential coefficient on its own.

$-2\dfrac{d\theta}{dt} = \dfrac{1}{3}e^{-2t} - t - 5$

$\dfrac{d\theta}{dt} = 2t + 10 - \dfrac{2}{3}e^{-2t}$

We can take two steps in one go; multiply by dt and write the integration signs.

$\displaystyle\int d\theta = \int \left(2t + 10 - \dfrac{2}{3}e^{-2t}\right) dt$

Integrate the three terms on the right then add the constant.

$\theta = t^2 + 10t + \tfrac{1}{3}e^{-2t} + A$ G.S.

Transpose for A and substitute the boundary condition.

$$A = \theta - t^2 - 10t - \tfrac{1}{3}e^{-2t}$$

$$A = \frac{\pi}{3} - (0)^2 - 10(0) - \frac{1}{3}e^{-2(0)} = \frac{\pi - 1}{3}$$

$$\theta = t^2 + 10t + \tfrac{1}{3}(e^{-2t} + \pi - 1) \quad \text{P.S.}$$

The value of A turns out to be quite neat.

The four examples which you just worked through gave the differential equation to begin with. There was no need for you to formulate it from a physical problem. Let's do that now.

EXAMPLE 13.3

Having carried out an experiment in which you measured the velocity of a body, you find the data obtained produces a curve that fits the equation $v = at^b$. From the graph, you have also obtained the values of the constants as $a = 4$ and $b = 1.2$. You also note that the initial condition was $(0,0)$. Form a differential equation and solve it to find the relationship between position and time.

Velocity is the rate of change of distance over time; it is a rate of change which allows you to write a differential equation.

$$v = 4t^{1.2}$$

$$\frac{ds}{dt} = 4t^{1.2}$$

This is a first-order differential equation of the type $\frac{dy}{dx} = f(x)$ which you can solve by direct integration.

$$\int ds = 4\int t^{1.2}\,dt$$

$$s = \frac{4}{2.2}t^{2.2} + A \quad \text{G.S.}$$

You can transpose the general solution to solve for A because you have the boundary condition $(0,0)$.

$$A = s - \frac{4}{2.2}t^{2.2} \quad \text{with } (t,s) = (0,0)$$

$$A = 0$$

Express the result to 3 s.f. which is a suitable accuracy for most practical applications.

$$s = 1.82t^{2.2} \quad \text{P.S.}$$

EXAMPLE 13.4

A further experiment with a moving mass indicates the following. When time, $t = 16\,\text{s}$, the velocity of the mass, $v = 20\,\text{m s}^{-1}$ and at $t = 20\,\text{s}$, the velocity has increased to $v = 40\,\text{m s}^{-1}$. You know the force acting on the body is a constant and the mass of the body is also a constant. Formulate a differential equation then find the equation which links velocity with time.

Newton's second law implies that with force and mass as constants, acceleration must be constant. Call it k. We tend to use the middle letters of the alphabet when constants are not constants of integration.

$$F = ma \quad \text{so} \quad a = \frac{F}{m}$$

$$a = k$$

$$\frac{dv}{dt} = k$$

$$\int dv = k \int dt$$

$$v = kt + A$$

$$v = at + A \quad \text{G.S.}$$

The general solution is easily found but yields two unknown constants. OK, we know that $k = a$. But we still have two unknowns.

Luckily, we have two sets of boundary conditions so we can solve simultaneously.

$$40 = 20a + A$$

$$20 = 16a + A$$

Subtracting the second equation from the first eliminates A, so we can solve for a.

$$20 = 4a$$

$$a = 5$$

Transposing the general solution and substituting for a leads to the value of the constant of integration, A.

$$A = v - at$$

$$A = 20 - 5(16) = -60$$

$$v = 5t - 60 \quad \text{P.S.}$$

The particular solution models the motion of the mass. We see it has a constant acceleration of $5\,\mathrm{ms^{-2}}$ and an initial velocity of $-60\,\mathrm{ms^{-1}}$.

PRACTICE EXERCISE 13.4

(a) Find the general solution of $\frac{dy}{dx} = 2x$. Then find the particular solution given that when $x = 0, y = 3$.

(b) Find the particular solution of $\frac{dy}{dx} = 3x^3 - 2x^2 - x + 5$ if it is known that its graph passes through the point $(1, 2)$.

(c) Solve, $\frac{dy}{dx} = x^2 + 3x$.

(d) Find the distance s of an object at time $t = 12\,\mathrm{s}$, given that its speed $\frac{ds}{dt} = -kt$ and initially its distance is $10\,\mathrm{m}$, then when $t = 2\,\mathrm{s}$ its distance is $4\,\mathrm{m}$.

Solutions

(a) $y = x^2 + 3$; (b) $y = \frac{3}{4}x^4 - \frac{2}{3}x^3 - \frac{1}{2}x^2 + 5x - \frac{31}{12}$; (c) $y = \frac{1}{3}x^3 + \frac{3}{2}x^2 + A$
(d) $-206\,\text{m}$

13.3 Differential equations of the form $dy/dx = f(y)$

This form of differential equation is quite common. It relates the rate of change of a function to the value of the function so its solution is always an exponential, as you will see.

EXAMPLE 13.5

Solve $\dfrac{dy}{dx} = ky$

The first step is to separate the differentials as before. y cannot be integrated with respect to x so we divide by y. Or multiply by its reciprocal, it amounts to the same thing.

$$\frac{dy}{dx} = ky$$

$$dy = ky\,dx$$

$$\frac{1}{y}\,dy = k\,dx$$

Now it is possible to integrate both sides. Remember that on the right we are integrating 1 with respect to x. We get to a general solution which gives us the natural log of the function of x. This is rather inconvenient so it should not be left as it is.

$$\int \frac{1}{y}\,dy = k\int dx$$

$$\ln y = kx + A$$

A reminder about logs. If:

$$\log_b N = x \quad \text{then} \quad N = b^x$$

taking antilogs,

$$b^{\log_b N} = b^x \quad \text{confirms that} \quad N = b^{\log_b N}$$

In this case we are dealing with natural logs which, as you know, are $\log_e$ so,

$$e^{\log_e y} = y \quad \text{means that} \quad e^{\ln y} = y$$

That deals with the LHS. On the right we can apply the rule of indices which states that:

$$b^m \times b^n = b^{m+n}$$

$$e^{\ln y} = e^{kx+A}$$

$$\therefore \quad y = e^{kx+A}$$

$$y = e^{kx} e^A$$

Finally, because A is a constant, e^A is just some other constant, say, B.

$$y = Be^{kx} \quad \text{G.S.}$$

Example 13.5 was dealt with in some detail in order to give you the opportunity to revise your knowledge of logs and exponentials. In time, this will not be necessary, you will be able to streamline the method. But, for the moment, you should concentrate on reinforcing your knowledge, so show the details of what you are doing.

Now for a particular solution.

EXAMPLE 13.6

Solve the differential equation (D.E.) $\dfrac{dy}{dx} - 3y = 0$ given that $y = 8$ when $x = 1.5$.

Rearrange to get the derivative on the LHS. Put dx on the RHS and y on the LHS. Integrate, as usual to obtain a general solution which is in log form and not very convenient.

$$\frac{dy}{dx} - 3y = 0$$

$$\frac{dy}{dx} = 3y$$

$$dy = 3y \, dx$$

$$\int \frac{1}{y} \, dy = \int 3 \, dx$$

$$\ln y = 3x + A \quad \text{G.S.}$$

Now we have a choice, we can use the boundary conditions to evaluate A then go on to get rid of the log. Or get rid of the log first and then evaluate the constant.

Let's do it both ways.

Method 1: we transpose for A, substitute the boundary conditions and get $A = -2.42$.

$$A = \ln y - 3x$$

$$= \ln 8 - 3(1.5) = -2.42$$

$$\ln y = 3x - 2.42 \quad \text{P.S.}$$

The particular solution is in log form so we take antilogs of both sides.

$$y = e^{3x - 2.42} = e^{3x} e^{-2.42}$$

This gives us the constant $e^{-2.42}$ which is 0.0889 correct to 3 significant figures.

$$y = 0.0889\,e^{3x}$$

Now for method 2. Convert the general solution from log to exponential form. This leads to some other constant B as in the previous example. Transpose for B and evaluate. The result is a particular solution which is exactly the same.

$$y = e^{3x+A} = e^{3x}\,e^A$$

$$y = B\,e^{3x} \quad \text{Better G.S.!}$$

$$B = y\,e^{-3x} = 8\,e^{-3(1.5)} = 0.0889$$

$$y = 0.0889\,e^{3x} \quad \text{Same P.S.!}$$

Which is the better method? In this example my preference is the second method; convert your G.S. to exponential form then look for the P.S. However, there may be circumstances which make the first method easier; find the particular solution in log form then take antilogs.

EXAMPLE 13.7

Use the 'second method' to find the particular solution of

$$\frac{3}{4}\frac{dy}{dx} + \frac{7}{8}y = y$$

given that $y = 3$ when $x = 2$.

Collect-up the y-terms. Collect-up the constants on the RHS. You don't have to, but it is neater that way. Divide by y and integrate both sides.

$$\frac{3}{4}\frac{dy}{dx} = y - \frac{7}{8}y = \frac{1}{8}y$$

$$\frac{dy}{dx} = \frac{4}{24}y = \frac{1}{6}y$$

$$dy = \frac{1}{6}y\,dx$$

$$\int \frac{1}{y}\,dy = \int \frac{1}{6}\,dx$$

$$\ln y = \tfrac{1}{6}x + A$$

Take antilogs to get a G.S. in exponential form. Transpose and substitute the boundary conditions to evaluate B. Note, in transposing for B, we multiply by the reciprocal of the exponential term.

$$y = e^{\frac{x}{6}+A} = e^{\frac{x}{6}}\,e^A$$

$$y = B\,e^{\frac{x}{6}} \quad \text{G.S.}$$

$$B = y\,e^{-\frac{x}{6}} = 3\,e^{-\frac{2}{6}} = 2.15$$

$$y = 2.15\,e^{\frac{x}{6}} \quad \text{P.S.}$$

EXAMPLE 13.8

Now use our 'first method' to find the particular solution of the same differential equation.

$$\ln y = \tfrac{1}{6}x + A \quad \text{G.S.}$$

Since you are going for a P.S. there is a good argument for leaving the G.S. in log form.

$$A = \ln y - \tfrac{1}{6}x = \ln 3 - \tfrac{2}{6} = 0.765$$

$$\ln y = \tfrac{1}{6}x + 0.765$$

$$y = e^{\frac{x}{6} + 0.765} = e^{\frac{x}{6}}e^{0.765}$$

$$y = 2.15\, e^{\frac{x}{6}} \quad \text{P.S.}$$

It's up to you to decide which method to use. In this case neither method is quicker, each requires the same number of steps. But remember, if you are just going for a G.S. you should take antilogs and present the solution in exponential form.

Now it's time to do a little modelling. The form of differential equation we are dealing with crops up quite often in engineering. Take an example which involves heat transfer.

EXAMPLE 13.9

Working for a firm which specializes in the manufacture of medical supplies, you have designed and built a prototype of an ice-box that will be used to carry transplant organs. In a carefully controlled experiment you froze a quantity of meat to exactly 0°C and placed it in the ice-box. Exactly 1 hour later you opened the box and measured the temperature of the meat and found it to be 1.32°C.

Now you have a good idea that the rate of temperature transfer is proportional to temperature difference. Starting with that idea, formulate a differential equation, find its solution and use it to predict the temperature of an item frozen to 0°C and then stored in the box for a period of 8 hours.

You are starting from scratch. You need some algebraic symbols. Let the temperature of the contents be θ. That makes the rate at which temperature changes over time the differential coefficient of θ with respect to t.

This, you believe to be proportional to the difference between the temperature contents, θ and the ambient temperature which you assume to be 25°C. Multiplying by a constant of proportionality (say k) leads to a differential equation.

$$\frac{d\theta}{dt} \propto 25 - \theta$$

$$\frac{d\theta}{dt} = k(25 - \theta)$$

Separate the differentials and divide by $(25 - \theta)$.

$$d\theta = k(25 - \theta)\, dt$$

$$\frac{1}{25 - \theta}\, d\theta = k\, dt$$

Now you are ready to integrate.

You have not seen this yet but please accept for the moment that:

$$\int \frac{1}{u} \, dx = -\ln u + C \quad \text{when } u = a - x$$

here, $a = 25$ and $x = \theta$

$$\therefore \quad \int \frac{1}{25 - \theta} \, d\theta = \int k \, dt$$

$$-\ln(25 - \theta) = kt + A \quad \text{G.S.}$$

You must now keep the G.S. in log form. If you antilog you will be faced with having to solve an exponential to evaluate the constants. That would involve converting back to log form again!

Transpose for A and substitute the initial conditions which we know to be 0°C when t is 0 s. Four significant figures is a reasonable degree of accuracy for this kind of problem.

$$A = -\ln(25 - \theta) - kt$$

with $\theta = 0$ when $t = 0$

$$A = -\ln(25 - 0) - k(0) = -3.219$$

With a value for A, substitute for k and use the boundary condition provided by the test data, $\theta = 1.32$ at $t = 3600$. Note we are sticking with SI so time is in seconds not hours.

$$kt = -\ln(25 - \theta) - A$$

$$k = \frac{-\ln(25 - \theta) - A}{t}$$

with $\theta = 1.32$ when $t = 3600$

$$k = \frac{-\ln(25 - 1.32) + 3.219}{3600} = 15.10 \times 10^{-6}$$

We substitute the values of the constants into the general solution which now needs to be converted into exponential form.

$$-\ln(25 - \theta) = 15.1t \times 10^{-6} - 3.219$$

$$\ln(25 - \theta) = 3.219 - 15.1t \times 10^{-6}$$

Taking antilogs gets rid of the log on the LHS and gives us the exponential on the right. The constant $e^{3.219}$ turns out to be 25.00. Factorizing 25 tidies things up.

$$25 - \theta = e^{3.219 - \frac{15.1t}{10^{6}}} = e^{3.219} \, e^{\frac{-15.1t}{10^{6}}}$$

$$= 25 \, e^{\frac{-15.1t}{10^{-6}}}$$

$$-\theta = 25 \, e^{\frac{-15.1t}{10^{6}}} - 25$$

$$\theta = 25 - 25 \, e^{\frac{-15.1t}{10^{-6}}}$$

$$\theta = 25 \left(1 - e^{\frac{-15.1t}{10^{6}}} \right) \quad \text{P.S.}$$

We now have the particular solution in an exponential form. The form should be familiar to you.

Finally you wished to know the temperature of the contents after 8 hours.

With $t = 8\,\text{h} = 28.8 \times 10^3\,\text{s}$,

$$\theta = 25\left(1 - e^{\frac{-15.1 \times 28.8 \times 10^3}{10^6}}\right) = 8.82°C$$

Your model predicts that the contents of your particular ice-box will have risen to almost 9°C after 8 hours of storage.

Suppose you were told that 5°C was a critical temperature at which the donor organs would perish. How much time would this box give you to transport the organs?

Go back to the log form of the particular solution and you can solve for t.

$$\ln(25 - \theta) = 3.219 - 15.1t \times 10^{-6}$$

$$15t \times 10^{-6} = 3.219 - \ln(25 - \theta)$$

$$t = \frac{3.219 - \ln(25 - \theta)}{15 \times 10^{-6}}$$

$$t = \frac{3.219 - \ln(25 - 5)}{15 \times 10^{-6}} = 14.88 \times 10^3\,\text{s}$$

$$= 4.13\,\text{h}$$

The model indicates that the critical time is just under 4 hours and 8 minutes.

There are some niggling details. Heat transfer that involves a change of state uses latent heat energy so it is likely that the temperature will increase at a slower rate than our model predicts. Different masses of donor organs will also have an effect. But don't be disappointed, the model provides a good first approximation!

EXAMPLE 13.10

A fundamental law of electric circuits is Kirchhoff's voltage law which leads us to conclude that at any instant the sum of the potential differences in an electric circuit is zero. Formulate a suitable differential equation that relates voltage with time in a circuit containing capacitance and resistance in series. Solve this D.E. to show how capacitor voltage depends on time in general.

Kirchhoff's law gives us an equation in which we can (according to Ohm's law) substitute for v_R. We do this because current is a rate of change of charge.

$$v_R + v_C = 0$$

$$iR + v_C = 0$$

$$i = \frac{dq}{dt}$$

We also know that charge stored in a capacitor is proportional to capacitance C and voltage v. This means that the rate of change of charge is proportional to the constant of capacitance C and the rate of change of voltage.

$$q = Cv$$

$$\frac{dq}{dt} = C\frac{dv}{dt}$$

We replace the rate of change of charge with the symbol i because this is current in coulombs per second.

$$i = C\frac{dv}{dt}$$

Now we can go back to the second equation we wrote and replace current with the term,

$$C\frac{dv}{dt}$$

This gives us a differential equation which provides an expression for rate of change of capacitor voltage.

$$\therefore \quad C\frac{dv}{dt}R + v = 0$$

$$CR\frac{dv}{dt} = -v$$

$$\frac{dv}{dt} = \frac{-v}{CR}$$

It is of the general form

$$\frac{dy}{dx} = f(y)$$

which we solve in the usual way.

$$\frac{1}{v}\frac{dv}{dt} = -\frac{1}{CR}$$

$$\int \frac{1}{v}\,dv = -\frac{1}{CR}\int dt$$

$$\ln v = -\frac{t}{CR} + A$$

$$v = e^{-\frac{t}{CR}+A} = e^{-\frac{t}{CR}}e^{A}$$

$$v = Be^{-\frac{t}{CR}} \quad \text{G.S.}$$

The general solution reveals that capacitor voltage decays exponentially with time.

PRACTICE EXERCISE 13.5

Solve:
(a) $\frac{dy}{dx} = -3y$; (b) $\frac{dy}{dx} = \frac{1}{2}y$; (c) $4\frac{ds}{dt} = 12s$; (d) $\frac{dQ}{dt} = kQ$; (e) $5\frac{dy}{dx} + y = 3y$

(f) $\frac{dy}{dx} = -13y$ with $(0, 4)$; (g) $\frac{di}{dt} = 6.25i$ with $(3, 2)$; (h) $\frac{dy}{dx} = ky$ with $(1, 3)$ and $(2, 7)$

Answers

(a) $y = Be^{-3x}$; (b) $y = Be^{\frac{x}{2}}$; (c) $s = Be^{3t}$; (d) $Q = Be^{kt}$; (e) $y = Be^{0.4x}$

(f) $y = 4e^{-13x}$; (g) $i = 14.4 \times 10^{-9}e^{6.25t}$; (h) $y = 1.29e^{0.847x}$

13.4 Differential equations of the form $dy/dx = f(x)g(y)$

This form of D.E. is known as **variable separable** since its solution is obtained by, first, separating the function of x from the function of y.

There is nothing difficult about this method. You simply apply some algebra to get the equation into the form:

$$\frac{1}{g(y)}\, dy = f(x)\, dx$$

Then you are in a position to integrate both sides:

$$\int \frac{1}{g(y)}\, dy = \int f(x)\, dx$$

EXAMPLE 13.11

Find the general solution of:

$$\frac{dy}{dx} = xy$$

Dividing by y is sufficient to separate the variables.

$$\frac{1}{y}\frac{dy}{dx} = x$$

Integrate in the usual manner.

$$\int \frac{1}{y}\, dy = \int x\, dx$$

$$\ln y = \tfrac{1}{2}x^2 + A$$

Antilog by expressing each side as a power of e.

$$y = e^{\frac{x^2}{2} + A}$$

Simplify.

$$y = B e^{\frac{x^2}{2}}$$

EXAMPLE 13.12

Find the general solution of:

$$\frac{dy}{dx} = \frac{1}{xy^2}$$

$$y^2 \frac{dy}{dx} = \frac{1}{x}$$

$$\int y^2\, dy = \int \frac{1}{x}\, dx$$

$$\tfrac{1}{3}y^3 = \ln x + A$$

Slightly different here. We get the log of the x-term so there is no need for antilogs.

Note that multiplying by 3 changes the value of the constant A to $3A$. But this is some other constant B.

$$y = \sqrt[3]{3\ln x + B}$$

$$y = (3\ln x + B)^{\frac{1}{3}}$$

EXAMPLE 13.13

Solve:

$$\frac{dy}{dx} = \frac{y}{x}$$

This one provides an opportunity to play with logs. If we wish, we can say that the constant A is equal to the log of some other constant, say B. This allows me to apply the law of logs which states that:

$$\log(mn) = \log m + \log n$$

$$\therefore \quad \frac{1}{y}\frac{dy}{dx} = \frac{1}{x}$$

$$\int \frac{1}{y}\,dy = \int \frac{1}{x}\,dx$$

$$\ln y = \ln x + A$$

$$\ln y = \ln x + \ln B$$

$$\ln y = \ln(Bx)$$

$$e^{\ln y} = e^{\ln(Bx)}$$

$$y = Bx$$

Otherwise we can proceed as before and use the laws of indices to simplify the solution.

$$e^{\ln y} = e^{\ln x + A}$$

$$y = e^{\ln x}e^{A}$$

$$y = xe^{A}$$

$$y = Bx$$

It turns out to be a simple linear example. Compare it with the next example.

EXAMPLE 13.14

Solve:

$$\frac{dy}{dx} = -\frac{x}{y}$$

Here, as in Example 13.12, the x- and y-factors are so related that the solution contains no logarithms (or exponentials). Note the negative possibility when we take a square root.

$$y\frac{dy}{dx} = -x$$

$$\int y\,dy = \int -x\,dx$$

$$\tfrac{1}{2}y^2 = -\tfrac{1}{2}x^2 + A$$

$$y^2 = 2A - x^2$$

$$y = \pm\sqrt{B - x^2}$$

This solution happens to be the equation of a circle with a radius that is $\sqrt{B}$.

EXAMPLE 13.15

Find the general solution of:

$$\frac{dy}{dx} = \frac{y + xy}{xy - x}$$

Before we can separate variables it is necessary to express the RHS as x- and y-factors. We can then multiply and divide by the y-factors. In order to be able to integrate we must then express the compound fractions as partial fractions. This is easily done here because y and x are common denominators.

$$\frac{dy}{dx} = \frac{y(1 + x)}{x(y - 1)}$$

$$\left(\frac{y - 1}{y}\right)\frac{dy}{dx} = \frac{1 + x}{x}$$

$$\left(1 - \frac{1}{y}\right)\frac{dy}{dx} = \frac{1}{x} + 1$$

$$\int\left(1 - \frac{1}{y}\right)dy = \int\left(\frac{1}{x} + 1\right)dx$$

$$y - \ln y = \ln x + x + A \quad \text{G.S.}$$

$$e^{y - \ln y} = e^{\ln x + x + A}$$

$$y^{-1}e^{y} = Bx\,e^{x} \quad \text{G.S.}$$

The last example shows that there are times when the logarithmic and exponential forms provide solutions that are equally awkward. In both cases we cannot get an expression for y. In which case it is a matter of individual choice whether to leave it in logarithmic form or not.

EXAMPLE 13.16

Find the particular solution of:

$$x\frac{dy}{dx} + 2y = 0$$

if $x = 0.25$ makes $y = 20$.

Subtract 2y and then divide by x to get the derivative on its own. Now you can separate the variables by dividing by y.

$$x\frac{dy}{dx} = -2y$$

$$\frac{dy}{dx} = \frac{-2y}{x}$$

$$\frac{1}{y}\frac{dy}{dx} = \frac{-2}{x}$$

Integrate and you have the general solution in log form. Remember, you are going for a particular solution so don't bother taking antilogs, go for the constant first.

$$\int\frac{1}{y}\,dy = -2\int\frac{1}{x}\,dx$$

$$\ln y = -2\ln x + A \quad \text{G.S.}$$

Substitute the constant into your general solution.

with $x = 0.25$, $y = 20$,

$$A = \ln 20 + 2\ln 0.25 = 0.223$$

$$\ln y = -2\ln x + 0.223$$

Here, we have a negative coefficient with the log term so we use the law,

$$\log(m^n) = n\log m$$

and raise x to the power of -2.

$$\ln y = \ln(x^{-2}) + 0.223$$

Now raising the terms of the equation as a power of e gives us the antilogs of the y and x terms and a value of 1.25 for the constant.

$$e^{\ln y} = e^{\ln x^{-2} + 0.223}$$

$$y = x^{-2}e^{0.223}$$

$$y = 1.25x^{-2}$$

Finally, we take the reciprocal of x^{-2}.

$$y = \frac{1.25}{x^2} \quad \text{P.S.}$$

EXAMPLE 13.17

Find the particular solution of:

$$\frac{ds}{dt} = 2e^{s-t}$$

if the curve of $s = f(t)$ passes through the point $(2, 1)$.

Applying the laws of indices allows us to write the RHS as a product.

$$\frac{ds}{dt} = 2e^s e^{-t}$$

Now we can separate the variables and integrate.

$$e^{-s}\frac{ds}{dt} = 2e^{-t}$$

$$\int e^{-s}\,ds = \int 2e^{-t}\,dt$$

$$-e^{-s} = -2e^{-t} + A$$

We evaluate A and substitute it into the general solution. Then multiply by -1, take logs and obtain an expression for s.

$$A = 2e^{-t} - e^{-s} = 2e^{-2} - e^{-1}$$

$$= -97.2 \times 10^{-3}$$

$$e^{-s} = 2e^{-t} + 97.2 \times 10^{-3}$$

$$s = -\ln(2e^{-t} + 97.2 \times 10^{-3})$$

You have probably realized by now that a particular solution is easily checked. Just substitute the boundary conditions and see if they satisfy your solution. A suitable check of the last solution would be:

$$s = -\ln(2e^{-2} + 97.2 \times 10^{-3}) = 1.0000$$

A 5 significant figure accuracy is good enough.

EXAMPLE 13.18

'The rate at which the pressure of a gas falls with the volume of the gas is directly proportional to pressure and inversely proportional to volume'. Use this statement to formulate a suitable differential equation and find the general solution that satisfies the differential equation.

Let pressure be p and volume be V.

The rate at which pressure changes with volume is the differential coefficient of pressure with respect to volume dp/dV.

The two proportionalities can be written separately.

$$\frac{dp}{dV} \propto p \quad \text{and} \quad \frac{dp}{dV} \propto \frac{1}{V}$$

A *fall* in pressure must be interpreted as a rate of change that is *negative* so we make the constant of proportionality $-k$.

$$\frac{dp}{dV} = -k\frac{p}{V}$$

The rest, you have seen before: separate variables, integrate and tidy up.

$$\frac{1}{p}\frac{dp}{dV} = -\frac{k}{V}$$

$$\int \frac{1}{p}\,dp = -k\int\frac{1}{V}\,dV$$

$$\ln p = -k \ln V + A$$

$$p = BV^{-k} = \frac{B}{V^k}$$

$$pV^k = B$$

The outcome is interesting and important.

Multiplying by V^k shows that the product of the pressure and volume of a gas is a constant. This implies that if we change the volume of a gas we must bring about a change in pressure which maintains the value of the constant B. What of the constant k? Well this is the constant which relates to a particular gas.

EXAMPLE 13.19

A mass of 20 kg is suspended by a spring having a stiffness of 2400 N m^{-1}. Initially the spring is extended by 200 mm and its speed is zero. Sketch a force diagram and use it to formulate a differential equation and solve it to find the particular equation that expresses the speed of the spring as a function of its extension.

Always start by identifying the variables and constants, assign symbols and jot down what is known.

Let F be the downward force of the mass.
Let T be the upward tension of the spring.
Let m be the mass.
Let x be the extension of the spring.
Let λ be the stiffness of the spring.

A diagram is essential in this case.

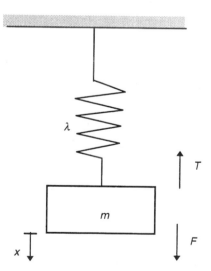

Start with Hooke's law, from which we know that the force in a spring is proportional to its stiffness and extension.

$$T = \lambda x$$

Newton's second law tells us that force in general is the product of mass and acceleration.

$$F = ma$$

Here we have a net force which is the downward force caused by acceleration due to gravity minus the upward force of the spring. This leads us to an equation in force.

$$F = mg - T$$

$$F = mg - \lambda x$$

We divide by m which makes the LHS acceleration.

$$\frac{F}{m} = g - \frac{\lambda x}{m}$$

$$a = g - \frac{\lambda x}{m}$$

Acceleration is the rate of change of speed with respect to time, dv/dt. This gives us a differential equation. However, we have saddled ourselves with three variables: velocity (v), displacement (x) and time (t).

$$\frac{dv}{dt} = g - \frac{\lambda x}{m}$$

We use the function of a function rule to write dv/dt as a product of dv/dx and dx/dt. Now dx/dt is speed so we can replace it with v. This gives us a differential equation in two variables v and x.

$$\frac{dv}{dt} = \frac{dv}{dx}\frac{dx}{dt} = \frac{dv}{dx}v$$

$$v\frac{dv}{dx} = g - \frac{\lambda x}{m}$$

The variables are already separated so we can integrate to get a general solution.

$$\int v\,dv = \int \left(g - \frac{\lambda x}{m} \right) dx$$

$$\frac{1}{2}v^2 = gx - \frac{\lambda x^2}{2m} + A \quad \text{G.S.}$$

Transpose for A, substitute the boundary conditions and the values of the constants. Take g as $9.81\,\mathrm{m\,s^{-2}}$, stiffness is $2400\,\mathrm{N\,m^{-1}}$ and the mass is $20\,\mathrm{kg}$.

$$A = \frac{1}{2}v^2 - gx + \frac{\lambda x^2}{2m}$$

With $x = 0.2$, $v = 0$

$$A = -9.81(0.2) + \frac{2400(0.2)^2}{2(20)} = 0.438$$

$$\frac{1}{2}v^2 = gx - \frac{\lambda x^2}{2m} + 0.438$$

Multiply by 2 and take the square root of the equation to get an expression for v. Note again, that in taking the square root of a number we must allow for the positive or negative possibility.

$$v = \pm\sqrt{2gx - \frac{\lambda x^2}{m} + 0.876}$$

$$v = \pm \sqrt{2gx - 120x^2 + 0.876} \quad \text{P.S.}$$

Since it is neater and more precise than its decimal approximation we leave g as it is.

PRACTICE EXERCISE 13.6

Solve:

(a) $\dfrac{dy}{dx} = \dfrac{x-2}{3y}$; (b) $(1-y)x\dfrac{dy}{dx} = y(x-1)$; (c) $\dfrac{ds}{dt} = e^{2s}\cos t$; (d) $\dfrac{dy}{dx} = \dfrac{-\sin x}{\cos y}$

(e) $\dfrac{1+u}{v^2}\dfrac{dv}{du} = 1$; (f) $\dfrac{dy}{dx} = \dfrac{3\sin 3x}{y}$ with $y = 2$ when $x = 1$

(g) $v\dfrac{dv}{dt} = 14t - 8$, with $v = 5$, $t = 2$; (h) $2x\dfrac{dy}{dx} - 3y = 0$ with $y = 2$, $x = 1$

(i) $\dfrac{d\theta}{dt} = \dfrac{1}{t^2 \sin\theta}$ with $\theta = \dfrac{\pi}{2}$, $t = 2$; (j) $\dfrac{dy}{dx} = 4e^{x-2y}$ with $y = 1$, $x = 0$

Solutions

(a) $y = \pm\sqrt{\frac{1}{3}(x^2 - 4x + C)}$; (b) $\dfrac{y}{e^y} = \dfrac{Be^x}{x}$; (c) $s = -\frac{1}{2}\ln(B - 2\sin t)$

(d) $y = \arcsin(\cos x + A)$; (e) $v = \dfrac{1}{B - \ln(1 + u)}$; (f) $y = \pm\sqrt{2.02 - 2\cos 3x}$

(g) $v = \pm\sqrt{14t^2 - 16t + 1}$; (h) $y = 2x^{\frac{3}{2}}$; (i) $\theta = \arccos\left(\dfrac{1}{t} - \dfrac{1}{2}\right)$

(j) $y = \frac{1}{2}\ln(8e^x - 0.611)$

13.5 Summary

That concludes your introduction to first-order differential equations. The following are the main points.

- Differential equations are classified by their order and degree. The order of a D.E. is the highest derivative, the degree is the power of the highest derivative. This will be important later on.
- The general solution of a D.E. is one that satisfies the D.E. and gives rise to a family of curves whose number is infinite. It will contain constants of integration that are equal in number to the degree of the equation.
- The particular solution of a D.E. is one that satisfies the D.E. and gives rise to one unique curve. A particular solution is possible if one or more sets of boundary conditions are known. A useful type of boundary condition is often an initial condition.
- D.E.s represent rates of change so they are very important in modelling dynamic relationships in science and engineering.

- The form $\dfrac{dy}{dx} = f(x)$ is solved by direct integration.

- The form $\dfrac{dy}{dx} = f(y)$ is solved by direct integration and always yields an exponential.

- The form $\dfrac{dy}{dx} = f(x)g(y)$ is solved by separation of variables prior to integration.

Assignment XI

In this assignment you will use differential equations to solve four engineering problems. You should aim to satisfy the following performance criteria.

- The engineering system is represented in the form of a differential equation.
- The variables of a differential equation are separated.
- Functions are integrated to obtain a general solution.
- Boundary conditions are identified for a given engineering problem.
- Boundary conditions are substituted to obtain a particular solution.
- Engineering systems are modelled using differential equations.

Start by seeking a full understanding of each problem situation.

PROBLEM 1

A chemical reaction's rate of propagation is thought to be indirectly proportional to the level of that reaction, R.

$$\frac{dR}{dt} \propto \frac{1}{R}$$

If, at the commencement of the reaction, R has a value of 5000 and after 5 s have elapsed the value of R has risen to 5500, predict the value of R when 50 s have elapsed.

Note: you are expected to find both the general and particular solutions of the differential equation as a means of determining the required answer.

PROBLEM 2

A hydraulic servo system's displacement-related change of exerted force, F, is proportional to the square of the error displacement, e.

$$\frac{dF}{de} \propto e^2$$

If the exerted force is zero when the error displacement is zero and is 500 N when the error displacement is 50 mm, determine what the exerted force will be when the error displacement is 60 mm.

Note: you are expected to find both the general and particular solutions of the differential equation as a means of determining the required answer.

PROBLEM 3

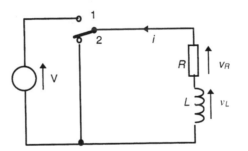

In the circuit shown, when the switch is moved to position 2, energy stored in the inductor will drive a current through resistor R. According to Kirchhoff's voltage law the sum of voltages around the loop will equal zero, so:

$$v_L + v_R = 0$$

From Ohm's law we know that:

$$v_R = iR$$

and the laws of electromagnetic induction give:

$$v_L = L\frac{di}{dt}$$

Use this information to form a differential equation and find the particular solution in the form of $i = f(t)$ given that initially, when $t = 0$, $i = 10$ A. You are not expected to evaluate the constants R and L.

PROBLEM 4

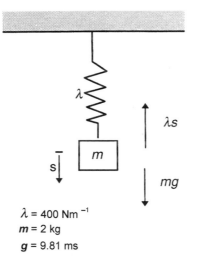

$\lambda = 400 \text{ Nm}^{-1}$
$m = 2 \text{ kg}$
$g = 9.81 \text{ ms}$

A spring of stiffness λ supports a mass m as shown. The extension of the spring is s. The upward force is the tension of the spring λs, the downward force is the mass accelerating due to gravity, mg. So the net force acting on the mass is:

$$F = mg - \lambda s$$

Applying Newton's second law and the function of a function rule for differentiation leads to:

$$v \frac{dv}{ds} = g - \frac{\lambda s}{m}$$

Find the particular equation which links v as a function of s given that initially the spring is extended by 0.1 m and has zero velocity.

chapter

14 Additional trigonometry

We start this chapter with a review of what has been covered. If you are unsure about something, turn back and revise the topic before proceeding any further. How confident are you with the following?

- the three trigonometric ratios, sine, cosine and tangent,
- Pythagoras' theorem,
- the number of degrees in a triangle,
- the area of a triangle by two methods, $\frac{1}{2}bh$ and $\frac{1}{2}ab\sin C$,
- the sine rule and the cosine rule,
- the three reciprocal ratios, cosecant, secant and cotangent,
- two useful identities, $\tan A = \sin A/\cos A$ and $\sin^2 A + \cos^2 A = 1$,
- the radian,
- decimals of a degree and minutes and seconds of a degree,
- trigonometric ratios for angles greater than $90°$,
- the properties of a circle.

There is rather a lot in the list but don't be put off, it is all quite basic. Do the exercise which follows. You will probably find that you remember more than you think.

PRACTICE EXERCISE 14.1

(a) Solve the triangle shown in Fig. 14.1.

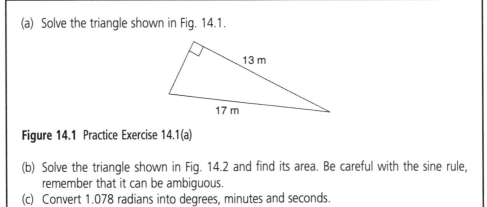

13 m

17 m

Figure 14.1 Practice Exercise 14.1(a)

(b) Solve the triangle shown in Fig. 14.2 and find its area. Be careful with the sine rule, remember that it can be ambiguous.
(c) Convert 1.078 radians into degrees, minutes and seconds.
(d) Express 137°32′33″ in radians.
(e) Express in degrees the principal angles whose tangent is 0.8391.

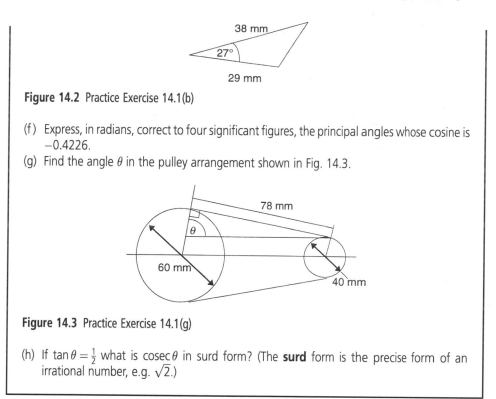

Figure 14.2 Practice Exercise 14.1(b)

(f) Express, in radians, correct to four significant figures, the principal angles whose cosine is −0.4226.

(g) Find the angle θ in the pulley arrangement shown in Fig. 14.3.

Figure 14.3 Practice Exercise 14.1(g)

(h) If $\tan \theta = \frac{1}{2}$ what is $\operatorname{cosec} \theta$ in surd form? (The **surd** form is the precise form of an irrational number, e.g. $\sqrt{2}$.)

The name **surd** has a curious origin. It is derived from the Latin for 'deaf' which was a mistranslation of the original Greek word meaning 'mute'. It's rather endearing to think of a number that can't speak its name.

Answers

(a) 11.0 m, 50°, 40°;   (b) 18 mm, 47°, 106°, 250 mm²;   (c) 61°45′54″
(d) 2.401 rad;   (e) 40°, −140°;   (f) 2.007 rad, −2.007 rad;   (g) 83°;   (h) $\sqrt{5}$

14.1 Properties of the circle

Contrary to popular opinion, it was Archimedes, not Pythagoras, who stated that for any circle the ratio of the circumference to its diameter is always equal to around 3.14. Later mathematicians gave this ratio a name so today we know it as π, which turns out to be irrational.

$$\frac{c}{d} = \pi = 3.141592654\ldots$$

You must remember that for any circle, area $= \pi r^2$. However, because diameter is easier to measure than radius, engineers often prefer the alternative:

$$\text{area} = \frac{\pi d^2}{4}$$

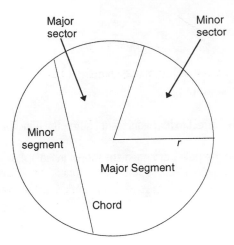

Figure 14.4 Parts of a circle

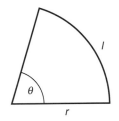

Figure 14.5 Minor sector of a circle

Now look at the minor sector shown in Figs. 14.4 and 14.5. The angle θ subtended by the arc l of a circle having a radius r is defined as the ratio:

$$\theta = \frac{l}{r} \text{ radians}$$

In SI units the length of the arc and the radius are in units of metres, so what is the unit of the radian?

The radian is just a ratio.

Multiplying the radian equation by r gives us a quick way of calculating the length of an arc of a sector of a circle.

Length of an arc of a sector:

$$\boldsymbol{l = r\theta \text{ metres}}$$

which confirms that for a full circle when $\theta = 2\pi$, the length of the arc is the circumference:

$$C = 2\pi r$$

Now, the ratio of θ to 2π gives us a sector as a fraction of a circle. Multiplying the area of a circle by this fraction gives the proportion of area occupied by the sector.

Area of a sector:

$$A = \pi r^2 \frac{\theta}{2\pi}$$

which simplifies to:

$$A = \tfrac{1}{2}r^2\theta$$

So we have simple ways of calculating lengths of arcs and areas of sectors *providing we work in radians.*

EXAMPLE 14.1

A 2 km tunnel is to be dug under a hill. Initially the tunnel is cut by a boring machine which cuts a circular profile having a radius of 4 m. The floor of the tunnel is then laid to a maximum depth of 1.5 m as shown in Fig. 14.6. Calculate the surface area of the tunnel wall and the volume of earth material to be removed.

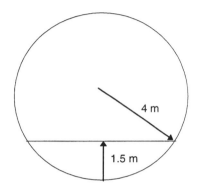

Figure 14.6 Tunnel and floor – see Example 14.1

Solution

Start by sketching the situation, putting in additional information. The radii form an isosceles triangle with the chord which represents the floor of the tunnel (see Fig. 14.7).

The isosceles triangle bisects to form two right-angled triangles of sides 4 and 2.5 m.

We can use the cosine ratio to find angle A.

Work in radians because this will make life easier, as you will see.

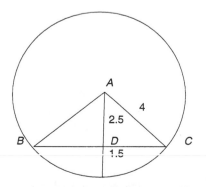

Figure 14.7 Sketch of the situation in Fig. 14.6, showing the isosceles triangle

$$\widehat{DAC} = \arccos\frac{2.5}{4} = 0.8957\,\text{rad}$$

$$\widehat{BAC} = 2 \times 0.8957 = 1.7914\,\text{rad}$$

Angle subtended by major arc $BC = 2\pi - 1.7914 = 4.4918\,\text{rad}$

Length of major arc $BC, l = r\theta = 4 \times 4.4918 = 17.9671\,\text{m}$

Surface area of tunnel walls $= 17.9671 \times 2000 = 35\,934\,\text{m}^2$

$$DC = \sqrt{4^2 - 2.5^2} = 3.1225\,\text{m}$$

Area of triangle $ABC = \frac{1}{2}bh = 3.1225 \times 2.5 = 7.8062\,\text{m}^2$

Area of major sector $ABC = \frac{1}{2}r^2\theta = \frac{1}{2}(4)^2 4.4918 = 35.9344\,\text{m}^2$

Total area of tunnel opening $= 35.9344 + 7.8062 = 43.7406$

Volume of material to be removed $= 43.7406 \times 2000 = 87\,481\,\text{m}^3$

Now try some exercises on your own.

PRACTICE EXERCISE 14.2

(a) Find the angle of lap, in degrees minutes and seconds, if the length of the belt drive in contact with the pulley shown in Fig. 14.8 is 150 mm and the diameter of the pulley is 200 mm.

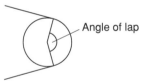

Figure 14.8 Pulley

(b) Find the required length of the belt drive shown in Fig. 14.9.

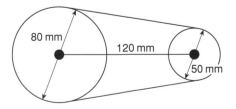

Figure 14.9 Belt drive

(c) Find the area of the shaded segment shown in Fig. 14.10.
(d) A steel bar having a diameter of 25 mm is to have a flat surface machined along its length to a depth of 5 mm. Find the remaining volume of steel if the length of the bar is 200 mm.

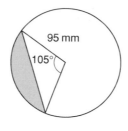

Figure 14.10 Shaded area

Answers

(a) $85°56'37''$; (b) $446\,\text{mm}$; (c) $3910\,\text{mm}^2$; (d) $84.2 \times 10^3\,\text{mm}^3$

14.2 Two useful triangles

Take a square having sides 1 unit long and bisect it to form two triangles as shown. Each of the triangles obtained is an isosceles triangle (this is sometimes known as the '45-45-90' triangle). The sides are: 1, 1 and (by Pythagoras' theorem) $\sqrt{2}$ (see Fig. 14.11). So we have some useful ratios:

$$\sin 45° = \frac{1}{\sqrt{2}}$$

$$\cos 45° = \frac{1}{\sqrt{2}}$$

$$\tan 45° = 1$$

The discovery of $\sqrt{2}$ in this way, gave the ancient mathematicians some headaches. As you know, this is an irrational number which cannot be formed as a simple quotient of two integers. Today, we accept it as a value

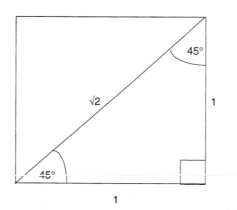

Figure 14.11 The 45–45–90 triangle

that can be expressed to some desired degree of accuracy but one that can never be known precisely. You must recognize it, since it crops up frequently in engineering.

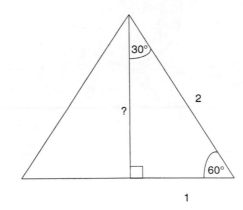

Figure 14.12 The 30–60–90 triangle

A second useful triangle is formed when we bisect an equilateral triangle having sides 2 units long. What is the length of the bisector?

$$\sqrt{3}$$

So here we have what is often called the '30-60-90' triangle (see Fig. 14.12). This also gives us some useful ratios:

$$\sin 30° = \cos 60° = \tfrac{1}{2}$$

$$\sin 60° = \cos 30° = \frac{\sqrt{3}}{2}$$

$$\tan 30° = \cot 60° = \frac{1}{\sqrt{3}}$$

$$\tan 60° = \cot 30° = \sqrt{3}$$

EXAMPLE 14.2

In electrical engineering, the addition of two voltage phasors in a three phase system leads to the result:

$$V_L = 2V_P \cos 30°$$

Use the 30-60-90 triangle to simplify this equation.

Solution

$$\cos 30° = \frac{\sqrt{3}}{2}$$

$$\therefore \quad V_L = 2V_P \frac{\sqrt{3}}{2}$$

$$V_L = \sqrt{3}V_P$$

14.3 Trigonometric identities

Let's remind ourselves that an **identity** is not an **equality**. An equality holds true for only certain values of the variable, while an identity is true for any value of the variable.

However, you will often see an identity written as an equality.

For example: $2x + 3 = 9$ is an equality since the only value of x which satisfies the equality is 3. On the other hand, you may remember that:

$$\tan A \equiv \frac{\sin A}{\cos A}$$

is an identity because it is satisfied by any value of A.

EXAMPLE 14.3

By referring to Fig. 14.13, see if you can remember how to prove the identities:

$$\tan A \equiv \frac{\sin A}{\cos A}$$

and $\sin^2 A + \cos^2 A \equiv 1.$

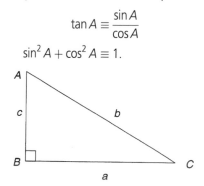

Figure 14.13 Right-angled triangle

Solution

$$\frac{\sin A}{\cos A} = \frac{\frac{a}{b}}{\frac{c}{b}} = \frac{ab}{bc} = \frac{a}{c} = \tan A$$

$$\therefore \quad \tan A = \frac{\sin A}{\cos A}$$

$$\sin^2 A + \cos^2 A = \frac{a^2}{b^2} + \frac{c^2}{b^2} = \frac{a^2 + c^2}{b^2}$$

$$\text{but, } b^2 = a^2 + c^2$$

$$\text{so, } \sin^2 A + \cos^2 A = \frac{a^2 + c^2}{a^2 + c^2}$$

$$\therefore \quad \sin^2 A + \cos^2 A = 1$$

The two identities you have just proved are important to us, so you should ensure that you know them well.

Now, let me bring in the reciprocal ratios and derive another couple of identities.

The identity, $\sec^2 A = 1 + \tan^2 A$

Taking an identity which we already know, we divide by $\cos^2 A$.

$$\sin^2 A + \cos^2 A = 1$$

$$\frac{\sin^2 A}{\cos^2 A} + \frac{\cos^2 A}{\cos^2 A} = \frac{1}{\cos^2 A}$$

The first term becomes the square of the tangent.

The second term is just 1.

The term on the right is the reciprocal of the cosine squared.

$$\tan^2 A + 1 = \sec^2 A$$

$$\therefore \quad \sec^2 A = 1 + \tan^2 A$$

The identity, $\operatorname{cosec}^2 A = 1 + \cot^2 A$

This time we divide by $\sin^2 A$.

$$\sin^2 A + \cos^2 A = 1$$

$$\frac{\sin^2 A}{\sin^2 A} + \frac{\cos^2 A}{\sin^2 A} = \frac{1}{\sin^2 A}$$

The first term becomes 1 and the second term is the reciprocal of the tangent squared.

The right is the reciprocal of the sine squared.

$$1 + \cot^2 A = \operatorname{cosec}^2 A$$

$$\therefore \quad \operatorname{cosec}^2 A = 1 + \cot^2 A$$

EXAMPLE 14.4

Show that $\cos^2 A - \sin^2 A = 1 - 2\sin^2 A$.

Solution

$$\sin^2 A + \cos^2 A = 1$$

$$\cos^2 A = 1 - \sin^2 A$$

$$\cos^2 A - \sin^2 A = 1 - \sin^2 A - \sin^2 A$$

$$\therefore \quad \cos^2 A - \sin^2 A = 1 - 2\sin^2 A$$

EXAMPLE 14.5

Show that:

$$\cot A = \frac{1 + \cot A}{1 + \tan A}$$

Solution

$$\frac{1 + \cot A}{1 + \tan A} = \frac{1 + \dfrac{\cos A}{\sin A}}{1 + \dfrac{\sin A}{\cos A}} = \frac{\dfrac{\sin A + \cos A}{\sin A}}{\dfrac{\cos A + \sin A}{\cos A}}$$

$$= \left(\frac{\sin A + \cos A}{\sin A}\right)\left(\frac{\cos A}{\cos A + \sin A}\right) = \frac{\cos A}{\sin A} = \cot A$$

$$\therefore \quad \cot A = \frac{1 + \cot A}{1 + \tan A}$$

There are other identities which we can derive in a similar manner, but that is not necessary for now. Just concentrate on four of the most important ones.

Four important identities

$$\tan A = \frac{\sin A}{\cos A}$$

$$\sin^2 A + \cos^2 A = 1$$

$$\sec^2 A = 1 + \tan^2 A$$

$$\operatorname{cosec}^2 A = 1 + \cot^2 A$$

The following exercise will give you practice in applying these identities to the solution of trigonometric equations.

PRACTICE EXERCISE 14.3

Find, in degrees, the principal angles that satisfy the following equations:

(a) $7 \sin y - 4 = 0$; (b) $2 \operatorname{cosec} y = -3.5$; (c) $2 \tan^2 x - 3 = 0$
(d) $5 \sin^2 \theta + \sin \theta = 2$; (e) $\sec \theta + 2 = \sec^2 \theta$; (f) $3 \cos^2 \phi + 2 \sin \phi = 3$

Find, in radians, the principal angles that satisfy the following equations:

(g) $5 \sec \theta + \tan^2 \theta = 4$; (h) $6 \cot x + 7 = 2 \operatorname{cosec}^2 x$

Answers

(a) $34.8°$ or $145.2°$; (b) $-34.8°$ or $145.2°$; (c) $50.8°$ or $-129.2°$, $129.2°$ or $-50.8°$
(d) $-47.8°$ or $-132.2°$, $32.7°$ or $147.3°$; (e) $180°$, $60°$ or $-60°$; (f) $0°$, $41.8°$ or $138.2°$
(g) 1.7424 rad or -1.7424 rad; (h) 0.2654 or -2.8762 rad, -0.9740 or 2.1675 rad

14.4 Compound angle formulae

You know, by now, that periodic functions such as those shown below can be used to model the behaviour of an electric current or an undamped vibrating spring. Angles such as $(\omega t + \phi)$ are known as **compound angles**. They are simply the sum of two angles. There are circumstances where it is useful to have some algebraic means of dealing with such angles.

$$a = A\sin(\omega t + \phi)$$

$$x = A\cos(t + \theta)$$

Work through the following analysis of a compound angle which is the sum of A and B and has the general form shown.

A compound angle:

$$\sin(A + B)$$

Figure 14.14 shows a line r which forms the compound angle $A + B$. First, take the angle A. It appears more than once in the diagram. It is also $\widehat{WZY}$ and $\widehat{VUZ}$.

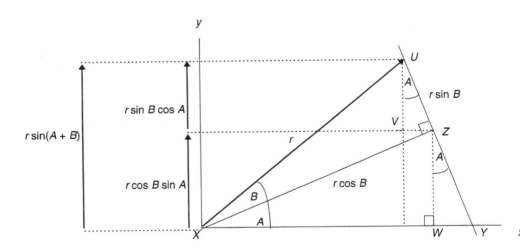

Figure 14.14 Diagram to prove the compound angle formula

In triangle XWZ: $\qquad\qquad A = 90 - \widehat{WZX}$

and, $\qquad\qquad\qquad \widehat{WZY} = 90 - \widehat{WZX}$

$$\therefore \quad \widehat{WZY} = A$$

and by similar triangles:

$$\widehat{VUZ} = \widehat{WZY} = A$$

Now look at the y-component of r. It is the side opposite angle $(A + B)$ which projects a length $r\sin(A + B)$ along the y-axis.

In the triangle XZU the side adjacent to angle B is $r\cos B$ and the side opposite is $r\sin B$.

In triangle XWZ the side opposite angle A is $r \cos B \sin A$. This is the vertical component of $r \cos B$ which projects the length $r \cos B \sin A$ along the y-axis as shown.

Finally, turning to the small triangle VZU, the projection on the y-axis of the length $r \sin B$ is $r \sin B \cos A$.

We can now compare the lengths of the projections on the y-axis.

It turns out that:

$$r \sin(A + B) = r \cos B \sin A + r \sin B \cos A$$

Dividing by r:

$$\sin(A + B) = \cos B \sin A + \sin B \cos A$$

Writing this in a more logical order gives us a **compound angle formula**:

$$\sin(A + B) = \sin A \cos B + \sin B \cos A$$

Continuing a similar analysis would give three other compound angle formulae which are:

$$\sin(A - B) = \sin A \cos B - \sin B \cos A$$

$$\cos(A + B) = \cos A \cos B - \sin A \sin B$$

$$\cos(A - B) = \cos A \cos B + \sin A \sin B$$

Taking all four it is easy to confuse them, but they can be summarized in two distinct forms.

Take careful note of the reversal of the $\pm$ sign in the second of the two formulae.

$$\mathbf{\sin(A \pm B) = \sin A \cos B \pm \sin B \cos A}$$

$$\mathbf{\cos(A \pm B) = \cos A \cos B \mp \sin A \sin B}$$

EXAMPLE 14.6

Use appropriate trigonometric identities and compound angle formulae to derive a compound angle formula for $\tan(A + B)$.

Solution

$$\tan(A + B) = \frac{\sin(A + B)}{\cos(A + B)} = \frac{\sin A \cos B + \sin B \cos A}{\cos A \cos B - \sin A \sin B}$$

dividing by $\cos A \cos B$:

$$\tan(A + B) = \frac{\dfrac{\sin A \cos B}{\cos A \cos B} + \dfrac{\sin B \cos A}{\cos A \cos B}}{\dfrac{\cos A \cos B}{\cos A \cos B} - \dfrac{\sin A \sin B}{\cos A \cos B}} = \frac{\tan A + \tan B}{1 - \tan A \tan B}$$

Now try one on your own.

PRACTICE EXERCISE 14.4

Apply the same methods used in Example 14.6 to derive a formula expressing $\tan(A - B)$ in terms of tangents of A and B.

Solution

$$\tan(A - B) = \frac{\tan A - \tan B}{1 + \tan A \tan B}$$

First time around, this sort of thing can be quite tiring so just take a note of the six compound angle formulae which have been derived. You can return to them later when you do some practice exercises.

Six compound angle formulae

$$\sin(A \pm B) = \sin A \cos B \pm \sin B \cos A$$

$$\cos(A \pm B) = \cos A \cos B \mp \sin A \sin B$$

$$\tan(A \pm B) = \frac{\tan A \pm \tan B}{1 \mp \tan A \tan B}$$

PRACTICE EXERCISE 14.5

(a) Prove the identity $\cos \theta = \sin(90° - \theta)$
(b) Show that $\sin \theta = \cos(90° - \theta)$
(c) Simplify, without using a calculator, $y = \sin(60 - x) + \cos(x + 30)$.
(d) Evaluate $\sin 165°$ without using a calculator.
(e) Evaluate $\tan 15°$ without a calculator.
(f) If $\sin A = \frac{4}{5}$ and $\cos B = \frac{5}{13}$, evaluate $\sin(A - B)$ without using a calculator.
(g) Now if $\sin A = \frac{4}{5}$ and $\cos B = \frac{5}{13}$, evaluate $\tan(A + B)$ without using a calculator.
(h) Evaluate $\sin 75°$ without a calculator.

Answers

(c) $y = \sqrt{3} \cos x - \sin x$; (d) $\dfrac{\sqrt{3} - 1}{\sqrt{8}}$; (e) $\dfrac{\sqrt{3} - 1}{1 + \sqrt{3}}$

(f) $-\frac{16}{65}$; (g) $-\frac{56}{33}$; (h) $\dfrac{1 + \sqrt{3}}{\sqrt{8}}$

14.5 Double angle formulae

A double angle is just a special case of a compound angle. So to develop expressions for double angles we can use the compound angles we know.

A double angle:

$$\sin(2A)$$

EXAMPLE 14.7

Apply a compound angle formula to derive an expression for the double angle $\sin(2A)$.

$$\sin 2A = \sin(A + A)$$
$$= \sin A \cos A + \sin A \cos A$$
$$\sin 2A = 2 \sin A \cos A$$

EXAMPLE 14.8

Apply a compound angle formula to derive an expression for the double angle $\cos(2A)$.

$$\cos 2A = \cos(A + A)$$
$$= \cos A \cos A - \sin A \sin A$$
$$\cos 2A = \cos^2 A - \sin^2 A$$

EXAMPLE 14.9

Apply a compound angle formula to derive an expression for the double angle $\tan(2A)$.

$$\tan 2A = \tan(A + A)$$
$$= \frac{\tan A + \tan A}{1 - \tan A \tan A}$$
$$\tan 2A = \frac{2 \tan A}{1 - \tan^2 A}$$

The double angle formulae can be very useful, as you will see in a moment.

Three double angle formulae

$$\mathbf{\sin 2A = 2 \sin A \cos A}$$

$$\mathbf{\cos 2A = \cos^2 A - \sin^2 A}$$

$$\mathbf{\tan 2A = \frac{2 \tan A}{1 - \tan^2 A}}$$

Focus your interest on the second of these. It opens up possibilities. Do you remember our table of the standard integrals? If not, look back and see if there is an integral that can deal with squares of trigonometric functions.

There isn't one. You might remember also that in finding the r.m.s. value of a function it was necessary to make a trigonometric substitution before we were able to integrate. Where did the substitution come from? Follow the next example.

EXAMPLE 14.10

Derive an expression for $\sin^2 A$ which it is possible to integrate directly.

Start with the second of the double angle formulae, it contains the term $\sin^2 A$.

$$\cos 2A = \cos^2 A - \sin^2 A$$

Remember the identity which can be transposed so that we can substitute for $\cos^2 A$.

$$\sin^2 A + \cos^2 A = 1$$

$$\cos^2 A = 1 - \sin^2 A$$

$$\cos 2A = (1 - \sin^2 A) - \sin^2 A$$

The double angle formula now only contains one trig. term which is squared.

$$\cos 2A = 1 - 2\sin^2 A$$

Some further transposition and you have an expression for $\sin^2 A$ which is easily integrated.

$$2\sin^2 A = 1 - \cos 2A$$

$$\sin^2 A = \tfrac{1}{2}(1 - \cos 2A)$$

PRACTICE EXERCISE 14.6

Find: $\displaystyle\int \cos^2\left(\theta - \frac{\pi}{2}\right)\,d\theta$

Solution

$\tfrac{1}{2}\theta - \tfrac{1}{4}\sin 2\theta + C$

14.6 The form $R\sin(\omega t \pm \alpha)$

It is possible (and often convenient) to express the sum of a sine and a cosine function in compound angle form.

$$a\sin \omega t + b\cos \omega t = R\sin(\omega t \pm \alpha)$$

Study the next part carefully.

Let: $\qquad a\sin \omega t + b\cos \omega t = R\sin(\omega t + \alpha)$

Then the right-hand-side can be re-written using the compound angle formula and re-arranged.

$$a \sin \omega t + b \cos \omega t = R(\sin \omega t \cos \alpha + \sin \alpha \cos \omega t)$$

$$a \sin \omega t + b \cos \omega t = R \cos \alpha \sin \omega t + R \sin \alpha \cos \omega t$$

We can then equate the coefficients of $\sin \omega t$ and the coefficients of $\cos \omega t$.

$$a = R \cos \alpha$$

$$b = R \sin \alpha$$

Squaring both equations and adding them gives:

$$a^2 = R^2 \cos^2 \alpha \quad \text{and} \quad b^2 = R^2 \sin^2 \alpha$$

$$a^2 + b^2 = R^2 \cos^2 \alpha + R^2 \sin^2 \alpha$$

Now, factorizing R^2 on the RHS reveals an identity which eliminates the trig. term and leads to a familiar expression for R^2.

$$a^2 + b^2 = R^2 (\cos^2 \alpha + \sin^2 \alpha)$$

$$R^2 = a^2 + b^2$$

Go back to:
$$R \sin \alpha = b \quad \text{and} \quad R \cos \alpha = a$$

Dividing the sine by the cosine function leads us to a simple solution for α.

$$\frac{R \sin \alpha}{R \cos \alpha} = \frac{b}{a} \qquad \therefore \quad \tan \alpha = \frac{b}{a}$$

$$\alpha = \arctan \frac{b}{a}$$

Summary

to transform: $a \sin \omega t + b \cos \omega t$

to the general form of a sinusoid: $R \sin(\omega t + \alpha)$

where: R is *amplitude* **and** α is *phase angle*

take: $R^2 = a^2 + b^2$

and: $\alpha = \arctan \dfrac{b}{a}$

But you do need to be careful. Remember that there are always two angles in the range between $-\pi$ and π which have the same tangent. The next example will show you how to establish which of the two angles is the correct value for α.

EXAMPLE 14.11

Express $3 \sin 2t + 4 \cos 2t$ in the form $R \sin(\omega t + \alpha)$.

ω is $2 \, \mathrm{rad \, s^{-1}}$ so we proceed.

Apply the compound angle formula.

$$3\sin 2t + 4\cos 2t = R\sin(2t + \alpha)$$
$$= R(\sin 2t \cos \alpha + \sin \alpha \cos 2t)$$
$$3\sin 2t + 4\cos 2t = R\cos \alpha \sin 2t + R\sin \alpha \cos 2t$$

Evaluate R. Note that it cannot be negative.

$$R = \sqrt{a^2 + b^2} = \sqrt{3^2 + 4^2} = 5$$

Equating coefficients of $\sin 2t$ indicates a positive cosine narrowing α down to 1st or 2nd quadrants.

$$R\cos \alpha = 3 \quad (+\text{ve so 1st or 4th quadrant})$$

A positive tangent puts α into the 1st quadrant.

$$\tan \alpha = \tfrac{4}{3} \quad (+\text{ve so 1st or 3rd quadrant}) \text{ so in the 1st quadrant}$$

Work in radians.

$$\alpha = \arctan \tfrac{4}{3} = 0.9273 \,\text{rad}$$
$$3\sin 2t + 4\cos 2t = 5\sin(2t + 0.9273)$$

Equating coefficients allowed us to test for one of two quadrants. The tangent then told us which of these to take.

In testing for quadrants, we chose to equate coefficients of $\sin 2t$ but you should see that we could just as easily have chosen to equate the coefficients of $\cos 2t$.

EXAMPLE 14.12

Express:

$$2\sin 2t - 4\cos 2t$$

in the form:

$$R\sin(\omega t + \alpha)$$

This time we test for quadrants by equating coefficients of $\cos 2t$.

$$\text{Let, } 2\sin 2t - 4\cos 2t = R\sin(2t + \alpha)$$
$$= R(\sin 2t \cos \alpha + \sin \alpha \cos 2t)$$
$$2\sin 2t - 4\cos 2t = R\cos \alpha \sin 2t + R\sin \alpha \cos 2t$$
$$R = \sqrt{a^2 + b^2} = \sqrt{2^2 + 4^2} = 4.47$$
$$R\sin \alpha = -4 \quad (-\text{ve sin})$$
$$\tan \alpha = \frac{b}{a} = \frac{-4}{2} \quad (-\text{ve tan})$$

A negative sine and negative tangent indicates that α lies in the 4th quadrant.

$$\alpha = -\arctan\tfrac{4}{2} = -1.107 \,\text{rad}$$

$$2\sin 2t - 4\cos 2t = 4.47\sin(2t - 1.107)$$

By choosing to express angles between the limits of $-\pi$ and π, an angle in the 4th quadrant must be negative.

Referring back to the last example, what would you state as the value of α if we chose to express angles between the limits of 0 and 2π?

$$\alpha = 2\pi - \arctan\tfrac{4}{2} = 5.176 \,\text{rad}$$

Whether you choose α so that $-\pi > \alpha \leq \pi$ or $0 < \alpha \leq 2\pi$ is up to you, but the general preference is for the former.

PRACTICE EXERCISE 14.7

(a) Express $12\sin \omega t + 18\cos \omega t$ as a compound angle. Check your result.
(b) Find the sum of $0.3\sin 3t$ and $0.3\cos 3t$.
(c) Transform $5.5\cos 5t - 9\sin 5t$ into $R\sin(\omega t + \alpha)$ form.
(d) Find the amplitude, frequency, periodic time and phase angle of the resultant of the two electric currents: $i_1 = 3\sin 377t$ and $i_2 = \cos 377t$.
(e) The solution of a differential equation predicts the position of an undamped vibrating mass as $s = 0.1\sin 10t + 0.3\cos 10t$. Find the amplitude of the vibrations, the initial position of the mass and its position 100 ms after commencement of the vibrations.
(f) On the same axes, plot the graphs of $2\sin 10\pi t$ and $-3\cos 10\pi t$ over one complete cycle. Plot the sum of the two graphs and express the sum in the form $R\sin(\omega t + \alpha)$.
(g) Draw the phasors of the three graphs you plotted in (f) and express them in polar form.

Answers

(a) $21.6\sin(\omega t + 0.983)$;   (b) $0.424\sin(3t + 0.785)$;   (c) $10.6\sin(5t + 2.59)$
(d) $3.16\,\text{A}$, $60\,\text{Hz}$, $16.7\,\text{ms}$, $0.322\,\text{rad}$;   (e) $316\,\text{mm}$, $300\,\text{mm}$, $246\,\text{mm}$

The last two problems should have reinforced your understanding of the link between phasor and waveform representations of circular functions.

Assignment XII

This assignment requires you to use trigonometry to solve engineering problems. The performance criteria which must be met for satisfactory completion are:

- engineering problems are represented using trigonometry,
- trigonometric ratios and identities are selected to solve engineering problems,
- angular properties of circles are calculated,
- graphs of trigonometric ratios and identities are sketched over the range 0 to 360°.

Exercise your skills and show what you know!

PROBLEM 1

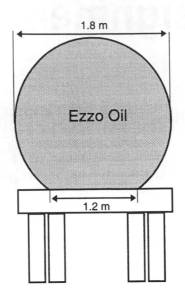

A new fleet of road tankers is being built for the Ezzo Oil Company. The cheapest and safest construction for the tanks is the shape of a cylinder with a flat base 1.2 m wide. The tanks must fit onto trailers that are 1.8 m wide and 10 m long, as shown. No overhang is allowed. The accountants want to know how much the tankers will be capable of carrying. Calculate the capacity of the tanks to the nearest 10 litres.

PROBLEM 2

Find, in degrees, all the principal angles that satisfy the equation: $2 \cot^2 \theta + 5 \csc \theta = 8$.

PROBLEM 3

You have misplaced your calculator. How do you find: $\sec 15°$?

PROBLEM 4

Find $\int (\cos^2 \theta) \, d\theta$

PROBLEM 5

The solution of a differential equation tells you that the position of a vibrating mass is described by $s = 0.45 \cos 12.6t - 0.78 \sin 12.6t$. Find the amplitude and phase angle of the vibrations.

PROBLEM 6

As chief engineer for the Gunless Brewery Company, you are responsible for training the Company's new graduate engineers. One of your trainees is not happy with the proof of the identity $\sin^2 A + \cos^2 A \equiv 1$. Plot graphs to convince her that this identity is true.

chapter

Vectors revisited

It is time to return to the topic that was briefly introduced in Chapter 5. There, we used trigonometry to solve simple vector and phasor problems. Although phasors are very similar to vectors, they are quite distinct from vectors so we will set phasors and their **complex number** notation aside for later.

As usual we need to get familiar with definitions and ground rules.

15.1   Vectors and scalars

You are quite familiar with the mathematics of **scalars**. Scalars are one-dimensional quantities that can be fully specified by one real number. They include, probably the most important scalar, time.

Vectors are quantities that require two dimensions to be fully specified, **magnitude** and **direction**. For example, a weather forecast which predicted a wind of 40 kn would not be very useful. We normally want to know the direction of the wind as well.

Perhaps wind is not a particularly good example of a vector quantity because the weather people specify the direction from which it blows, e.g. 40 kn north-east means a speed of 40 kn blowing *from* the north-east. In engineering, it is normal to state the direction in which a vector is acting. So, for example, *weight* might be specified as 75 newtons downwards. Downwards being the direction *toward* which the force of weight is acting.

PRACTICE EXERCISE 15.1

From the following list, identify which is a vector and which is a scalar. With each vector state its magnitude.

(a) Temperature;   (b) Speed;   (c) Velocity;   (d) Mass;   (e) Distance
(f) Displacement;   (g) Acceleration;   (h) Population;   (i) Area;   (j) Energy
(k) Momentum;   (l) Magnetic flux

Answers

(a) Temperature is a scalar
(b) Speed is a scalar
(c) Velocity is a vector having speed and direction
(d) Mass is a scalar
(e) Distance is a scalar
(f) Displacement is a vector having distance and direction
(g) Acceleration is a vector because it is rate of change of velocity
(h) Population is a scalar
(i) Area is a scalar
(j) Energy is a scalar
(k) Momentum is a vector because it is a function of velocity
(l) Magnetic flux is a vector having field strength and direction

With vectors it is important to be clear. Try the following exercise. You may find the result surprising.

PRACTICE EXERCISE 15.2

The earth's orbit around the sun is periodic. The periodic time is one year. Does this mean that the earth's motion around the sun is at a constant velocity and its acceleration is zero?

Acceleration is the rate of change of velocity over time. The speed of the earth is constant but its direction is not. It is changing continuously. Because direction is changing, velocity is not constant. If velocity is not constant then acceleration is not zero. Relative to the sun, the earth is undergoing constant acceleration due to the gravitational force of the sun.

In fact, according to Newton's second law any object which has a net force acting on it must be accelerating. But you should know that by now!

So it is important to remember that a vector quantity changes if either its *magnitude* or *direction* changes. A change in velocity is brought about by either a change in speed or a change in direction (or both of course).

Graphical representation

This is quite straightforward. The magnitude of a vector is represented by a line whose length is proportional to the magnitude according to some specified scale. The direction of a vector is represented by an arrow pointing in the direction in which the vector is acting (Fig. 15.1). Usually the arrow is placed at the 'nose' of the vector. However, you will also come across vectors where the arrow is drawn in the middle (Fig. 15.2).

Figure 15.1 Graphical representation of a vector

Figure 15.2 Another representation of a vector

Direction actually consists of two properties: **orientation**, which is the slope of the line and **sense** which is the direction of the arrowhead.

It is important that in a graphical representation of a particular vector a scale must be specified. This of course does not apply when dealing with a general case as in the figures above.

Algebraic representation

This is not quite so straightforward because there are a number of alternatives in use, all equally accepted.

Most books use bold lettering to distinguish a vector from a scalar. This is not possible to do in hand-written form so underlining is also used.

Another method uses ordered pairs of upper case letters with an arrow or just a plain line above. The order of the letters specifies the 'tail' and 'nose' of the vector. Ordered pairs are normally used to represent displacement vectors.

$$\mathbf{a}$$
$$\underline{a}$$
$$\overrightarrow{AB}$$
$$\overline{AB}$$
$$A \longrightarrow B$$

The algebraic representation of 'magnitude of' is a pair of vertical lines either side of the vector symbol. Mathematicians tend to call it **modulus** instead of 'magnitude'.

$$\text{modulus of } \mathbf{a} = |\mathbf{a}|$$

The mathematical term for direction is **argument** for which the abbreviation 'arg **a**' is used.

$$\text{argument of } \mathbf{a} = \arg \mathbf{a}$$

The algebraic representation of direction is often the 'angle' symbol $\angle\theta$ where the value of θ is taken as either a positive or negative angular displacement from the x-axis of the Cartesian coordinate system.

$$\mathbf{a} = |\mathbf{a}|\angle\theta$$

However, the $\angle\theta$ symbol to represent direction is more commonly used with complex numbers rather than vectors. With vectors the direction can be specified as an angular displacement from some known reference. The vertical, the horizontal and the points of the compass are all used, depending on the situation.

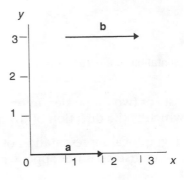

Figure 15.3 Equality

Equality

Since vectors are specified *entirely* by their magnitude and direction any two vectors having equal magnitude and direction are equal vectors. They need not occupy the same position in space.

The vectors **a** and **b** shown in Fig. 15.3 are equal because they are parallel and of the same length even though the tail of vector **b** does not start from the origin.

Scaling

Multiplying a vector by a scalar produces another vector which has a different magnitude but the same direction. This is called **scaling**.

For example, in vector form, the displacement vector of an object travelling at constant velocity is the product of time and velocity where time is a scalar (Fig. 15.4).

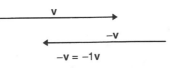

Figure 15.4 Scaling

Inverse

Inverse vectors are those that have equal magnitude and opposite sense. Multiplying a vector by the scalar -1 produces an inverse vector (Fig. 15.5).

Figure 15.5 Inverse

Zero vector

The zero vector, **0**, has a magnitude of zero and therefore no specified direction. It is created when we scale by zero.

$$0\mathbf{a} = \mathbf{0}$$

EXAMPLE 15.1

(a) If $\mathbf{a} = \overline{PQ}$ what is $\overline{QP}$?

(b) The velocities of two aircraft are $\mathbf{u}$ and $\mathbf{v}$. If velocity $\mathbf{u}$ is $350\,\text{km h}^{-1}$ east and the velocity of $\mathbf{v}$ is $500\,\text{km h}^{-1}$ east, sketch the velocity vectors and write down velocity $\mathbf{u}$ in terms of $\mathbf{v}$.

(c) If another aircraft is travelling west at velocity $\mathbf{w}$ and $|\mathbf{w}| = 250\,\text{km h}^{-1}$ write down $\mathbf{w}$ in terms of $\mathbf{v}$.

(d) If $\mathbf{v}$ is the velocity of a wind of 20 kn from the north-west what is the velocity of a wind of 60 kn from the south-east?

(e)

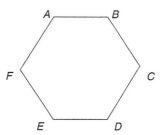

Figure 15.6 Regular hexagon

If Fig. 15.6 is a regular hexagon what are the algebraic relationships between the following?

(i) $\overline{AB}$ and $\overline{ED}$

(ii) $\overline{AF}$ and $\overline{DC}$

(iii) $\overline{AA}$ and $\mathbf{0}$

(f) Refer to Fig. 15.6 again. Find the displacement vector that is equal to:

$$\overline{AB} + \overline{BC} + \overline{CD} + \overline{DE} + \overline{EF}$$

hence find the zero vector for the hexagon.

Solutions

(a) $\overline{QP} = -\mathbf{a}$

(b)

$$\frac{|\mathbf{u}|}{|\mathbf{v}|} = \frac{350}{500} \quad |\mathbf{u}| = \frac{35}{50}|\mathbf{v}|$$

$$|\mathbf{u}| = 0.7|\mathbf{v}|$$

Since directions are equal, $\mathbf{u} = 0.7\mathbf{v}$

(c)

$$\frac{|\mathbf{w}|}{|\mathbf{v}|} = \frac{250}{500} \quad |\mathbf{w}| = \frac{1}{2}|\mathbf{v}|$$

Opposite directions give the vectors opposite sense, so $\mathbf{w} = -0.5\mathbf{v}$

(d) The wind is opposite and scaled by a factor of 3 so it is: $-3\mathbf{v}$

(e)

(i) $\overline{AB} = \overline{ED}$

(ii) $\overline{AF} = -\overline{DC}$

(iii) $\overline{AA} = \mathbf{0}$

(f) Following the vectors from tail to nose gives the same displacement (position) as following the single vector, $\overline{AF}$, so:

$$\overline{AB} + \overline{BC} + \overline{CD} + \overline{DE} + \overline{EF} = \overline{AF}$$

$$\overline{AB} + \overline{BC} + \overline{CD} + \overline{DE} + \overline{EF} - \overline{AF} = \mathbf{0}$$

15.2 Addition of vectors

Two vectors can be added together by applying one of two rules: the triangle rule and the parallelogram rule. Let's look at them again.

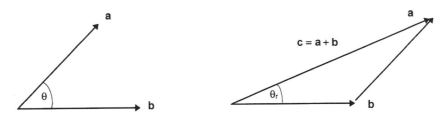

The triangle rule for vector addition

Adding two vectors by the triangle rule requires one vector to be drawn 'nose to tail' with the other vector. It does not matter which way around this is done. The resultant of the two vectors is obtained by drawing a line which forms a triangle. The direction of the resultant must be such that the sum of all three is the zero vector, i.e.

$$\mathbf{a} + \mathbf{b} = \mathbf{c}$$

$$\therefore \quad \mathbf{a} + \mathbf{b} - \mathbf{c} = \mathbf{0}$$

The triangle rule can be expanded to carry out the addition of any number of vectors in which the resultant vector will complete a closed figure called a polygon. In all cases the resultant is the vector which joins the tail of the first vector to the nose of the last vector. The direction of the resultant is always opposite to the chain formed by its components – clockwise or anti-clockwise.

EXAMPLE 15.2

Find the sum of the five vectors shown in Fig. 15.7.

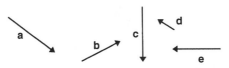

Figure 15.7 Five vectors

The first step is to join up the vectors, nose to tail to form a chain. We want the sum of the vectors so that the resultant is:

$$r = a + b + c + d + e$$

We choose to form a chain that is clockwise so the resultant must be anticlockwise. But the solution will be equally valid if the chain was anticlockwise with the resultant drawn in the clockwise direction. And, as with any addition, the order does not matter:

$$b + a + c + e + d = r$$

as well. Try it.

Solution

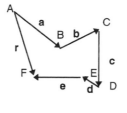

Or:

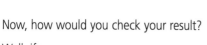

Now, how would you check your result?

Well, if:

$$r = a + b + c + d + e$$

then,

$$a + b + c + d + e - r = 0$$

This is the zero vector. Which has no magnitude. It goes nowhere! Figure 15.8 shows that $-r$ takes us back to the point of origin, A.

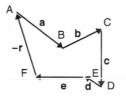

Figure 15.8 Checking the result

This also introduces the principle of subtracting vectors.

**To find the difference of two vectors we simply add the
negative of the vector:**

$$\mathbf{a} - (+\mathbf{b}) = \mathbf{a} + (-\mathbf{b})$$

So the rule for the addition and subtraction of vectors is quite straightforward. The algebra is the same. The triangle rule helps us to represent the procedure graphically. But how would we obtain numerical solutions?

One method, which you have seen before, would be to get some graph paper, a protractor, a ruler, draw vectors to scale and measure off the resultant according to the triangle rule. Another, quicker and more accurate method would be to apply our knowledge of trigonometry. In which case, just a sketch of the vectors would be sufficient. For revision, work through the following examples.

EXAMPLE 15.3

Two downward forces are acting on a support as shown in Fig. 15.9. Find the resultant force vector.

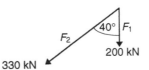

Figure 15.9 Two vectors acting on a support

The first step is to sketch the vectors according to the triangle rule.

We put F_1 on to the end of F_2 and close up the triangle with the resultant vector F_r.

For convenience we label the angles A, B and C.

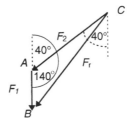

F_2 forms an angle of 40° with the vertical so angle A is 140°.

We now have a triangle for which we know two sides and the angle between. We can use the cosine rule to find the magnitude of F_r.

$$a^2 = b^2 + c^2 - 2bc \cos A$$

$$|\mathbf{F_r}| = \sqrt{330^2 + 200^2 - 2(330)(200)\cos 140}$$

$$= 500\,\text{kN}$$

We are then able to apply the sine rule to find angle C.

$$\frac{\sin C}{c} = \frac{\sin A}{a}$$

$$C = \arcsin\left(c\frac{\sin A}{a}\right) = \arcsin\left(200\frac{\sin 140}{500}\right)$$

Now, angle C is the difference between 40° and the angle of the resultant from the vertical, which we call angle θ.

$$C = 22.9° \quad \theta = 40 - 14.9 = 25.1°$$

What would be a suitable rough check?

A sketch of the triangle drawn roughly to scale.

EXAMPLE 15.4

A belt-driven pulley is attached to a shaft as shown in Fig. 15.10. Calculate the magnitude and direction of the resultant force acting on the shaft.

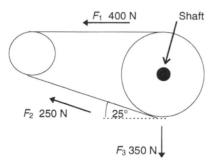

Figure 15.10 Belt-driven pulley system

This involves the addition of three vectors. A sketch, drawn roughly to scale, will provide a rough check.

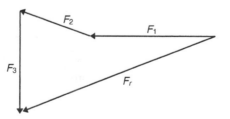

Then the problem needs to be tackled in two stages. We start by resolving F_1 and F_2 which gives us F_{r1}: the third side of triangle 1 whose angles are labelled A, B and C.

Triangle 1

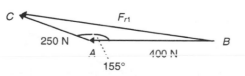

The angle of F_2 from the horizontal forms the difference between angle A and 180°. So angle A is $180 - 25 = 155°$.

Again, two sides and an angle between allow us to apply the cosine rule to find the magnitude of the resultant.

$$a = \sqrt{b^2 + c^2 - 2bc\cos A}$$

$$|F_{r1}| = \sqrt{250^2 + 400^2 - 2(250)(400)\cos 155°} = 635.4\,\text{N}$$

Then the sine rule can be used to find angle B which is the angle of the resultant vector above the horizontal.

$$B = \arcsin\left(b\frac{\sin A}{a}\right) = \arcsin 250\frac{\sin 155}{635.4} = 9.57°$$

To find the resultant of all three forces we construct a second triangle and resolve F_{r1} with F_3 to find the resultant of the three forces, F_r.

Triangle 2

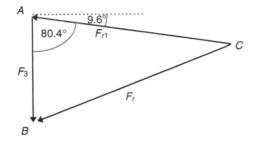

The approach is the same as for triangle 1.

$$|F_r| = \sqrt{635.4^2 + 350^2 - 2(635.4)(350)\cos 80.4°} = 672.5\,\text{N}$$

$$C = \arcsin\left(350\frac{\sin 80.4}{672.4}\right) = 30.8° \quad \theta = 30.9 - 9.6 = 21.3°$$

Resultant:

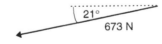

Now, you may think that using trigonometry gets to be somewhat tedious when resolving more than two vectors, and rightly so. There is another, faster and more elegant method for dealing with multi-vector problems. It involves expressing the vectors in terms of their **Cartesian components**. We will come to it shortly. First another quick look at the other rule for vector addition.

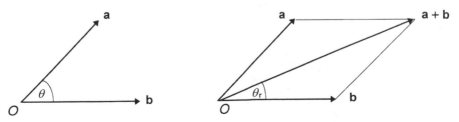

The parallelogram rule for vector addition

The difference here is simply that we draw both vectors from a common origin, O, then draw lines parallel and equal in length to the two vectors. The resultant is the diagonal which acts in a direction away from the origin, O.

Essentially, this is no different to the triangle rule except that graphically we can only deal with two vectors at a time. It is not possible to construct a polygon of vectors as it is when we extend the triangle rule to the addition of more than two vectors.

15.3 Cartesian components

Unit vector

When we multiply a scalar quantity by its reciprocal that scalar becomes 1 or **unity**.

In the same way, multiplying any (non-zero) vector by the reciprocal of its magnitude will cause its magnitude to become 1. Now the reciprocal of the magnitude is a scalar and multiplication by a scalar is called **scaling** which, you know, does not change the direction of the vector. A vector which has a magnitude of 1 and a known direction is called a **unit vector**.

This is all a bit wordy so look at the algebra, it's clearer.

For any scalar (other than zero),

$$\frac{1}{a}a = 1$$

For any vector (other than the zero vector),

$$\frac{1}{|\mathbf{a}|}\mathbf{a} = 1\mathbf{a}$$

$1\mathbf{a}$ is a vector with the magnitude of 1 and the direction of $\mathbf{a}$.

But what is the point!

The point is that it provides us with a convenient means of expressing vectors in terms of **Cartesian components**. By convention we use $\mathbf{i}$ to represent the unit vector which points in the positive direction of the x-axis and $\mathbf{j}$ to represent the unit vector which points in the positive direction of the y-axis. $\mathbf{i}$ and $\mathbf{j}$ are called the Cartesian unit vectors (Fig. 15.11).

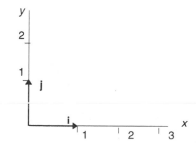

Figure 15.11 Cartesian unit vectors

Now we have a means of expressing any vector having the same direction as either of the Cartesian unit vectors, as a product of a scalar and its corresponding unit vector, **i** or **j**.

PRACTICE EXERCISE 15.3

Express the vectors shown in Fig. 15.12 in terms of their Cartesian unit vectors.

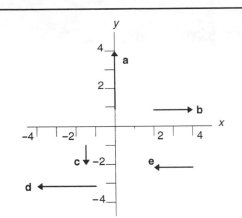

Figure 15.12 Vectors in Practice Exercise 15.3

Answers

(a) = 3**j**;   (b) = 2**i**;   (c) = −**j**;   (d) = −3**i**;   (e) = −2**i**

But what happens if the direction of our vector does not correspond to either the *x*- or *y*-axis of the Cartesian coordinate system?

Look at Fig. 15.13. **a** represents any vector in the *xy*-plane. It projects a horizontal line along the *x*-axis and a vertical line along the *y*-axis. We call these projections the **Cartesian components** (or just **components** for

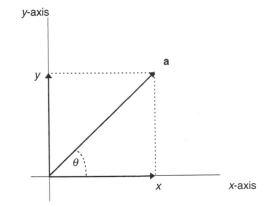

Figure 15.13 Vector in the *xy*-plane

short) of the vector. The general way of expressing a vector in **component form** is:

$$\mathbf{a} = x\mathbf{i} + y\mathbf{j}$$

where x and y are scalars; the Cartesian components of **a**.

Once we have expressed a vector in component form the operations of addition, subtraction and scaling can be carried out according to the rules of ordinary real numbers because x and y are just ordinary real numbers.

Note: Try not to confuse *components of a vector* with *component vectors*. When two vectors **a** and **b** are added we say the resultant vector (**a** + **b**) has component vectors **a** and **b**!

EXAMPLE 15.5

If:

$$\mathbf{a} = 3\mathbf{i} - 3\mathbf{j}, \quad \mathbf{b} = -\mathbf{i} + 5\mathbf{j} \quad \text{and} \quad \mathbf{c} = -\mathbf{j}$$

Find the resultant vector:

$$\mathbf{r} = \mathbf{a} + \mathbf{b} + \mathbf{c}$$

We simply apply the rules of ordinary algebra and collect-up like terms. No more triangles or parallelograms!

Solution

$$\mathbf{r} = (3\mathbf{i} - 3\mathbf{j}) + (-\mathbf{i} + 5\mathbf{j}) + (-\mathbf{j})$$
$$= 3\mathbf{i} - \mathbf{i} - 3\mathbf{j} + 5\mathbf{j} - \mathbf{j}$$
$$\mathbf{r} = 2\mathbf{i} + \mathbf{j}$$

Cartesian and polar form

In Chapter 5 we saw how it is possible to apply trigonometry to convert the position of a point in a plane from polar to Cartesian form and vice-versa.

The same ideas apply when representing vectors in a plane. Figure 15.14 serves as a reminder.

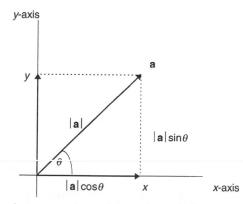

Figure 15.14 Converting between polar and Cartesian coordinates

If we know the magnitude and direction of a vector we have its polar form:

$$\mathbf{a} = |\mathbf{a}|\angle\theta$$

To go into Cartesian form:

$$\mathbf{a} = x\mathbf{i} + y\mathbf{j}$$

we take:

$$\mathbf{a} = |\mathbf{a}|(\cos\theta)\mathbf{i} + |\mathbf{a}|(\sin\theta)\mathbf{j}$$

In Fig. 15.14 the projection of $\mathbf{a}$ on the x-axis is the side *adjacent* to θ and its projection on the y-axis is equal to the side *opposite* θ.

EXAMPLE 15.6

A vessel is moving at a speed of 12 kn on a course of 30° from north. A tidal current is running at a speed of 4 kn in a direction of 130° from north. Find the velocity vector of the vessel in Cartesian form.

Let the vessel's velocity vector be $\mathbf{v}$ and the current vector $\mathbf{u}$. Then it is useful to sketch the vectors against a set of Cartesian coordinates. A compass bearing of 30° forms an angle θ_v of 60° and the current is running at an angle $\theta_u = -40°$.

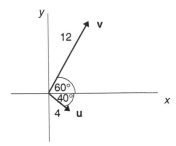

The procedure is quite straightforward. Once the vectors are in Cartesian form they can be added according to the normal rules of algebra.

$$\mathbf{v} = |\mathbf{v}|(\cos\theta_v)\mathbf{i} + |\mathbf{v}|(\sin\theta_v)\mathbf{j}$$
$$= (12\cos 60)\mathbf{i} + (12\sin 60)\mathbf{j}$$
$$\mathbf{v} = 6\mathbf{i} + 10.39\mathbf{j}$$
$$\mathbf{u} = |\mathbf{u}|(\cos\theta_u)\mathbf{i} + |\mathbf{u}|(\sin\theta_u)\mathbf{j}$$
$$= (4\cos -40)\mathbf{i} + (4\sin -40)\mathbf{j}$$
$$\mathbf{u} = 3.06\mathbf{i} - 2.57\mathbf{j}$$
$$\mathbf{v} + \mathbf{u} = 6\mathbf{i} + 10.39\mathbf{j} + 3.06\mathbf{i} - 2.57\mathbf{j}$$
$$= 9.06\mathbf{i} + 7.82\mathbf{j}$$

To go from Cartesian to polar form is simple trigonometry again. The magnitude of the vector is the hypotenuse of a right-angled triangle.

And, given the sides opposite and adjacent to the angle of interest we can use the tangent ratio. But to get the angle right you must look at the unit vectors to see which quadrant is defined.

In general for any vector **a**,

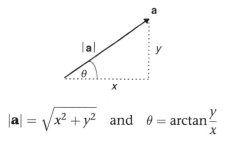

$$|\mathbf{a}| = \sqrt{x^2 + y^2} \quad \text{and} \quad \theta = \arctan\frac{y}{x}$$

if θ lies in the 1st quadrant.

EXAMPLE 15.7

Find the magnitude and direction of the resultant velocity of the vessel in Example 15.6.

Pythagoras' theorem helps us find the magnitude of the velocity which is the speed. Both unit vectors are positive so angle θ lies in the first quadrant of the Cartesian system. To convert θ to a compass bearing we must subtract it from 90°.

Solution

$$\mathbf{r} = \mathbf{v} + \mathbf{u} = 9.06\mathbf{i} + 7.82\mathbf{j}$$

$$|\mathbf{r}| = \sqrt{x^2 + y^2}$$

$$= \sqrt{9.06^2 + 7.82^2} = 12.0\,\text{kn}$$

$$\theta = \arctan\frac{y}{x} = \arctan\frac{7.82}{9.06} = 41°$$

Speed is 12 kn and direction is 49° from N.

Note that the Cartesian component form of a vector is sometimes called the **rectangular form** and sometimes the **algebraic form**. Different names for the same thing.

PRACTICE EXERCISE 15.4

With reference to Fig. 15.15:

(a) Write the vectors in component form.
(b) Find, in algebraic form, the resultant of $\mathbf{a} + \mathbf{b} - \mathbf{c}$.
(c) Find the rectangular form of $\frac{3}{4}\mathbf{b} - 5\mathbf{d}$.
(d) Find the modulus and argument of vector $\mathbf{d} - \mathbf{f} + 2\mathbf{b}$.

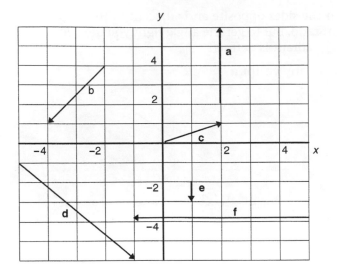

Figure 15.15 Vectors for Practice Exercise 15.4

Solutions

(a) $\mathbf{a} = 4\mathbf{j}$; $\mathbf{b} = -2\mathbf{i} - 3\mathbf{j}$; $\mathbf{c} = 2\mathbf{i} + \mathbf{j}$; $\mathbf{d} = 4\mathbf{i} - 5\mathbf{j}$; $\mathbf{e} = -\mathbf{j}$; $\mathbf{f} = -6\mathbf{i}$

(b) $\mathbf{r} = -4\mathbf{i}$; (c) $\mathbf{r} - 18.5\mathbf{i} + 22.75\mathbf{j}$; (d) $|\mathbf{r}| = 12.5$, $\arg \mathbf{r} = -61°$

EXAMPLE 15.8

Figure 15.16 is a force vector diagram of the forces acting on part of a structure which supports a floor in a building. If point m is in equilibrium, the vector sum of the forces is zero. Given that:

$$\mathbf{F}_1 = 3\mathbf{i} + 4\mathbf{j},$$
$$\mathbf{F}_2 = 4\mathbf{j},$$
$$\mathbf{F}_4 = -10.4\mathbf{i} - 6\mathbf{j},$$
$$\mathbf{F}_5 = 10.4\mathbf{i} - 6\mathbf{j}$$

find the x- and y-components of $\mathbf{F}_3$.

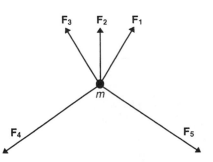

Figure 15.16 Force vector diagram

Solution

In equilibrium, $\mathbf{F_1 + F_2 + F_3 + F_4 + F_5 = 0}$ so $\mathbf{F_3 = -F_1 - F_2 - F_4 - F_5}$

$$\mathbf{F_3} = -(3\mathbf{i} + 4\mathbf{j}) - (4\mathbf{j}) - (-10.4\mathbf{i} - 6\mathbf{j}) - (10.4\mathbf{i} - 6\mathbf{j})$$

$$= -3\mathbf{i} - 4\mathbf{j} - 4\mathbf{j} + 10.4\mathbf{i} + 6\mathbf{j} - 10.4\mathbf{i} + 6\mathbf{j}$$

$$= -3\mathbf{i} + 4\mathbf{j}$$

$$\therefore \quad x_3\mathbf{i} + y_3\mathbf{j} = -3\mathbf{i} + 4\mathbf{j}$$

Equating coefficients of $\mathbf{i}$ and $\mathbf{j}$ gives: $x_3 = -3$, $y_3 = 4$

15.4 Vectors in three dimensions

Forces and motions of objects are not necessarily constrained to move in a plane. The microprocessor which controls the plotter of a CAD machine needs to process data in the xy-plane but what about a robot that is programmed to spot weld car bodies? The robot has three degrees of freedom. It must be capable of finding a position in space. For this, it will need to be programmed with data that define a displacement vector in three dimensions.

So it is necessary to add a third axis to the Cartesian coordinate system, the z-axis.

The three-dimensional (3-D) coordinate system follows the convention of the **right hand rule**. If the direction of the thumb represents the positive direction of the x-axis, then the first finger represents the y-axis and the second finger the z-axis. This coordinate system is known as the **right-handed system**. It is important to follow the convention because a left-handed system *is not the same*! Try it.

We can now define a unit vector in the direction of the z-axis. We call it $\mathbf{k}$. See Fig. 15.17. With three unit vectors we can represent three-dimensional

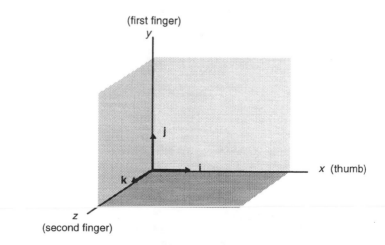

Figure 15.17 Three-dimensional coordinate system

vectors algebraically.

$$\mathbf{a} = x\mathbf{i} + y\mathbf{j} + z\mathbf{k}$$

The operations of addition and scaling of three-dimensional vectors are carried out in exactly the same way as they are for two-dimensional vectors; we simply scale and add the components of the vectors. We can state this more formally.

Given two vectors **a** and **b** having components a_1, a_2, a_3 and b_1, b_2, b_3 respectively, the sum of the vectors is the sum of their components.

If $\mathbf{a} = a_1\mathbf{i} + a_2\mathbf{j} + a_3\mathbf{k}$ and $\mathbf{b} = b_1\mathbf{i} + b_2\mathbf{j} + b_3\mathbf{k}$, then

$$\mathbf{a} + \mathbf{b} = (a_1 + b_1)\mathbf{i} + (a_2 + b_2)\mathbf{j} + (a_3 + b_3)\mathbf{k}$$

If m is a scaling factor and the vector is **a**, the product $m\mathbf{a}$ is the product of m and the components of **a**.

$$m\mathbf{a} = m(a_1\mathbf{i} + a_2\mathbf{j} + a_3\mathbf{k}) = ma_1\mathbf{i} + ma_2\mathbf{j} + ma_3\mathbf{k}$$

So, as before, the operations of addition (also subtraction) and scaling of three-dimensional vectors are carried out according to the same rules as for ordinary real numbers, providing the vectors are expressed in Cartesian component form.

Just one more thing. How do we find the magnitude and direction of a three-dimensional vector? Let's see.

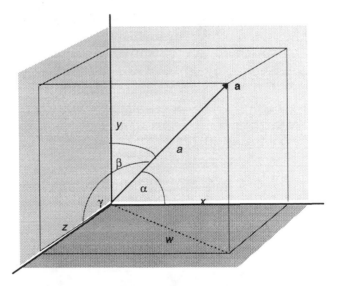

Figure 15.18 Modulus and argument of a 3-D vector

Given:

$$\mathbf{a} = x\mathbf{i} + y\mathbf{j} + z\mathbf{k}$$

then,

$$w^2 = x^2 + z^2$$

and

$$a^2 = w^2 + y^2$$

so

$$a^2 = x^2 + z^2 + y^2$$

therefore,

$$|\mathbf{a}| = \sqrt{x^2 + y^2 + z^2}$$

So finding the magnitude of a three-dimensional vector requires Pythagoras' theorem to be applied twice. Quite simple.

Finding the direction requires finding three angles. These are labelled α, β and γ (pronounced *alpha*, *beta* and *gamma*) in Fig. 15.18. The idea is to take what are called the **direction cosines** of the vector **a**. So:

$$\cos \alpha = \frac{x}{|\mathbf{a}|}, \quad \cos \beta = \frac{y}{|\mathbf{a}|}, \quad \cos \gamma = \frac{z}{|\mathbf{a}|}$$

EXAMPLE 15.9

A satellite GPS (global positioning system) tracks the movement of a missile. With reference to the centre of the globe, the displacement vector of the missile is $\mathbf{r} = a\mathbf{i} + b\mathbf{j} + c\mathbf{k}$. At t_0 the vector components are recorded as: $a_0 = -551\,267.0$, $b_0 = 3\,556\,230.0$ and $c_0 = 122\,236.0$ metres.

If 36 s later at t_1, the vector components of the missile are: $a_1 = 1210.0$, $b_1 = 3\,912\,341.0$ and $c_1 = -568\,371.0$ metres, find the average speed of the missile and its direction cosines.

Solution

The first thing to do is to make a sketch. 3-D sketches are difficult so a 2-D representation will do.

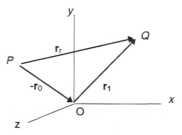

Displacement vectors are usually represented by the letter **r**. Let the displacement vector at t_0 be $\mathbf{r}_0$ and at t_1, $\mathbf{r}_1$. Note that these are displacements from the origin; in this case chosen to be the centre of the earth.

Now, let the vector representing the displacement of the missile from $\mathbf{r}_0$ to $\mathbf{r}_1$ be $\mathbf{r}_r$.

By the triangle rule, $\mathbf{r}_r = \mathbf{r}_1 - \mathbf{r}_0$.

$$\mathbf{r}_r = \mathbf{r}_1 - \mathbf{r}_0$$
$$= (a_1\mathbf{i} + b_1\mathbf{j} + c_1\mathbf{k}) - (a_0\mathbf{i} + b_0\mathbf{j} + c_0\mathbf{k})$$
$$= (a_1 - a_0)\mathbf{i} + (b_1 - b_0)\mathbf{j} + (c_1 - c_0)\mathbf{k}$$

If you find this difficult, try to appreciate that a movement of the missile from P to Q is equivalent to it moving from P to O then from O to Q.

In Cartesian component form we need only to add the components of the vectors. The numbers are awkward here so we keep them out for as long as we can.

Average velocity is total displacement over time taken. We take the difference of each component in turn. Then divide by the time taken to obtain the velocity vector in Cartesian form.

$$\mathbf{v} = \frac{\mathbf{r}_r}{t}$$

$$= \frac{1}{t}[(a_1 - a_0)\mathbf{i} + (b_1 - b_0)\mathbf{j} + (c_1 - c_0)\mathbf{k}]$$

$$a_1 - a_0 = 1210 - (-551\,267) = 552\,477$$

$$b_1 - b_0 = 3\,912\,341 - 3\,556\,230 = 356\,111$$

$$c_1 - c_0 = 568\,371 - (-122\,236) = -690\,607$$

$$\mathbf{v} = \tfrac{1}{36}(552\,477\mathbf{i} + 356\,111\mathbf{j} - 690\,607\mathbf{k})$$

$$= 15\,346.6\mathbf{i} + 9892.0\mathbf{j} - 19\,183.5\mathbf{k}$$

Speed is just the magnitude of velocity so we calculate the modulus of the velocity vector using Pythagoras' theorem.

Speed,

$$|\mathbf{v}| = \sqrt{15\,346.6^2 + 9892^2 + 19\,183.5^2}$$

$$= 26\,484\,\mathrm{m\,s^{-1}}$$

Our result is correct to $\pm 1\,\mathrm{m\,s^{-1}}$.

$$\cos\alpha = \frac{a_r}{|\mathbf{v}|} = \frac{15\,347}{26\,484} = 0.579$$

$$\cos\beta = \frac{b_r}{|\mathbf{v}|} = \frac{9892}{26\,484} = 0.374$$

$$\cos\gamma = \frac{c_r}{|\mathbf{v}|} = \frac{-19\,184}{26\,484} = -0.724$$

The direction cosines indicate that the average motion of the missile relative to the x, y, z coordinates is inclined approximately $55°$, $68°$ and $-44°$, respectively.

Now practise by doing the next exercise.

PRACTICE EXERCISE 15.5

Given that, $\mathbf{a} = 5\mathbf{i} - 2\mathbf{j} + 3\mathbf{k}$, $\mathbf{b} = -7\mathbf{i} - \mathbf{j} + 4\mathbf{k}$ and $\mathbf{c} = 3\mathbf{j} - 6\mathbf{k}$, evaluate:

(a) $-3\mathbf{a}$; (b) $\mathbf{b} - 2\mathbf{c}$; (c) $3\mathbf{a} - 2\mathbf{b} + \tfrac{1}{2}\mathbf{c}$; (d) $|\mathbf{a} + 4\mathbf{b}|$
(e) modulus of $\mathbf{a} + \mathbf{b} - 3\mathbf{c}$; (f) the direction cosines of $11\mathbf{a} - 8\mathbf{b}$.

Solutions

(a) $-15\mathbf{i} + 6\mathbf{j} - 9\mathbf{k}$; (b) $-7\mathbf{i} - 7\mathbf{j} + 16\mathbf{k}$; (c) $29\mathbf{i} - \frac{5}{2}\mathbf{j} - 2\mathbf{k}$
(d) 30.4; (e) 27.8; (f) 0.992, −0.125, 0.009

There is one last thing.

15.5 Matrix form of a vector

When dealing with 3-D vectors we can end up with problems that involve arrays. **Matrix algebra** is a convenient mathematical technique of dealing with arrays.

The matrix form of a vector expresses its components as a **column matrix**. In general, any vector in the component form:

$$\mathbf{a} = a_1\mathbf{i} + a_2\mathbf{j} + a_3\mathbf{k}$$

may be written as a column matrix:

$$\mathbf{a} = \begin{bmatrix} a_1 \\ a_2 \\ a_3 \end{bmatrix}$$

The point of this will become clear in the next chapter.

Assignment XIII

Problems in more than one dimension

In this assignment you will use vectors to solve four engineering problems. The performance criteria for assessment are as follows.

- Represent engineering systems using vectors.
- Define properties of vectors.
- Perform operations on vectors.
- Solve engineering problems involving vectors using appropriate techniques.

Get interested in the problems.

PROBLEM 1

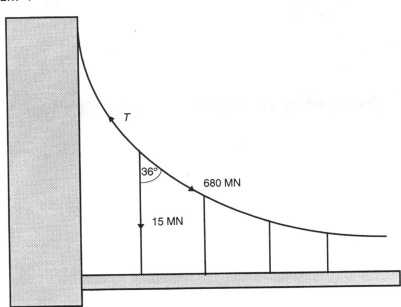

The tensions in two of the support cables of a suspension bridge are 680 MN and 15 MN as shown. Apply the *triangle rule* for vector addition to find the magnitude of the tension, T.

PROBLEM 2

Four forces are acting on a mass. Find, in Cartesian form, the force vector required to keep the acceleration vector of the mass at $\mathbf{0}$ given that the four forces are:

$$\mathbf{F}_1 = 12\mathbf{i} - 3\mathbf{j} + 7\mathbf{k}$$
$$\mathbf{F}_2 = 9\mathbf{i} + 6\mathbf{j}$$
$$\mathbf{F}_3 = -5\mathbf{i} + 18\mathbf{j} + 3\mathbf{k}$$
$$\mathbf{F}_4 = -\mathbf{i} - 8\mathbf{j} - \mathbf{k}$$

PROBLEM 3

A robot is programmed to start cutting when the position coordinates, in millimetres, of its cutting head are $(5, 3, -2)$ and finish the cut when the coordinates of the head are $(1, -2, 6)$. Find the length of the cut.

PROBLEM 4

A docking manoeuvre between a space station and a space shuttle is being planned. The space station has a mass of 12.61 tonnes and moves at a velocity of $(\mathbf{i} + 3\mathbf{j} + \mathbf{k})$. The mass of the shuttle is 1.36 tonnes. Calculate the required speed and the direction cosines of the shuttle to ensure that, after docking, the combined momentum of the two vehicles is $(14\,000\mathbf{i} + 36\,000\mathbf{j} + 12\,000\mathbf{k})$.

Remember that momentum

$$\mathbf{p} = m\mathbf{v}\,\mathrm{kg\,m\,s}^{-1}$$

is a conserved quantity, i.e. it does not vary with time. This means that total momentum before a collision is equal to total momentum after a collision.

16 Revision II

Assignment XIV

Additional maths

This summative exercise will require you to use differential equations, trigonometry and vectors to solve engineering problems. The following performance criteria will apply:

- Engineering systems are represented in the form of differential equations.
- The variables of a differential equations are separated.
- Functions are integrated to obtain a general solution.
- Boundary conditions are identified for a given engineering system.
- Boundary conditions are substituted to obtain a particular solution.
- Engineering systems are modelled using differential equations.
- Engineering systems are represented using vectors.
- Properties of vectors are defined.
- Operations are performed on vectors.
- Engineering problems involving vectors are solved using appropriate techniques.
- Engineering problems are represented using trigonometry.
- Trigonometric ratios and identities are selected to solve engineering problems.
- Angular properties of circles are calculated.
- Graphs of trigonometric ratios and identities are sketched over the range 0 to 360°.

Read the criteria, revise the topics and, when ready, tackle the assignment. The format will be familiar to you by now. There are 20 multiple choice questions which you should answer in 2 hours under normal examination conditions.

Good luck!

QUESTION 1

$$\frac{d^2y}{dx^2} - 3\left(\frac{dy}{dx}\right)^2 + 7 = 0$$

The equation above is a differential equation of:

(a) the first order and second degree,
(b) the second order and first degree,
(c) first order and first degree,
(d) second order and second degree.

QUESTION 2

The general solution of a differential equation gives rise to:

(a) an equation that is the unique solution to a problem,
(b) an equation in which all constants are known,
(c) a set of boundary conditions,
(d) a family of curves that is infinite in number.

QUESTION 3

Select from the following the one statement about solving differential equations which is *false*:

(a) Boundary conditions allow us to evaluate the constant of integration.
(b) The particular solution to a differential equation is a numerical value.
(c) Initial conditions are a type of boundary condition so they lead to a particular solution.
(d) *Separation of variables* is often the first step in solving a differential equation.

QUESTION 4

Experimental data indicate that the rate at which current changes in an inductive circuit is proportional to current. The differential equation which correctly models the behaviour of the current in the circuit is:

(a) $\dfrac{di}{dt} = kt$

(b) $\dfrac{di}{dt} = k$

(c) $\dfrac{di}{dt} = i$

(d) $\dfrac{di}{dt} = ki$

QUESTION 5

$$\frac{dy}{dx} = 2x$$

The correct solution to the differential equation above is:

(a) $y = x^2 + C$
(b) $y = 4x^2 + C$
(c) $y = A e^x + C$
(d) $y = 2 \ln x + C$

QUESTION 6

$$\frac{dy}{dx} = 3y$$

For the equation above, select the one option which is *incorrect*.

(a) $\ln y = 3x + A$
(b) $y = B e^{3x}$
(c) $y = \ln 3x + A$
(d) $y = e^{3x} e^A$

QUESTION 7

It is found that the velocity of an object moving in a straight line is described by the differential equation shown below.

$$\frac{ds}{dt} = 2.5t$$

If the position s of the object is 10 at $t = 2$, the particular equation which gives the position of the object is:

(a) $s = 2.5t^2 + C$
(b) $s = 1.25t^2 + 5$
(c) $s = 1.25t^2 - 125$
(d) $s = 5t^2 - 10$

QUESTION 8

For the equation shown below, select the one option which is *incorrect*:

$$\frac{dy}{dx} = y(2x - 5)$$

(a) $\ln y = x^2 - 5x + A$
(b) $y = e^{x^2 - 5x + A}$
(c) $y = e^{x^2} - 5x + A$
(d) $y = B e^{x^2 - 5x}$

QUESTION 9

Which one of the following is *not* a vector quantity?

(a) Angular velocity.
(b) Force.
(c) Energy.
(d) Magnetic flux.

QUESTION 10

Which one of the following is *not* a scalar quantity?

(a) Displacement.
(b) Speed.
(c) Time.
(d) Distance.

QUESTION 11

There are a number of ways of representing vectors using algebraic symbols. Which one of the following is *not* a vector?

(a) $\mathbf{a}$
(b) $\bar{a}$
(c) $\vec{a}$
(d) $|a|$

QUESTION 12

The magnitude and direction of the vector

$$\mathbf{a} = 12\mathbf{i} - 8\mathbf{j}$$

is:

(a) -14.4, $33.7°$
(b) 14.4, $-33.7°$
(c) 14.4, $56.3°$
(d) 14.4, $-56.3°$

QUESTION 13

The three forces, shown, are acting upon an object in space.

$$\mathbf{a} = 3\mathbf{i} - 4\mathbf{j} + 2\mathbf{k}$$

$$\mathbf{b} = -4\mathbf{k}$$

$$\mathbf{c} = -2\mathbf{i} + 6\mathbf{j} + 6\mathbf{k}$$

The resultant of the forces, $\mathbf{a} + \mathbf{b} + \mathbf{c}$, is:

(a) $\mathbf{i} + 2\mathbf{j} + 4\mathbf{k}$
(b) $5\mathbf{i} + 10\mathbf{j} + 12\mathbf{k}$
(c) $-6\mathbf{i} - 24\mathbf{j} - 48\mathbf{k}$
(d) $6\mathbf{i} - 10\mathbf{j}$

QUESTION 14

The magnitude of the momentum

$$\mathbf{p} = 27.4\mathbf{i} - 13.4\mathbf{j} + 8.72\mathbf{k}$$

is:

(a) $23.9 \, \text{kg m s}^{-1}$
(b) $22.7 \, \text{kg m s}^{-1}$
(c) $31.7 \, \text{kg m s}^{-1}$
(d) $49.5 \, \text{kg m s}^{-1}$

QUESTION 15

If $\cos\theta = \frac{1}{2}$, then:

(a) $\operatorname{cosec}\theta = \dfrac{2}{\sqrt{3}}$

(b) $\operatorname{cosec}\theta = \sqrt{3}$
(c) $\operatorname{cosec}\theta = \sqrt{5}$

(d) $\operatorname{cosec}\theta = \dfrac{1}{\sqrt{3}}$

QUESTION 16

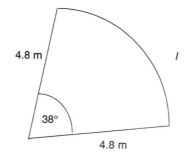

Figure 16.1 Sector of a circle—see Questions 16 and 17

The area of the sector shown in Fig. 16.1 is:

(a) $438 \, \text{m}^2$
(b) $7.64 \, \text{m}^2$
(c) $3.18 \, \text{m}^2$
(d) $1.59 \, \text{m}^2$

QUESTION 17

The length of the arc, l shown in Fig. 16.1, is:

(a) $182 \, \text{m}$
(b) $7.92 \, \text{m}$
(c) $7.24 \, \text{m}$
(d) $3.18 \, \text{m}$

QUESTION 18

Which one of the following identities is *correct*:

(a) $\cot\theta\sin\theta \equiv \sin\theta$
(b) $\cot\theta\sin\theta \equiv \sec\theta$
(c) $\cot\theta\sin\theta \equiv \cos\theta$
(d) $\cot\theta\sin\theta \equiv \cos^2\theta$

QUESTION 19

The solution which *does not* satisfy the trigonometric equation

$$3\cos^2\theta - 7\sin\theta = -1$$

is:

(a) $28.4°$
(b) $-151.6°$
(c) 0.495 rad
(d) 2.647 rad

QUESTION 20

The expression

$$2\sin\omega t + 3\cos\omega t$$

is the compound angle expansion of:

(a) $\sqrt{13}\sin(\omega t + 0.983)$
(b) $5\sin(\omega t + 0.983)$
(c) $\sqrt{13}\sin(\omega t - 0.983)$
(d) $3.61\sin(\omega t + 2.159)$

End

chapter

17 Matrices

Matrices were the idea of Arthur Cayley, an English lawyer who became a mathematician. In 1858 he demonstrated the use of matrices in the solution of simultaneous linear equations, a purpose for which they are still widely used.

The word *matrix* has a curious origin. It is derived from the Latin for *womb*: a place where something develops. In general it means any array of cells arranged in rows and columns.

In mathematics it is a rectangular array of **elements** written in rows and columns. The elements of a matrix may be algebraic terms or they may be any number, real or complex. If you have not come across matrices before then, to begin with, you may find it difficult to accept that a matrix cannot be reduced to a single numerical value.

For instance, the matrices **A** and **B** *cannot* be 'evaluated'. The arrays of elements in the brackets are not reducible in any way.

$$\mathbf{A} = \begin{bmatrix} -1 & 5 & a \\ \cos\theta & -6 & 5 \\ 0.2 & 12 & \frac{3}{7} \end{bmatrix}$$

$$\mathbf{B} = \begin{bmatrix} 2 & 3 & -7 \\ \pi & 0.5 & 2 \\ 67 & -1 & 9 \end{bmatrix}$$

Think of a matrix as a data store.

Note the symbol which represents the matrix. Symbols save a lot of writing when we carry out arithmetic or algebraic operations on matrices.

$$\mathbf{A} + \mathbf{B}$$

is better than writing

$$\begin{bmatrix} -1 & 5 & a \\ \cos\theta & -6 & 5 \\ 0.2 & 12 & \frac{3}{7} \end{bmatrix} + \begin{bmatrix} 2 & 3 & -7 \\ \pi & 0.5 & 2 \\ 67 & -1 & 9 \end{bmatrix}$$

Note also the convention of using bold lettering, similar to vector symbols. In hand-written form it is usual to put a wavy line above or below the letter, thus, $\tilde{A}$.

It is, of course, essential to nest the elements of a matrix in brackets. Square brackets are the most common but round brackets are also used.

17.1 Introduction to matrices

The order of a matrix

The **order** of a matrix defines the size of the array in terms of the number of rows and columns it contains. It is expressed as $m \times n$, where m is the number of rows and n is the number of columns. So the matrices **A** and **B**, above, are both order 3×3. Any order is possible.

$$\mathbf{D} = [x \quad y], \quad \text{or} \quad \mathbf{D} = (x, y)$$

$$\mathbf{E} = \begin{bmatrix} 3 \\ 2 \\ 5 \end{bmatrix} \quad \mathbf{F} = [7]$$

$$\mathbf{G} = \begin{bmatrix} a & b & c & d \\ e & f & g & h \end{bmatrix} \quad \mathbf{H} = \begin{bmatrix} 3 & 5 \\ 18 & 4 \\ -9 & 3 \\ 2 & -6 \end{bmatrix}$$

Matrix **D** is known as a **line matrix** (of order 1×2). You may recognize it as an ordered pair which represents the coordinates of a point in a plane. Matrix **E** is a **column matrix** (having order 3×1) which is one method of representing a 3-D vector.

QUESTION

State the order of matrices: **F**, **G** and **H** shown above.

ANSWER

$$1 \times 1, 2 \times 4, 4 \times 2$$

Transpose of a matrix

When the rows of a matrix are interchanged with their corresponding columns we obtain what is called the **transpose** of the matrix. The transpose of a matrix is identified by an apostrophe or by the superscript ^{T}, as shown. See how row 1 becomes column 1 and row 2 becomes column 2.

If,

$$\mathbf{A} = \begin{bmatrix} 1 & -3 & 0 \\ 2 & -1 & 5 \end{bmatrix}$$

then,

$$\mathbf{A}^{\mathrm{T}} \text{ or } \mathbf{A}' = \begin{bmatrix} 1 & 2 \\ -3 & -1 \\ 0 & 5 \end{bmatrix}$$

A special case of a transpose is a column matrix written as a line matrix indicated by curly brackets. Matrix **B** is an example. The purpose of this is simply to save writing space!

$$\mathbf{B} = \begin{bmatrix} 1 \\ 2 \\ 3 \end{bmatrix} = \{1 \quad 2 \quad 3\}$$

QUESTION

What is the order of matrix **C**?

$$\mathbf{C} = \{a \quad b \quad c\}$$

$$3 \times 1$$

Zero matrix

The zero matrix is defined as one in which all the elements are zero and, just like the zero vector, its symbol is **0**. The example shown is an order 2×2 zero matrix.

$$\mathbf{0} = \begin{bmatrix} 0 & 0 \\ 0 & 0 \end{bmatrix}$$

Identity matrix

This is the matrix equivalent of the ordinary number 1. It is one in which all the main diagonal elements have the value of 1 and all the rest zero. The identity matrix is usually represented by the letter **I**.

$$\mathbf{I} = \begin{bmatrix} 1 & 0 & 0 \\ 0 & 1 & 0 \\ 0 & 0 & 1 \end{bmatrix}$$

Equality

Two matrices are equal if all their corresponding elements are equal.

If,

$$\begin{bmatrix} 2 & 4 \\ -2 & a \\ b & -1 \end{bmatrix} = \begin{bmatrix} 2 & 4 \\ -2 & 5 \\ 3 & -1 \end{bmatrix}$$

then, $a = 5$ and $b = 3$

Matrix notation

The generalized form of matrix notation is: $[a_{ij}]$ where an element a is specified by its unique 'address'. We say that a is the element in the ith row and the jth column and the highest values of i and j are m and n which specify the order of the matrix.

Double subscript notation:

$$[a_{ij}] = \begin{bmatrix} a_{11} & a_{12} & a_{13} & a_{14} \\ a_{21} & a_{22} & a_{23} & a_{24} \\ a_{31} & a_{32} & a_{33} & a_{34} \end{bmatrix}$$

Specifying an item of data by its address is how a modern stored program control makes a computer find and manipulate data. So matrix methods are useful to computer software engineers.

I wonder if back in the 1850s, Arthur Cayley had any idea that this might become an application of his idea?

Addition and subtraction of matrices

Matrices can only be added or subtracted if they are of the *same order*. The process is quite straightforward. We simply find the sum or difference of each of the corresponding elements.

RULE

$$\begin{bmatrix} a & b \\ c & d \end{bmatrix} + \begin{bmatrix} e & f \\ g & h \end{bmatrix} = \begin{bmatrix} a+e & b+f \\ c+g & d+h \end{bmatrix}$$

EXAMPLE 17.1

If,

$$A = \begin{bmatrix} 3 & -2 \\ 11 & 7 \end{bmatrix}, \quad B = \begin{bmatrix} -2 & -3 \\ 9 & 2 \end{bmatrix}$$

find

$$C = A + B \quad \text{and} \quad D = A - B$$

Solution

$$C = \begin{bmatrix} 3 & -2 \\ 11 & 7 \end{bmatrix} + \begin{bmatrix} -2 & -3 \\ 9 & 2 \end{bmatrix} = \begin{bmatrix} 1 & -5 \\ 20 & 9 \end{bmatrix}$$

$$D = \begin{bmatrix} 3 & -2 \\ 11 & 7 \end{bmatrix} - \begin{bmatrix} -2 & -3 \\ 9 & 2 \end{bmatrix} = \begin{bmatrix} 5 & 1 \\ 2 & 5 \end{bmatrix}$$

17.2 Matrix multiplication

Multiplication by a scalar

Multiplication by a scalar, i.e. a single element matrix, is simple. Think of the scalar as being a factor outside brackets so that each element of the matrix must be multiplied by that factor.

IN GENERAL

$$k \begin{bmatrix} a & b \\ c & d \end{bmatrix} = \begin{bmatrix} ka & kb \\ kc & kd \end{bmatrix}$$

The process is reversible. We can factorize a matrix.

EXAMPLE 17.2

Find the product $k\mathbf{A}$, given that:

$$k = 3 \quad \mathbf{A} = \begin{bmatrix} 4 & -2 & 7 \\ x & 1 & \frac{1}{3} \end{bmatrix}$$

Solution

$$k\mathbf{A} = 3 \begin{bmatrix} 4 & -2 & 7 \\ x & 1 & \frac{1}{3} \end{bmatrix} = \begin{bmatrix} 12 & -6 & 21 \\ 3x & 3 & 1 \end{bmatrix}$$

EXAMPLE 17.3

Factorize:

$$\begin{bmatrix} 14 & 42 & -21 \\ 21 & 7 & 84 \\ -49 & 28 & -7 \end{bmatrix}$$

Solution

$$7 \begin{bmatrix} 2 & 6 & -3 \\ 3 & 1 & 12 \\ -7 & 4 & -1 \end{bmatrix}$$

Multiplication of two matrices

This is not quite so straightforward. Two matrices can be multiplied only if the number of columns of the first is equal to the number of rows of the second. This is called **conformability**. Two matrices are conformable for multiplication if their orders are:

$$l \times m \quad \text{and} \quad m \times n$$

and their product will be a matrix of order:

$$l \times n$$

IN GENERAL

If $\mathbf{A} = \begin{bmatrix} a & b \\ c & d \end{bmatrix}$ and $\mathbf{B} = \begin{bmatrix} e & f \\ g & h \end{bmatrix}$

$$\mathbf{AB} = \begin{bmatrix} a & b \\ c & d \end{bmatrix} \begin{bmatrix} e & f \\ g & h \end{bmatrix}$$

$$= \begin{bmatrix} ae + bg & af + bh \\ ce + dg & cf + dh \end{bmatrix}$$

The procedure is to multiply, in turn, each column of the second matrix by each row of the first matrix. It takes a little practice.

EXAMPLE 17.4

If,

$$A = \begin{bmatrix} 3 & -2 \\ 1 & 5 \end{bmatrix}$$

$$B = \begin{bmatrix} 2 & 4 \\ 6 & -1 \end{bmatrix}$$

$$C = \begin{bmatrix} 3 \\ -4 \\ 2 \end{bmatrix}$$

$$D = \begin{bmatrix} -1 & 3 & 2 \\ -5 & x & 7 \end{bmatrix}$$

$$E = \begin{bmatrix} 0.5 & -1 & 2 \end{bmatrix}$$

$$F = \begin{bmatrix} 1 \\ 2 \end{bmatrix}$$

$$G = \{1 \quad 2 \quad 3\}$$

$$H = \begin{bmatrix} 2 & -4 \\ 6 & 8 \\ 10 & 4 \end{bmatrix}$$

Find:

(a) **AB**
(b) **BA**
(c) **CE**
(d) **FE**
(e) **DH**
(f) **HD**
(g) **FH**
(h) **AF**
(i) **EG**
(j) **CD**

Solutions

(a) $AB = \begin{bmatrix} 3 & -2 \\ 1 & 5 \end{bmatrix}\begin{bmatrix} 2 & 4 \\ 6 & -1 \end{bmatrix} = \begin{bmatrix} 6-12 & 12+2 \\ 2+30 & 4-5 \end{bmatrix} = \begin{bmatrix} -6 & 14 \\ 32 & -1 \end{bmatrix}$

(b) $BA = \begin{bmatrix} 2 & 4 \\ 6 & -1 \end{bmatrix}\begin{bmatrix} 3 & -2 \\ 1 & 5 \end{bmatrix} = \begin{bmatrix} 6+4 & -4+20 \\ 18-1 & -12-5 \end{bmatrix} = \begin{bmatrix} 10 & 16 \\ 17 & -17 \end{bmatrix}$

(c) $\mathbf{CE} = \begin{bmatrix} 3 \\ -4 \\ 2 \end{bmatrix} [0.5 \quad -1 \quad 2] = \begin{bmatrix} 1.5 & -3 & 6 \\ -2 & 4 & -8 \\ 1 & -2 & 4 \end{bmatrix}$

(d) $\mathbf{FE} = \frac{1}{2}[0.5 \quad -1 \quad 2] = [0.25 \quad -0.5 \quad 1]$

(e) $\mathbf{DH} = \begin{bmatrix} -1 & 3 & 2 \\ -5 & x & 7 \end{bmatrix} \begin{bmatrix} 2 & -4 \\ 6 & 8 \\ 10 & 4 \end{bmatrix}$

$= \begin{bmatrix} -2+18+20 & 4+24+8 \\ -10+6x+70 & 20+8x+28 \end{bmatrix} = \begin{bmatrix} 36 & 36 \\ 60+6x & 48+8x \end{bmatrix}$

(f) $\mathbf{HD} = \begin{bmatrix} 2 & -4 \\ 6 & 8 \\ 10 & 4 \end{bmatrix} \begin{bmatrix} -1 & 3 & 2 \\ -5 & x & 7 \end{bmatrix}$

$= \begin{bmatrix} -2+20 & 6-4x & 4-28 \\ -6-40 & 18+8x & 12+56 \\ -10-20 & 30+4x & 20+28 \end{bmatrix} = \begin{bmatrix} 18 & 6-4x & -24 \\ -46 & 18+8x & 68 \\ -30 & 30+4x & 48 \end{bmatrix}$

(g) $\mathbf{FH} = \frac{1}{2} \begin{bmatrix} 2 & -4 \\ 6 & 8 \\ 10 & 4 \end{bmatrix} = \begin{bmatrix} 1 & -2 \\ 3 & 4 \\ 5 & 2 \end{bmatrix}$

(h) $\mathbf{AF} = \begin{bmatrix} 3 & -2 \\ 1 & 5 \end{bmatrix} \frac{1}{2} = \begin{bmatrix} 1.5 & -1 \\ 0.5 & 2.5 \end{bmatrix}$

(i) $\mathbf{EG} = [0.5 \quad -1 \quad 2] \begin{bmatrix} 1 \\ 2 \\ 3 \end{bmatrix} = [0.5 - 2 + 6] = [3.5]$

(j) Not conformable for multiplication.

It is worth glancing back over the last set of examples and taking careful note of the following points.

- Examples (a) and (b) indicate that matrix multiplication is not commutative, i.e. $\mathbf{AB} \neq \mathbf{BA}$.
- Also, you may have realized that multiplication of square matrices of the same order is always conformable since the number of columns of the first will always equal the number of columns of the second.
- Example (c) confirms the general rule that a multiplication of matrices of orders $(l \times m)$ and $(m \times n)$ yields a product matrix which is $(l \times n)$. In this case $(3 \times 1)(1 \times 3)$ has the product (3×3).
- Example (d) shows that multiplication by a single element matrix is similar to expanding an ordinary pair of brackets. Not the same, of course. The brackets must remain.
- Examples (e) and (f) exercise your skill in the technique. Multiplying column 1 by row 1 yields an element which is a sum of three terms and so on.

- Example (h) shows that multiplying by a single element matrix is commutative.
- Example (i) depends on how alert you are! You were given the *transpose* of **G**.
- Finally example (j) was not possible to carry out. The product of an order (3×1) and an order (2×3) matrix does not exist.

Go through the next set of examples and note the conclusions that follow.

EXAMPLE 17.5

If,

$$A = \begin{bmatrix} 3 & 1 & 4 \\ 2 & 8 & 9 \\ 5 & 7 & 6 \end{bmatrix}$$

$$B = \begin{bmatrix} 2 & 1 & -3 \\ 6 & 3 & -9 \end{bmatrix}$$

$$C = \begin{bmatrix} 1 & 9 \\ 4 & -6 \\ 2 & 4 \end{bmatrix}$$

Find,

(a) **IA**
(b) **AI**
(c) **0A**
(d) **BC**

Solutions

(a) $\mathbf{IA} = \begin{bmatrix} 1 & 0 & 0 \\ 0 & 1 & 0 \\ 0 & 0 & 1 \end{bmatrix} \begin{bmatrix} 3 & 1 & 4 \\ 2 & 8 & 9 \\ 5 & 7 & 6 \end{bmatrix} = \begin{bmatrix} 3 & 1 & 4 \\ 2 & 8 & 9 \\ 5 & 7 & 6 \end{bmatrix}$

(b) $\mathbf{AI} = \begin{bmatrix} 3 & 1 & 4 \\ 2 & 8 & 9 \\ 5 & 7 & 6 \end{bmatrix} \begin{bmatrix} 1 & 0 & 0 \\ 0 & 1 & 0 \\ 0 & 0 & 1 \end{bmatrix} = \begin{bmatrix} 3 & 1 & 4 \\ 2 & 8 & 9 \\ 5 & 7 & 6 \end{bmatrix}$

(c) $\mathbf{0A} = \begin{bmatrix} 0 & 0 & 0 \\ 0 & 0 & 0 \\ 0 & 0 & 0 \end{bmatrix} \begin{bmatrix} 3 & 1 & 4 \\ 2 & 8 & 9 \\ 5 & 7 & 6 \end{bmatrix} = \begin{bmatrix} 0 & 0 & 0 \\ 0 & 0 & 0 \\ 0 & 0 & 0 \end{bmatrix}$

(d) $\mathbf{BC} = \begin{bmatrix} 2 & 1 & -3 \\ 6 & 3 & -9 \end{bmatrix} \begin{bmatrix} 1 & 9 \\ 4 & -6 \\ 2 & 4 \end{bmatrix}$

$= \begin{bmatrix} 2+4-6 & 18-6-12 \\ 6+12-18 & 54-18-36 \end{bmatrix} = \begin{bmatrix} 0 & 0 \\ 0 & 0 \end{bmatrix}$

Conclusions.

- Examples (a) and (b) show that multiplication by the identity matrix, **I**, is equivalent to multiplication by 1 in ordinary arithmetic. Note also that **IA** = **AI** so multiplication by the identity matrix is commutative.
- Example (c) shows that the zero matrix behaves like zero in ordinary arithmetic. Note also that multiplication by **0** is also commutative.
- Example (d) is surprising. In ordinary number algebra the result: $ab = 0$ leads us to conclude that either a or b is zero. Example (d) indicates that this is not necessarily the case with matrices.

17.3 Second-order square matrices

For the moment we will concentrate just on square, order 2, matrices. They are fairly easy to deal with and provide straightforward examples of the application of matrices to engineering problems. But first, you must practise with the techniques.

PRACTICE EXERCISE 17.1

If, $\mathbf{A} = \begin{bmatrix} 2 & 0 \\ 3 & -1 \end{bmatrix}$ $\mathbf{B} = \begin{bmatrix} -4 & -7 \\ 5 & 8 \end{bmatrix}$ $\mathbf{C} = \begin{bmatrix} 12 & -19 \\ -27 & 41 \end{bmatrix}$ and $\mathbf{D} = \begin{bmatrix} \dfrac{7}{x} & 2x \\ \sin x & e^x \end{bmatrix}$

find:

(a) $\mathbf{A} + \mathbf{B}$; (b) $\mathbf{A} - \mathbf{B}$; (c) $3\mathbf{A}$; (d) $2(\mathbf{B} - \mathbf{A})$; (e) $\mathbf{AB}$; (f) $\frac{1}{3}\mathbf{BA}$
(g) $\mathbf{I} + \mathbf{C}$; (h) $\mathbf{AD}$; (i) $\mathbf{B}^2$; (j) $(\mathbf{B} - \mathbf{C})^2$; (k) $\mathbf{D}^\mathsf{T}$; (l) $\mathbf{CC}^\mathsf{T}$

Solutions

(a) $\begin{bmatrix} -2 & -7 \\ 8 & 7 \end{bmatrix}$; (b) $\begin{bmatrix} 6 & 7 \\ -2 & -9 \end{bmatrix}$; (c) $\begin{bmatrix} 6 & 0 \\ 9 & -3 \end{bmatrix}$; (d) $\begin{bmatrix} -12 & -14 \\ 4 & 18 \end{bmatrix}$

(e) $\begin{bmatrix} -8 & -14 \\ -17 & -29 \end{bmatrix}$; (f) $\begin{bmatrix} -\frac{29}{3} & \frac{7}{3} \\ \frac{34}{3} & -\frac{8}{3} \end{bmatrix}$; (g) $\begin{bmatrix} 13 & -19 \\ -27 & 42 \end{bmatrix}$

(h) $\begin{bmatrix} \dfrac{14}{x} & 4x \\ \left(\dfrac{21}{x} - \sin x\right) & (6x - e^x) \end{bmatrix}$; (i) $\begin{bmatrix} -19 & -28 \\ 20 & 29 \end{bmatrix}$; (j) $\begin{bmatrix} 640 & -588 \\ -1568 & 1473 \end{bmatrix}$

(k) $\begin{bmatrix} \dfrac{7}{x} & \sin x \\ 2x & e^x \end{bmatrix}$; (l) $\begin{bmatrix} 505 & -1103 \\ -1103 & 2410 \end{bmatrix}$

Now for a few more definitions. Be patient, it all comes neatly together in time.

Determinant of a matrix

It is easy, at the beginning, to get confused between matrices and determinants. They are not the same thing at all. A matrix is composed of discrete elements, it cannot be reduced to a single numerical value. Determinants have a numerical value; their value can be 'determined'. To find the determinant of a square matrix the following rule is applied.

For any square, order 2, matrix, the determinant is the difference of the product of the diagonal elements.

The symbol which is used to represent a determinant is either:

$$\det \mathbf{A} \quad \text{or} \quad \Delta\mathbf{A} \quad \text{or} \quad |\mathbf{A}|$$

The elements of a determinant must be written between vertical lines (rather like the modulus of a vector) and never between brackets. Remember, the determinant of **A** is not the matrix **A**.

If,

$$\mathbf{A} = \begin{bmatrix} a & b \\ c & d \end{bmatrix}$$

then,

$$\det \mathbf{A} = \begin{vmatrix} a & b \\ c & d \end{vmatrix}$$

which is evaluated by,

$$\det \mathbf{A} = ad - cb$$

You might find the following a handy way of remembering what to do:

$$+ \searrow \quad \nearrow -$$

EXAMPLE 17.6

Find the determinant of the matrix:

$$\mathbf{A} = \begin{bmatrix} -3 & -2 \\ 5 & 4 \end{bmatrix}$$

Solution

$$\det \mathbf{A} = \begin{vmatrix} -3 & -2 \\ 5 & 4 \end{vmatrix} = -12 - (-10) = -2$$

Adjoint of a matrix

To find the **adjoint** of any matrix we first need the **minors** and **cofactors** of the matrix. Double subscript notation comes in useful here.

To find the minor of an element ignore the other elements in the same row and the same column. The remaining element is the minor.

Finding the minors is the first step towards finding the cofactors.

If,

$$A = \begin{bmatrix} a_{11} & a_{12} \\ a_{21} & a_{22} \end{bmatrix}$$

then,

the minor of a_{11} is a_{22}
the minor of a_{12} is a_{21}
the minor of a_{21} is a_{12}
the minor of a_{22} is a_{11}

To find the cofactors of a matrix the general rule is:

$$\text{cofactor } a_{ij} = (-1)^{i+j}(\text{minor of } a_{ij}).$$

The cofactor of each element is found by multiplying the minor of that element by -1 raised to the power defined by the ith row plus the jth column in which the element of interest resides.

If,

$$A = \begin{bmatrix} a_{11} & a_{12} \\ a_{21} & a_{22} \end{bmatrix}$$

then,

the cofactor of a_{11} is $(-1)^2 a_{22} = a_{22}$
the cofactor of a_{12} is $(-1)^3 a_{21} = -a_{21}$
the cofactor of a_{21} is $(-1)^3 a_{12} = -a_{12}$
the cofactor of a_{22} is $(-1)^4 a_{11} = a_{11}$

It isn't easy at first but if you look carefully at the working, you will find it is not as difficult as you think.

Now, having established how to find the values of the minors and the cofactors of our matrix we can write its **matrix of cofactors**.

$$\begin{bmatrix} a_{22} & -a_{21} \\ -a_{12} & a_{11} \end{bmatrix}$$

Finally we obtain the **adjoint** of matrix **A** by writing the **transpose of the matrix of cofactors**.

$$\text{adj } A = \begin{bmatrix} a_{22} & -a_{12} \\ -a_{21} & a_{11} \end{bmatrix}$$

In general, if:

$$A = \begin{bmatrix} a_{11} & a_{12} \\ a_{21} & a_{22} \end{bmatrix}$$

$$\text{adj } A = \begin{bmatrix} a_{22} & -a_{12} \\ -a_{21} & a_{11} \end{bmatrix}$$

Confused? Of course you are but the general procedure for finding the adjoint of a matrix simplifies when dealing with an order 2 matrix. Again, I think arrows and signs come in useful for this:

EXAMPLE 17.7

Find the adjoint of the matrix:

$$A = \begin{bmatrix} -3 & -2 \\ 5 & 4 \end{bmatrix}$$

Solution

Exchange elements a_{11} and a_{22}, and change the signs of elements a_{12} and a_{21}:

$$\text{adj } A = \begin{bmatrix} 4 & 2 \\ -5 & -3 \end{bmatrix}$$

Simple then! But you need to hold on to what you have just learned in order to deal with the next section.

To find the determinant of a matrix:

To find the adjoint of a matrix:

Inverse of a matrix

With matrices, the equivalent of the reciprocal is called the **inverse**. The inverse of a matrix is indicated in the same way we indicate the reciprocal of an ordinary number.

$$\frac{a}{b} = ab^{-1} \quad \text{and} \quad aa^{-1} = 1$$

Now the property of an inverse matrix is the same as the property of a reciprocal. Can you think what it is?

The product of any number and its reciprocal is unity.

Zero being the exception, of course.

The inverse of $\mathbf{A}$ is $\mathbf{A}^{-1}$, so

$$\mathbf{A}^{-1}\mathbf{A} = \mathbf{I} = \begin{bmatrix} 1 & 0 \\ 0 & 1 \end{bmatrix}$$

the identity matrix.

For example:

$$7^{-1} \times 7 = 1$$

And the matrix equivalent of unity, you know, is the identity matrix, **I**.

The remaining question is how do we find the inverse of a given matrix?

You divide the adjoint of the matrix
by the determinant of the matrix.

Note the rule carefully. It explains the purpose of what we have been doing over the preceding sections.

If,

$$\mathbf{A} = \begin{bmatrix} a & b \\ c & d \end{bmatrix}$$

$$\mathbf{A}^{-1} = \frac{\text{adj } \mathbf{A}}{\det \mathbf{A}} = \frac{1}{|\mathbf{A}|} \text{adj } \mathbf{A}$$

$$\mathbf{A}^{-1} = \frac{1}{ad - cb} \begin{bmatrix} d & -b \\ -c & a \end{bmatrix}$$

EXAMPLE 17.8

Find the inverse of:

$$\mathbf{A} = \begin{bmatrix} 3 & -2 \\ 1 & -4 \end{bmatrix}$$

then check the result.

Solution

$$\mathbf{A}^{-1} = \frac{1}{|\mathbf{A}|} \text{adj } \mathbf{A}$$

$$= \frac{1}{-12 + 2} \begin{bmatrix} -4 & 2 \\ -1 & 3 \end{bmatrix} = \frac{1}{-10} \begin{bmatrix} -4 & 2 \\ -1 & 3 \end{bmatrix}$$

$$\mathbf{A}^{-1} = \begin{bmatrix} 0.4 & -0.2 \\ 0.1 & -0.3 \end{bmatrix}$$

Check:

$$\mathbf{A}\mathbf{A}^{-1} = \begin{bmatrix} 3 & -2 \\ 1 & -4 \end{bmatrix} \begin{bmatrix} 0.4 & -0.2 \\ 0.1 & -0.3 \end{bmatrix}$$

$$= \begin{bmatrix} 1.2 - 0.2 & -0.6 + 0.6 \\ 0.4 - 0.4 & -0.2 + 1.2 \end{bmatrix} = \begin{bmatrix} 1 & 0 \\ 0 & 1 \end{bmatrix} = \mathbf{I}$$

The check indicates that the product of the matrix and its inverse is the identity matrix. This is correct.

There are a couple of other things. Study the next examples carefully.

EXAMPLE 17.9

If,

$$A = \begin{bmatrix} 6 & 8 \\ -1 & 2 \end{bmatrix}$$

find $A^{-1}A$.

$$A^{-1}A = AA^{-1}$$

Solution

$$A^{-1} = \frac{1}{12+8}\begin{bmatrix} 2 & -8 \\ 1 & 6 \end{bmatrix} = \begin{bmatrix} 0.1 & -0.4 \\ 0.05 & 0.3 \end{bmatrix}$$

$$A^{-1}A = \begin{bmatrix} 0.1 & -0.4 \\ 0.05 & 0.3 \end{bmatrix}\begin{bmatrix} 6 & 8 \\ -1 & 2 \end{bmatrix}$$

$$= \begin{bmatrix} 0.6+0.4 & 0.8-0.8 \\ 0.3-0.3 & 0.4+0.6 \end{bmatrix} = \begin{bmatrix} 1 & 0 \\ 0 & 1 \end{bmatrix} = I$$

This example illustrates that multiplication of a matrix by its inverse is commutative.

EXAMPLE 17.10

Find the inverse of:

$$A = \begin{bmatrix} -3 & 3 \\ -2 & 2 \end{bmatrix}$$

Solution

$$A^{-1} = \frac{1}{-6+6}\begin{bmatrix} 2 & -3 \\ 2 & -3 \end{bmatrix} = \frac{1}{0}\begin{bmatrix} 2 & -3 \\ 2 & -3 \end{bmatrix}$$

is irrational.

So a matrix with a determinant which is equal to zero has no inverse. This type of matrix is called a **singular matrix**.

PRACTICE EXERCISE 17.2

Find the inverse of each of the following:

(a) $\begin{bmatrix} 5 & 10 \\ 2 & -7 \end{bmatrix}$; (b) $\begin{bmatrix} 2 & -3 \\ -1 & 6 \end{bmatrix}$; (c) $\begin{bmatrix} -10 & 26 \\ 2 & -18 \end{bmatrix}$; (d) $\begin{bmatrix} 1 & 0 \\ 0 & 1 \end{bmatrix}$; (e) $\begin{bmatrix} 7 & 7 \\ -3 & -3 \end{bmatrix}$

Solutions

(a) $\begin{bmatrix} \frac{7}{55} & \frac{2}{11} \\ \frac{2}{55} & -\frac{1}{11} \end{bmatrix}$; (b) $\begin{bmatrix} \frac{2}{3} & \frac{1}{3} \\ \frac{1}{9} & \frac{2}{9} \end{bmatrix}$; (c) $\begin{bmatrix} -\frac{9}{64} & -\frac{13}{64} \\ -\frac{1}{64} & -\frac{5}{64} \end{bmatrix}$; (d) $\begin{bmatrix} 1 & 0 \\ 0 & 1 \end{bmatrix}$; (e) singular matrix

Practice Exercise 17.2 highlights one or two interesting points.

- It pays to work with fractions.
- It pays to check by multiplying the original matrix by its inverse.
- The inverse of the identity matrix is also an identity matrix.
- When any two rows or any two columns are the same, you are dealing with a singular matrix.

Finally, it must be pointed out that the short cuts for finding the determinant and adjoint of a matrix only apply to second-order square matrices. With higher-order matrices it is necessary to pay more attention to minors and cofactors.

17.4 Solution of simultaneous equations using matrices

In Chapter 4 you looked at two algebraic methods of solving simultaneous linear equations: the **substitution** and **elimination** methods. Glance back and refresh your knowledge. We started with the example from the Mud Assignment for which simultaneous equations were used to find the correct mixture of baryte and water to give the required mud density.

Here are the equations again.

$$37x + 11y = 30.5$$

$$x + y = 1$$

We can write them in matrix form. On each side of the equation is an order 2 column matrix.

$$\begin{bmatrix} 37x + 11y \\ x + y \end{bmatrix} = \begin{bmatrix} 30.5 \\ 1 \end{bmatrix}$$

On the right is a matrix of constants. On the left there is something more interesting. It is a matrix which is the product of two matrices, i.e. it can be factorized. The factors turn out to be a matrix of the coefficients and a matrix of the variables. Apply the rule of matrix multiplication to check that these are indeed factors of the original matrix.

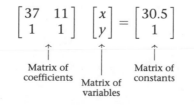

Now, life will be easier if we write out the equation of matrices in an abbreviated form. Let the matrix of coefficients be **A**, the matrix of the variables **X**, and the matrix of constants **B**.

Multiplying by the inverse of **A** turns **A** into the identity matrix, **I**. The identity matrix has the same property as 1 in ordinary numbers. The result is a solution for **X**, the matrix of variables. Simple.

$$\mathbf{AX} = \mathbf{B}$$

$$\mathbf{A}^{-1}\mathbf{AX} = \mathbf{A}^{-1}\mathbf{B}$$

$$\mathbf{IX} = \mathbf{A}^{-1}\mathbf{B}$$

$$\mathbf{X} = \mathbf{A}^{-1}\mathbf{B}$$

We need the inverse of the matrix of coefficients. To find the inverse the adj **A** and the |**A**| must be found.

$$\mathbf{A}^{-1} = \frac{1}{|\mathbf{A}|}\,\text{adj}\,\mathbf{A}$$

The solution of **X** is the product, **A**⁻¹**B**, so we write the matrices in full and multiply.

$$\begin{bmatrix} 37 & 11 \\ 1 & 1 \end{bmatrix}^{-1} = \frac{1}{37-11}\begin{bmatrix} 1 & -11 \\ -1 & 37 \end{bmatrix} = \frac{1}{26}\begin{bmatrix} 1 & -11 \\ -1 & 37 \end{bmatrix}$$

The result is an order 2 column matrix.

$$\mathbf{X} = \mathbf{A}^{-1}\mathbf{B}$$

$$\begin{bmatrix} x \\ y \end{bmatrix} = \frac{1}{26}\begin{bmatrix} 1 & -11 \\ -1 & 37 \end{bmatrix}\begin{bmatrix} 30.5 \\ 1 \end{bmatrix}$$

$$= \frac{1}{26}\begin{bmatrix} 30.5 - 11 \\ -30.5 + 37 \end{bmatrix} = \frac{1}{26}\begin{bmatrix} 19.5 \\ 6.5 \end{bmatrix}$$

$$= \begin{bmatrix} 0.75 \\ 0.25 \end{bmatrix}$$

Remember the law of matrix equality. If two matrices are equal their corresponding elements are equal.

So we get:

$$x = 0.75,\ y = 0.25$$

EXAMPLE 17.11

Solve:

$$4x = 3y + 1$$

$$x + 3y - 19 = 0$$

The procedure can be streamlined.

Re-arrange to get variables on the left, constants on the right. This gives you an equality of two column matrices.

$$\begin{bmatrix} 4x - 3y \\ x + 3y \end{bmatrix} = \begin{bmatrix} 1 \\ 19 \end{bmatrix}$$

Factorize the LHS to obtain the product of the matrix of coefficients and the matrix of variables.

$$\begin{bmatrix} 4 & -3 \\ 1 & 3 \end{bmatrix}\begin{bmatrix} x \\ y \end{bmatrix} = \begin{bmatrix} 1 \\ 19 \end{bmatrix}$$

Multiply the RHS by the inverse of **A**. This involves dividing by the determinant and multiplying by the adjoint of **A**. Using,

$$\mathbf{X} = \mathbf{A}^{-1}\mathbf{B} = \left(\frac{1}{|\mathbf{A}|}\operatorname{adj}\mathbf{A}\right)\mathbf{B}$$

$$\begin{bmatrix} x \\ y \end{bmatrix} = \frac{1}{12+3}\begin{bmatrix} 3 & 3 \\ -1 & 4 \end{bmatrix}\begin{bmatrix} 1 \\ 19 \end{bmatrix} = \frac{1}{15}\begin{bmatrix} 3+57 \\ -1+76 \end{bmatrix}$$

$$= \frac{1}{15}\begin{bmatrix} 60 \\ 75 \end{bmatrix}$$

$$= \begin{bmatrix} 4 \\ 5 \end{bmatrix}$$

$$x = 4, y = 5$$

Remember to check your results. Here, back-substitution gives:

$$4(4) - 3(5) = 1$$
$$4 + 3(5) = 19$$

Correct!

Perhaps, now, you can appreciate the value of all that hard work learning the definitions, the properties of standard matrices (unit, zero, singular) and the rules of arithmetical operations with matrices. The rules may have seemed rather complicated and odd. But remember, matrices were designed with the solution of simultaneous equations in mind. They work perfectly.

EXAMPLE 17.12

Solve:

$$2x - y = -5$$
$$-2x + y = -3$$

Solution

$$\begin{bmatrix} 2x - y \\ -2x + y \end{bmatrix} = \begin{bmatrix} -5 \\ -3 \end{bmatrix}$$

$$\begin{bmatrix} 2 & -1 \\ -2 & 1 \end{bmatrix}\begin{bmatrix} x \\ y \end{bmatrix} = \begin{bmatrix} -5 \\ -3 \end{bmatrix}$$

$$\begin{bmatrix} x \\ y \end{bmatrix} = \frac{1}{2-2}\begin{bmatrix} 1 & 1 \\ 2 & 2 \end{bmatrix}\begin{bmatrix} -5 \\ -3 \end{bmatrix}$$

$$= \frac{1}{0}\begin{bmatrix} 1 & 1 \\ 2 & 2 \end{bmatrix}\begin{bmatrix} -5 \\ -3 \end{bmatrix}$$

No solution

Here the determinant is zero. Dividing the adjoint by zero is irrational.

This means that the matrix of coefficients has no inverse. It is a **singular matrix** and no solution is possible.

Example 17.12 seems to indicate that the matrix method does not always work. Is there a flaw in the method? Try the substitution and the elimination methods on these equations.

These do not work either. The fact is, there is nothing wrong with any of the methods. This particular pair of simultaneous equations has no solution. The graphs of these equations have the same gradient, they will never cross, so it is impossible to find values of x and y which satisfy both equations. In this situation we say that the equations are **inconsistent**.

Now look at a practical problem.

EXAMPLE 17.13

On 15 August, a car assembly plant in Liverpool was charged £390 for the delivery of 50 gearboxes from Swansea and 20 which came from Dagenham.

On 21 August the charge for the delivery of 45 gearboxes from Swansea and 70 gearboxes from Dagenham was £715.

Determine the transportation cost of a gearbox from Swansea and the transportation cost of a gearbox from Dagenham.

Solution

You need to sort out the facts. Look for the variables. Eliminate the 'noise'.

Let the cost of transporting a gearbox from Swansea be £x and the cost of a gearbox from Dagenham £y. We see that £50x plus £20y made £390 and £45x plus £70y made £715.

That gives two simultaneous equations which can be solved by matrices. The dates are the 'noise' which we can ignore.

$$50x + 20y = 390$$

$$45x + 70y = 715$$

$$\begin{bmatrix} 50x & 20y \\ 45x & 70y \end{bmatrix} = \begin{bmatrix} 390 \\ 715 \end{bmatrix}$$

$$\begin{bmatrix} 50 & 20 \\ 45 & 70 \end{bmatrix}\begin{bmatrix} x \\ y \end{bmatrix} = \begin{bmatrix} 390 \\ 715 \end{bmatrix}$$

$$\begin{bmatrix} x \\ y \end{bmatrix} = \begin{bmatrix} 50 & 20 \\ 45 & 70 \end{bmatrix}^{-1} \begin{bmatrix} 390 \\ 715 \end{bmatrix}$$

$$= \frac{1}{3500 - 900} \begin{bmatrix} 70 & -20 \\ -45 & 50 \end{bmatrix} \begin{bmatrix} 390 \\ 715 \end{bmatrix}$$

$$= \frac{1}{2600} \begin{bmatrix} (70 \times 390) + (-20 \times 715) \\ (-45 \times 390) + (50 \times 715) \end{bmatrix}$$

$$= \frac{1}{2600} \begin{bmatrix} 13\,000 \\ 18\,200 \end{bmatrix} = \begin{bmatrix} 5 \\ 7 \end{bmatrix}$$

$$x = £5, y = £7$$

PRACTICE EXERCISE 17.3

Apply the matrix method to each of the following problems.

(a) Solve: $\begin{array}{l} 5x - 3y = 7 \\ -3x + 7y = 1 \end{array}$; (b) Solve: $\begin{array}{l} x_2 = 3x_1 - 5 \\ x_2 = 10 - x_1 \end{array}$; (c) Solve: $\begin{array}{l} 2F_1 + 5F_2 + 14.6 = 0 \\ 3.1F_1 + 1.7F_2 + 2.06 = 0 \end{array}$

(d) The total running time, from their last services, of two steam turbines, is 5800 hours. The last service of turbine 2 was 600 hours after the last service of turbine 1. What are the running hours of each turbine since it was last serviced?

Solutions

(a) $x = 2, y = 1$; (b) $x_1 = 3.75, x_2 = 6.25$; (c) $F_1 = 1.2, F_2 = -3.4$
(d) 3200 hours, 2600 hours

Assignment XV

This assignment requires you to use matrices to solve engineering problems. Note the performance criteria which must be met for satisfactory completion.

- Engineering problems are represented using matrix notation.
- Standard matrices are defined.
- Operations are performed on matrices.
- Engineering problems involving matrices are solved using appropriate techniques.

A word of warning, before you start. Simultaneous linear equations are prone to what is called **ill-conditioning**. This means that they can be very sensitive to small changes in the variables. For example, a slight rounding off in an early part of the calculation can produce results that are significantly incorrect.

Ill-conditioning occurs when the graphs of the equations are almost parallel. Just to see for yourself; use a ruler to draw a couple of lines on graph paper making them almost parallel. What happens at the point where they cross? It's fuzzy and a slight tilt of the gradient of one graph causes the crossing-point to move appreciably.

So it pays to be especially cautious with simultaneous equations: work to a degree of accuracy much greater than you would normally.

Be careful, work methodically,
show each step of your solutions clearly.

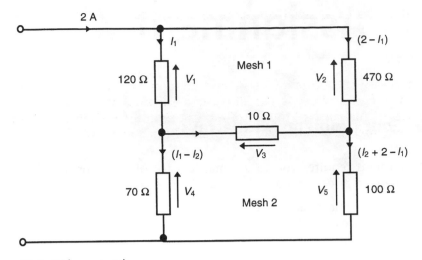

Figure 17.1 Bridge network

PROBLEM 1 – MESH ANALYSIS

The circuit in Fig. 17.1 is known as a *bridge network*. The bridge draws a supply current of 2 A, which divides between the various branches as shown.

Any closed loop in a network is known as a *mesh*. In mesh 1 of the network shown, there are three volt drops which, according to Kirchhoff's voltage law, have an algebraic sum of zero. So you can write: $V_1 - V_2 + V_3 = 0$. The same holds true for mesh 2, so: $V_4 - V_3 - V_5 = 0$. Now, in accordance with Ohm's law, the product of a resistance and the current it carries is volts drop, so the two voltage equations can be re-written as:

$$120I_1 - 470(2 - I_1) + 10I_2 = 0$$
$$270(I_1 - I_2) - 10I_2 - 100(I_2 + 2 - I_1) = 0$$

Use matrices to find the branch currents in the circuit of Fig. 17.1.

PROBLEM 2 – BATTERY ACID

At a given temperature, the required density of the electrolyte in a lead–acid battery is $1280 \, \mathrm{kg \, m^{-3}}$. The electrolyte is made up from a solution of sulphuric acid in distilled water. At the same temperature, the density of sulphuric acid is $1428 \, \mathrm{kg \, m^{-3}}$ and the density of the distilled water is $1002 \, \mathrm{kg \, m^3}$. Find the quantity of sulphuric acid and the quantity of distilled water required to make up 1 litre of electrolyte. Apply matrices to the solution of this problem; remember to show and justify the steps you take.

PROBLEM 3 – TYRES

Groundhog Ltd is a firm that makes two types of car tyre: *Longlife* and *Supergrip*. A *Longlife* tyre requires 4.22 kg of rubber, and 1.11 kg of nylon. A *Supergrip* tyre requires 2.02 kg of rubber, and 4.10 kg of nylon. An inventory shows that Groundhog has the following stock of raw materials: 3.76 tonnes of rubber, and 3.43 tonnes of nylon. How many of each type of tyre can be produced before stocks run out?

PROBLEM 4 – THE COST OF A VOYAGE

The total cost of 180 tonnes of fuel oil and 2.5 tonnes of lubricating oil over one voyage of a cargo vessel is £65 300. On a subsequent voyage the cost of 150 tonnes of fuel and 2.1 tonnes of lubricating oil is £54 432. Find the cost per tonne of fuel oil and the cost per tonne of lubricating oil.

chapter
18 Complex numbers

You may have forgotten by now but in Chapter 4, Exercise 6, you were finding the roots of equations and came across the following problem.

$$3x^2 + 12x = -15$$

A simple quadratic; or so it seems.

$$3x^2 + 12x + 15 = 0$$

$$x = \frac{-b \pm \sqrt{b^2 - 4ac}}{2a}$$

$$= \frac{-12 \pm \sqrt{144 - 4(3)(15)}}{6}$$

$$= -2 \pm \frac{\sqrt{-36}}{6}$$

Applying the formula method leads us to a term with $\sqrt{-36}$. Try as we might, it is impossible to find the root of a negative number.

So we do our best: $-36 = -1 \times 36$.

$$= -2 \pm \frac{\sqrt{(-1)36}}{6}$$

This gives us $\sqrt{36}$ over six which is 1. A simple solution seems tantalizingly close.

$$= -2 \pm \frac{\sqrt{-1}\sqrt{36}}{6}$$

$$= -2 \pm \sqrt{-1}$$

But what to do with $\sqrt{-1}$? A question first posed by the great Greek mathematicians in Alexandria at around 250 AD.

The answer to this question was given almost 1300 years later by Gerolamo Cardano and Rafaello Bombelli, two mathematicians working in Bologna, Italy.

They said, in effect, that we do absolutely nothing! The roots of this type of equation are just what they say they are.

$$x = -2 + \sqrt{-1} \quad \text{and} \quad -2 - \sqrt{-1}$$

By accepting the existence of $\sqrt{-1}$ Cardano and Bombelli demonstrated a classic example of lateral thinking. This is not to say that all the mathematicians of Bologna were comfortable with the idea. Far from it. The quirky nature of $\sqrt{-1}$ made them believe that they had discovered something mystical so they called it an 'imaginary' number and $\sqrt{-1}$ became 'i'.

Let $\sqrt{-1} = i$, so:

$$x = -2 + i, \; -2 - i$$

Today, we say that the roots of this type of equation are **complex numbers** that consist of a **real** part and an **imaginary part**. So, in our example, $-2 + i$ is a complex number in which -2 is real and i is imaginary.

The history of i is very revealing about the way mathematics was treated as some part of the physical world; a science waiting to be discovered by man. Today's idea is different. We now generally accept that mathematics is a man-made device, an artefact (or *art* for short). $\sqrt{-1}$, which we call i, exists because we say so!

But what has all this to do with engineering? Well, $\sqrt{-1}$ is so important to engineers that they have given it a symbol of its own. In order not to confuse i with electric current, they call it j.

18.1 The Cartesian form of a complex number

The general form of a complex number in Cartesian form expresses x as the real and y the imaginary component. If you look back to Chapter 5 you will recognize that we are dealing with a two-dimensional number which can, for example, describe the position of a point in the x, y-plane.

For

$$z = x + jy$$

we say

$$x = \operatorname{Re} z$$

and

$$y = \operatorname{Im} z$$

where

$$j = \sqrt{-1}$$

You may also remember that the Cartesian form can, if we wish, be converted to polar form, but we can come back to that later.

The Cartesian form of a complex number is sometimes called the **rectangular form** and sometimes the **algebraic form** because it allows us to carry out operations according to the normal rules of algebra, treating the real and imaginary parts as the terms of a binomial.

Powers of j

Operations with j can produce interesting results. Look at some examples.

Since, $j = \sqrt{-1}$

then, $j^2 = -1,$ real number

and $j^3 = j(j^2) = -j,$ imaginary

$j^4 = j(j^3) = -j^2 = -(-1) = 1,$ real again

$j^5 = j(j^4) = j,$ imaginary again

$j^6 = j(j^5) = j^2 = -1$

Taking increasing powers of j causes it to change in a regular pattern from j to -1 to $-j$ to 1 and back to j again and so on.

Notice also how the powers of j cause it to alternate between real and imaginary. You will see in a while that this property is very useful.

Addition and subtraction

For addition and subtraction we treat complex numbers as binomials and simply collect up the real and imaginary terms.

EXAMPLE 18.1

Evaluate:

(a) $(3 + j) + (-2 - j4)$
(b) $(4 - j5) - (3 - j2)$

If:

$$z_1 = 12 - j15, z_2 = -7 + j3, z_3 = 5 + j5$$

Find:

(c) $z_1 + z_2$
(d) $z_1 - z_2 + z_3$

Solution

(a) $(3 + j) + (-2 - j4) = 1 - j3$
(b) $(4 - j5) - (3 - j2) = 1 - j3$
(c) $z_1 + z_2 = (12 - j15) + (-7 + j3)$
$$= 5 - j12$$

(d) $z_1 - z_2 + z_3 = (12 - j15) - (-7 + j3) + (5 + j5)$
$$= 12 + 7 + 5 - j15 - j3 + j5$$
$$= 24 - j13$$

Multiplication

Multiplication is just as straightforward. It is the equivalent of a binomial expansion.

EXAMPLE 18.2

If:

$$z_1 = 2 - j5$$
$$z_2 = -1 + j3$$
$$z_3 = 5 + j$$
$$z_4 = 5 - j$$

Evaluate:

(a) $z_1 z_2$
(b) $3z_2 z_3$
(c) $z_1(z_2 + z_3)$
(d) $\frac{1}{2}(z_2 - z_3)^2$
(e) $z_3 z_4$

Solution

(a) $z_1 z_2 = (2 - j5)(-1 + j3) = -2 + j6 + j5 - j^2 15$
$\quad\quad = -2 + j11 - (-1)15 = -2 + j11 + 15$
$\quad\quad = 13 + j11$

(b) $3z_2 z_3 = 3(-1 + j3)(5 + j)$
$\quad\quad\quad = 3(-5 - j + j15 - 3) = 3(-8 + j14)$
$\quad\quad\quad = -24 + j42$

(c) $z_1(z_2 + z_3) = (2 - j5)(-1 + j3 + 5 + j)$
$\quad\quad\quad\quad = (2 - j5)(4 + j4) = 8 + j8 - j20 + 20$
$\quad\quad\quad\quad = 28 - j12$

(d) $\frac{1}{2}(z_2 - z_3)^2 = \frac{1}{2}(-1 + j3 - 5 - j)^2$
$\quad\quad\quad\quad = \frac{1}{2}(-6 + j2)^2 = \frac{1}{2}(-6 + j2)(-6 + j2)$
$\quad\quad\quad\quad = \frac{1}{2}(36 - j24 - 4) = 16 - j12$

(e) $z_3 z_4 = (5 + j)(5 - j) = 25 - j5 + j5 + 1 = 26$

The last problem is of special interest. The product of $(5 + j)$ and $(5 - j)$ is wholly real. We can use this.

Division

Division by an imaginary number is not possible but we can turn a complex number into a real number by applying the procedure you just saw.

Given a complex number in the general form, $a + jb$, see if you can formulate a rule to make it wholly real.

Multiply by $a - jb$

$a - jb$ is called the **complex conjugate** of $a + jb$. Any complex number multiplied by its complex conjugate becomes wholly real.

To form the conjugate of a complex number simply multiply the imaginary part by -1.

Now let's see how we can make this work for us when we look for the quotient of complex numbers.

EXAMPLE 18.3

Evaluate:

$$\frac{4 - j3}{2 + j}$$

The first step is to *rationalize the denominator* (make it a real number) by multiplying by its complex conjugate. In this case it is $2 - j$.

To ensure we are dealing with an equivalent fraction we must also multiply the numerator by $2 - j$.

The rest is clear.

Solution

$$\frac{4 - j3}{2 + j} = \frac{(4 - j3)(2 - j)}{(2 + j)(2 - j)}$$

$$= \frac{(4 - j3)(2 - j)}{4 - j2 + j2 + 1}$$

$$= \frac{(4 - j3)(2 - j)}{5}$$

$$= \frac{8 - j4 - j6 - 3}{5}$$

$$= \frac{5 - j10}{5} = 1 - j2$$

EXAMPLE 18.4

Evaluate:

(a) $\dfrac{10 - j6}{1 + j3}$

(b) $\dfrac{1}{4 - j3}$

(c) $\dfrac{3 + j5}{j2}$

(d) $\dfrac{1}{j}$

Solution

(a) $\dfrac{10 - j6}{1 + j3} = \dfrac{(10 - j6)(1 - j3)}{(1 + j3)(1 - j3)} = \dfrac{10 - j30 - j6 - 18}{1 + 9}$

$\quad = \dfrac{-8 - j36}{10} = -0.8 - j3.6$

(b) $\dfrac{1}{4 - j3} = \dfrac{4 + j3}{(4 - j3)(4 + j3)} = \dfrac{4 + j3}{16 + 9} = \dfrac{4 + j3}{25}$

$\quad = 0.16 + j0.12$

(c) $\dfrac{3 + j5}{j2} = \dfrac{(3 + j5) - j2}{j2(-j2)} = \dfrac{-j6 + 10}{4} = 2.5 - j1.5$

(d) $\dfrac{1}{j} = \dfrac{-j}{j(-j)} = \dfrac{-j}{1} = -j$

Problem (d) in the last set of examples leads to a useful rule that is reminiscent of one that applies to indices.

$$\frac{1}{j} = -j$$

and

$$\frac{1}{-j} = j$$

It also makes division by a number that is wholly imaginary a little easier. Take problem (c). We need only multiply by $-j$. Look at the next example.

EXAMPLE 18.5

Evaluate:

$$\frac{3 + j5}{j2}$$

by multiplying by $-j$.

Solution

$$\frac{3 + j5}{j2} = \frac{-j3 + 5}{2} = 2.5 - j1.5$$

So, when complex numbers are expressed in **Cartesian form**, otherwise known as **algebraic form**, we can carry out the normal operations of:

- addition and subtraction by collecting real and imaginary terms,
- multiplication by applying the rule of binomial expansion,
- division if we multiply by the **conjugate** of the denominator,
- taking powers remembering the all important property of $\sqrt{-1}$.

So much for the arithmetic. Now, in technology, $\sqrt{-1}$ is important enough that recently a contemporary scientist called it one of the seven wonders of the world!

Complex numbers are used in quantum physics, fluid mechanics, electronics and electrical engineering. Let's take a look at a simple electrical problem.

EXAMPLE 18.6

Impedance in electric circuits is the combined effect of resistance and reactance. Because power is only dissipated in the resistance and because reactance causes a 90° phase shift between voltage and current, we treat resistance as the real part and reactance as the imaginary part of the impedance.

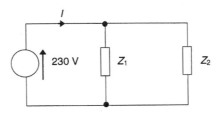

Figure 18.1 Circuit for Example 18.6

Two impedances, Z_1 and Z_2, are connected in parallel across a 230 V supply as shown in Fig. 18.1. Calculate the current drawn by the circuit given that:

$$Z_1 = 36\Omega + j21\Omega$$
$$Z_2 = 12\Omega - j45\Omega$$

The first step is to work out the combined impedance of the parallel branches. With just two impedances, this can be done by what is called the **product-over-sum** formula:

$$Z = \frac{Z_1 Z_2}{Z_1 + Z_2}$$

The second step is to use Ohm's law to find the circuit current:

$$I = \frac{V}{Z}$$

Solution

$$Z = \frac{Z_1 Z_2}{Z_1 + Z_2} = \frac{(36 + j21)(12 - j45)}{(36 + j21) + (12 - j45)}$$

$$= \frac{432 - j1620 + j252 + 945}{48 - j24} = \frac{1377 - j1368}{48 - j24}$$

$$= \frac{(1377 - j1368)(48 + j24)}{(48 - j24)(48 + j24)}$$

$$= \frac{66\,096 - j32\,616 + 32\,832}{2304 + 576}$$

$$= \frac{98\,928 - j32\,616}{2880} = 34.35 - j11.325$$

$$I = \frac{V}{Z} = \frac{230}{34.35 - j11.325}$$

$$= \frac{230(34.35 + j11.325)}{(34.35 - j11.325)(34.35 + j11.325)}$$

$$= \frac{7900.5 + j2604.75}{1179.992 + 128.256} = \frac{7900.5 + j2604.75}{1308.248}$$

$$I = 6.04 + j1.99\,A$$

The value obtained is in Cartesian form, so it is expressed in terms of its horizontal (real) and vertical (imaginary) components.

A couple of other points. You have seen that we can represent $\sqrt{-1}$ by the letter i (which stands for imaginary) or by the letter j preferred by engineers because of the possible confusion of i with electric current. So be prepared to see both.

Also, mathematicians tend not to worry about the position of i or j. For example, $3 + i2$ may be written as $3 + 2i$ just as correctly. So be prepared to see both of these too.

PRACTICE EXERCISE 18.1

Evaluate:

(a) $(3 + 5i) - (12 - 2i)$; (b) $(3 - i4)(4 + i7)$; (c) $(4 + i6)^2$; (d) $\frac{9j(1 - j)}{3j}$; (e) $\frac{15 - j4}{3 - j2}$

Given that: $z_1 = 20 - j15$, $z_2 = 3 + j$, $z_3 = -6j$, $z_4 = 6.53 + j(100\pi \times 25 \times 10^{-3})$ evaluate:

(f) $z_1 + z_4$; (g) $\frac{1}{z_1} + \frac{1}{z_2}$; (h) $\frac{z_2 z_3}{z_2 + z_3}$; (i) $\frac{12}{z_4}$; (j) $\frac{z_4}{z_3}$; (k) $\frac{z_3^3}{z_2^2}$; (l) $\frac{z_2}{z_3 z_4}$

Solutions

(a) $-9 + 7i$; (b) $40 + i5$; (c) $-20 + i48$; (d) $3 - j3$; (e) $4.08 + j1.38$
(f) $26.5 - j7.15$; (g) $(303 - j76) \times 10^{-3}$; (h) $3.18 - j0.706$; (i) $0.751 - j0.903$
(j) $-1.31 + j1.09$; (k) $130 + j173$; (l) $(48.1 + j18.7) \times 10^{-3}$

18.2 Argand diagram

In Chapter 1 you saw that we can represent the set of real numbers graphically by a **number line**.

The Swiss mathematician Jean Argand suggested that a similar graphical representation could be used for complex numbers. He gave us the **Argand diagram** on which a complex number of the form $x + jy$ can be represented as a point in a plane having the Cartesian coordinates (x, y).

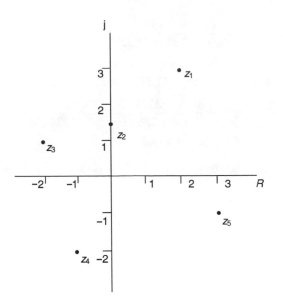

Figure 18.2 Argand diagram

The diagram consists of the x-axis, the real number line and a y-axis, the imaginary number line. It is usual to label these the R- and j-axes respectively.

The Argand diagram stresses the close similarity that exists between complex numbers and 2-D vectors. Complex numbers provide us with a means of expressing two-dimensional quantities although their engineering application is different to that of vectors.

PRACTICE EXERCISE 18.2

Express the points z_1 to z_5 shown in Fig. 18.2 in rectangular form. Remember, the 'rectangular', 'algebraic' and 'Cartesian' forms are all the same.

Solution

$$z_1 = 2 + j3, \quad z_2 = j1.5, \quad z_3 = -2 + j, \quad z_4 = -1 - j2, \quad z_5 = 3 - j$$

Phasors

When complex numbers are used to represent sinusoids (voltage and current in electrical systems or harmonic motion in mechanical systems) they are called **phasors**. We have dealt with them before but let's take a different approach.

When we draw a phasor on an Argand diagram we do not represent it as a point in the plane but a line with an arrowhead.

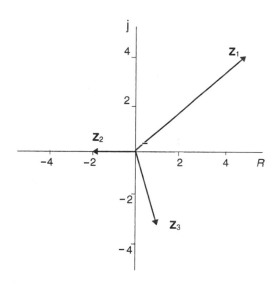

Figure 18.3 Three phasors

To distinguish phasors from other (real) number quantities we represent them by a bold letter, **Z**, like vectors and matrices. In hand-written form it is usual to put a wavy line above the letter $\tilde{Z}$.

The Argand diagram of Fig. 18.3 shows three phasors:

$$\mathbf{Z}_1 = 5 + j4,$$

$$\mathbf{Z}_2 = -2,$$

$$\mathbf{Z}_3 = 1 - j3$$

j-operator

Earlier, you saw the curious effect of multiplying a number repeatedly by j. It caused the number to alternate between real and imaginary:

$$j = j$$

$$j^2 = -1$$

$$j^3 = -j$$

$$j^4 = 1, \text{ etc}\ldots$$

The Argand diagram shows the graphical effect of this. In Fig. 18.4, repeated multiplication of the real number $z = 3$ by j causes it to change to j3, -3, $-j3$, 3 again and so on.

This anticlockwise rotation is called a 'j-operation', consequently j is often called the **j-operator**. Consider the phasor:

$$\mathbf{Z} = 2 + j4$$

A 'j-operation' gives:

$$j\mathbf{Z} = -4 + j2$$

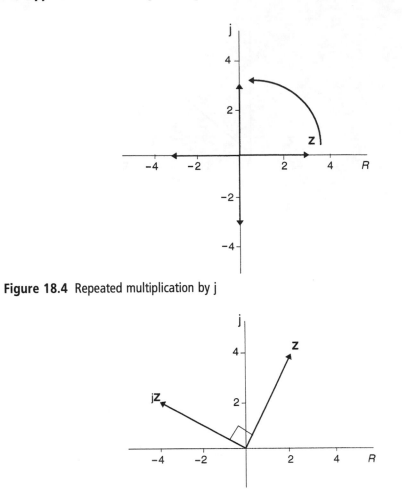

Figure 18.4 Repeated multiplication by j

Figure 18.5 j-operation

This rotates **Z** anticlockwise by 90° without changing the amplitude (length) of the phasor (see Fig. 18.5).

18.3 The polar form of a complex number

By some simple application of trigonometry, the Argand diagram can lead us to express a complex number in polar form.

For any phasor expressed in the form:

$$\mathbf{z} = x + \mathrm{j}y$$

we can take the magnitude of **z** to be the length r, and the angular displacement anticlockwise from the positive direction of the real axis to be θ.

Referring to Fig. 18.6, the real component x is related to r and θ by:

$$\cos\theta = \frac{x}{r} \quad \text{so} \quad x = r\cos\theta$$

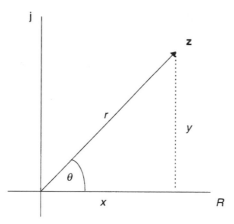

Figure 18.6 Polar form of a complex number

and the imaginary component:

$$\sin\theta = \frac{y}{r} \quad \text{so} \quad y = r\sin\theta$$

Substituting into the algebraic form of a complex number we find what is called the 'long' polar form of a complex number.

This is often abbreviated by engineers to the 'short' polar form. And in engineering the modulus is called the **amplitude** and the argument is called the **phase** of a phasor.

In long polar form: $\mathbf{z} = r\cos\theta + jr\sin\theta$ or $\mathbf{z} = r(\cos\theta + j\sin\theta)$

In short polar form: $\mathbf{z} = r\angle\theta$

where,

$$r = |\mathbf{z}|, \quad \text{the 'modulus' of } \mathbf{z}$$

and,

$$\theta = \arg \mathbf{z}, \quad \text{the 'argument' of } \mathbf{z}$$

You have seen this before so it should present no difficulty now.

Converting between rectangular and polar forms

It should be clear from Fig. 18.6 that the amplitude of a phasor is simply the hypotenuse of a right angled triangle.

To go from rectangular to polar form we can apply Pythagoras' theorem and use the tangent ratio to evaluate phase.

The only thing you need to be careful about, is to know into which quadrant your phasor extends. This you will know from the signs of the x- and y-components.

The method for taking a complex number from polar to rectangular form was shown in the previous section but, for convenience, here it is again.

To go from rectangular to polar form:

$$\mathbf{z} = x + jy \rightarrow \mathbf{z} = r\angle\theta$$

$$r = \sqrt{x^2 + y^2}$$

$$\theta = \arctan\frac{y}{x}$$

To go from polar to rectangular form:

$$\mathbf{z} = r\angle\theta \rightarrow \mathbf{z} = x + jy$$

$$x = r\cos\theta$$

$$y = r\sin\theta$$

EXAMPLE 18.7

Convert:

$$\mathbf{z} = 15.2 - j27.8$$

into polar form and back.

Solution

$$\mathbf{z} = r\angle\theta$$

$$= \sqrt{15.2^2 + 27.8^2}\angle - \arctan\frac{27.8}{15.2}$$

$$\mathbf{z} = 31.7\angle - 61.3°$$

$$\mathbf{z} = x + jy$$

$$= 31.7(\cos -61.3 + j\sin -61.3)$$

$$\mathbf{z} = 15.2 - j27.8$$

So it isn't difficult.

No doubt you will learn to use the appropriate calculator keys for this purpose. To go from 'x-plus-jy' to 'r-theta' form use R → P. To go from 'r-theta' to 'x-plus-jy' form use P → R.

Using the calculator is faster and you don't have to worry about which angular quadrant you are dealing with. However, you must appreciate by now how easy it is to make blunders when we rely entirely on a calculator. Sketching an Argand diagram is always a sensible rough check, whether you use the calculator to do your conversion or not.

PRACTICE EXERCISE 18.3

Convert the following phasors into polar form:

(a) $\mathbf{z}_1 = 5 + j3$; (b) $\mathbf{z}_2 = -3 - j7$; (c) $\mathbf{z}_3 = 140 - j2130$
(d) $\mathbf{z}_4 = (-27.5 + j31.2) \times 10^{-6}$

Convert the following phasors into Cartesian form:

(e) $z_5 = 5\angle -53.1°$; (f) $z_6 = 12.4\angle -131°$; (g) $z_7 = 230\angle 120°$

(h) $z_8 = 11.2\left[\cos\left(-\dfrac{2\pi}{3}\right) + j\sin\left(-\dfrac{2\pi}{3}\right)\right]$

Solutions

(a) $z_1 = 5.83\angle 31°$; (b) $z_2 = 7.62\angle -113°$; (c) $z_3 = 2135\angle -86°$

(d) $z_4 = 41.6 \times 10^{-6}\angle 131°$; (e) $z_5 = 3 - j4$; (f) $z_6 = -8.14 - j9.36$

(g) $z_7 = -115 + j199$; (h) $z_8 = -5.60 - j9.70$

If you had trouble with the last of these, the chances are that you forgot to put your calculator into radian mode. Not for the first time probably!

A complex number in polar form is a very suitable method of representing a phasor. Take the general form of a phasor:

$$a = A\sin(\omega t + \phi)$$

where A is amplitude and ϕ is phase angle, i.e. the modulus and argument of a complex number in polar form.

The reason why it is useful to have complex numbers expressed in alternative forms is because some operations are much more easily carried out in polar form and other operations can only be done in Cartesian form.

Addition and subtraction

This is only possible in Cartesian form. In Chapter 5 you learned how to add phasors by resolving horizontal and vertical components. Look back to Chapter 5, Example 5.8. Although you may not have been aware of it, in order to add the phasors, you first had to express them as complex numbers in Cartesian form then you converted the result back to polar form.

This is the normal procedure except that we now have a shorthand for complex numbers which enables us to do addition and subtraction more quickly. To add or subtract complex numbers expressed as phasors (or in polar form) convert them to Cartesian form, add their real and imaginary components and convert back to polar form.

EXAMPLE 18.8

Given the phasors:

$$v_1 = 35\sin(\omega t - 0.982)$$
$$v_2 = 24\sin(\omega t + 1.232)$$
$$v_3 = 43\sin(\omega t + 2.357)$$

Find:

$$v_r = v_1 + v_2 - v_3$$

Solution

First we express the phasors in short polar form. Convert to Cartesian form. Add. Convert back to polar form and write the resultant voltage phasor. Working in radians of course.

$$v_r = 35\angle - 0.982 + 24\angle 1.232 - 43\angle 2.357$$

$$= (35\cos -0.982 + j35\sin 0.982)$$

$$+ (24\cos 1.232 + j24\sin 1.232) - (43\cos 2.357 + j43\sin 2.357)$$

$$= (19.44 - j29.11) + (7.98 + j22.64) - (30.43 + j30.38)$$

$$= 57.85 - j36.85$$

$$= \sqrt{57.85^2 + 36.85^2} \angle - \arctan \frac{36.85}{57.85}$$

$$= 68.6\angle - 0.567$$

$$v_r = 68.6\sin(\omega t - 0.567)$$

Multiplication

This is interesting. Take any two complex numbers given in short polar form:

$$\mathbf{z}_1 = r_1 \angle \theta_1$$

$$\mathbf{z}_2 = r_2 \angle \theta_2$$

We first express them in long polar form. Then write the product:

$$\mathbf{z}_1 \mathbf{z}_2$$

Now there is some multiplication which amounts to no more than expanding brackets. We then collect the real and imaginary terms.

$$\mathbf{z}_1 = r_1 \angle \theta_1 = r_1(\cos \theta_1 + j \sin \theta_1)$$

$$\mathbf{z}_2 = r_2 \angle \theta_2 = r_2(\cos \theta_2 + j \sin \theta_2)$$

$$\mathbf{z}_1 \mathbf{z}_2 = r_1(\cos \theta_1 + j \sin \theta_1) r_2(\cos \theta_2 + j \sin \theta_2)$$

$$= r_1 r_2(\cos \theta_1 + j \sin \theta_1)(\cos \theta_2 + j \sin \theta_2)$$

$$= r_1 r_2(\cos \theta_1 \cos \theta_2 + j \cos \theta_1 \sin \theta_2 + j \sin \theta_1 \cos \theta_2 - \sin \theta_1 \sin \theta_2)$$

$$= r_1 r_2[(\cos \theta_1 \cos \theta_2 - \sin \theta_1 \sin \theta_2) + j(\cos \theta_1 \sin \theta_2 + \sin \theta_1 \cos \theta_2)]$$

Two compound angle formulae come to the rescue and we have the product of the two complex numbers in long polar form.

By the compound angle formulae:

$$\cos(A + B) = \cos A \cos B - \sin A \sin B$$

$$\sin(A + B) = \cos A \sin B + \sin A \cos B$$

We obtain:

$$\mathbf{z}_1\mathbf{z}_2 = r_1r_2[\cos(\theta_1 + \theta_2) + j\sin(\theta_1 + \theta_2)]$$

or,

$$\mathbf{z}_1\mathbf{z}_2 = r_1r_2\angle\theta_1 + \theta_2$$

The short polar form reveals the multiplication to be a very simple process indeed.

To multiply complex numbers in polar form, multiply the moduli and add the arguments:

$$r_1\angle\theta_1 \times r_2\angle\theta_2 = r_1r_2\angle\theta_1 + \theta_2$$

EXAMPLE 18.9

Evaluate:

$$(47\angle 37°)(22\angle -81°)$$

Solution

$$(47\angle 37°)(22\angle -81°) = 47 \times 22\angle 37 - 81$$
$$= 1034\angle -44°$$

Division

A similar analysis leads to the conclusion that division of numbers in polar form is just as easy.

To divide complex numbers in polar form, divide the moduli and subtract the arguments:

$$\frac{r_1\angle\theta_1}{r_2\angle\theta_2} = \frac{r_1}{r_2}\angle\theta_1 - \theta_2$$

EXAMPLE 18.10

If,

$$\mathbf{z}_1 = 20\angle 27°$$
$$\mathbf{z}_2 = 11\angle -49°$$

evaluate:

(a) $\mathbf{y} = \dfrac{1}{\mathbf{z}_1}$

(b) $\mathbf{z} = \dfrac{\mathbf{z}_1}{\mathbf{z}_2}$

Solution

(a) $\mathbf{y} = \dfrac{1}{\mathbf{z}_1} = \dfrac{1\angle 0}{20\angle 27} = \dfrac{1}{20}\angle 0 - 27 = 0.05\angle -27°$

(b) $z = \dfrac{z_1}{z_2} = \dfrac{20 \angle 27}{11 \angle -49} = \dfrac{20}{11} \angle 27 - (-49) = 1.82 \angle 76°$

Powers and roots

Here we look to the rule for multiplication in polar form. To take the square of a complex number we multiply it by itself, which requires us to square the modulus and multiply the argument by 2. Given:

$$\mathbf{z} = r\angle\theta$$

$$\mathbf{z}^2 = r\angle\theta \times r\angle\theta = rr\angle\theta + \theta$$

$$\mathbf{z} = r^2 \angle 2\theta$$

The general result for any complex number raised to the power of n is called **DeMoivre's theorem**.

As a mathematical theorem it is usual to express it in the long polar form.

In general:

$$\mathbf{z}^n = r^n \angle n\theta$$

or

$$\mathbf{z}^n = r^n (\cos n\theta + j \sin n\theta)$$

Now let's look to DeMoivre's theorem to see if it helps with roots of complex numbers.

Remember that the root of a number is just a fractional power. Take a square root. The theorem allows us to raise the modulus to the power of one half and multiply the argument by a half. Given:

$$\mathbf{z} = r\angle\theta$$

$$\sqrt{\mathbf{z}} = \sqrt{r\angle\theta}$$

$$\mathbf{z}^{\frac{1}{2}} = (r\angle\theta)^{\frac{1}{2}}$$

$$= r^{\frac{1}{2}} \angle \tfrac{1}{2}\theta$$

In other words, take the root of the modulus and divide the argument by 2.

DeMoivre's theorem works for any value of n. The method for finding powers and roots is exactly the same.

EXAMPLE 18.11

Find: $\sqrt[3]{\mathbf{z}}$
if, $\mathbf{z} = 27 \angle 120°$

Solution

$$\sqrt[3]{\mathbf{z}} = \mathbf{z}^{\frac{1}{3}} = 27^{\frac{1}{3}} \angle \tfrac{120}{3} = 3 \angle 40°$$

The root which we found in Example 18.11 is sometimes called the **principal root** because complex numbers have more than one root. But that is for a later study. For the moment we can confine ourselves to dealing with just the principal roots.

Now we know how to carry out all the operations of addition, subtraction, multiplication and taking powers and roots. The following summary is worth noting down:

- addition and subtraction is only possible in Cartesian form,
- multiplication and division is much easier in polar form,
- powers and roots are taken in polar form,
- in carrying out operations with complex numbers it is frequently necessary, and often desirable, to convert between polar and rectangular forms.

PRACTICE EXERCISE 18.4

Evaluate:

(a) $(240\angle 36°)(415\angle -85°)$;   (b) $\dfrac{(120\angle -72°)(78\angle 67°)}{(117\angle 6°)(415\angle -24°)}$;   (c) $\sqrt{(42\angle -152°)^3}$

Given: $i_1 = 27\sin\left(100\pi t - \dfrac{\pi}{3}\right)$, $i_2 = 38\sin\left(100\pi t - \dfrac{\pi}{6}\right)$, $i_3 = 31\sin\left(100\pi t + \dfrac{\pi}{4}\right)$

Evaluate:

(d) $i_1 + i_2$;   (e) $i_2 - i_3$;   (f) $i_1(i_2 + i_3)$;   (g) $\dfrac{i_1 - i_2}{i_3}$;   (h) $(i_1 + i_3)^2$

Given: $Z_1 = 27 - j15$, $Z_2 = 820 + j974$, $Z_3 = 12 - j0.33$

Evaluate and express the following in rectangular form:

(i) $Z = \dfrac{Z_1 Z_2}{Z_1 + Z_2}$;   (j) $Y = \dfrac{1}{Z_1} + \dfrac{1}{Z_2} + \dfrac{1}{Z_3}$;   (k) Z_1^3;   (l) $\sqrt[3]{Z_2}$

(m) If $z = 3(\cos 32° + j\sin 32°)$ find z^3 and express it in algebraic form.

(n) An electric circuit consists of a resistance, a capacitance and an inductance connected in series across a 230 V supply. Find the amplitude of the circuit current and its phase angle relative to the supply voltage given that:

resistance, $R = 36\,\Omega$, capacitive reactance, $X_C = \dfrac{1}{j0.08}\,\Omega$ and inductive reactance, $X_L = j27.4\,\Omega$

Solutions

(a) $99.6 \times 10^3 \angle -49°$;   (b) $193 \times 10^{-3} \angle 13°$;   (c) $272\angle 132°, 272\angle -48°$
(d) $62.8\sin(100\pi t - 0.74)$;   (e) $42.4\sin(100\pi t - 1.31)$;   (f) $1483\sin(100\pi t - 0.994)$
(g) $0.642\sin(100\pi t + 2.57)$;   (h) $1257\sin(100\pi t - 0.082)$;   (i) $27.2 - j14.3$
(j) $112 \times 10^{-3} + j17.4 \times 10^{-3}$;   (k) $29.5 \times 10^3 \angle -87.2°$;   (l) $10.8\angle 16.6°$
(m) $-2.82 + j26.9$;   (n) $5.90\,A\angle -22.5°$

Assignment XVI

Real and imaginary problems

This assignment requires you to use complex numbers to solve four engineering problems. Note the performance criteria which must be met for satisfactory completion.

- Numbers are represented on an Argand diagram.
- Engineering systems are represented using complex numbers.
- Calculations involving complex numbers are performed.

Do justice to $\sqrt{-1}$!

1. SIGNAL TRANSMISSION

A telecommunication signal launched on a transmission line is presented with an impedance that is called the **characteristic impedance** of the line. Characteristic impedance is determined in the following way:

$$Z_0 = \sqrt{\frac{R + j\omega L}{G + j\omega C}}\ \Omega$$

where:

R is the distributed resistance of the conductors in ohms per metre
G is the distributed conductance of the dielectric in siemens per metre
L is the distributed inductance of the conductors in henrys per metre
C is the distributed capacitance of the dielectric in farads per metre

and are called the primary constants of the line.

And $\omega = 2\pi f$ is the angular frequency in radians per second, as usual.

Like any impedance, the characteristic impedance of a transmission line determines the relationship between the signal current and the signal voltage.

Another important measure of the performance of a transmission line under signal conditions is called the **propagation constant** because it determines the speed at which a signal travels along the line. The propagation constant is also a function of the primary constants of the line, it is given by:

$$\gamma = \sqrt{(R + j\omega L)(G + j\omega C)}$$

A particular twin copper line has the following primary constants:

$$R = 55\,\text{m}\Omega\,\text{m}^{-1}$$
$$G = 39\,\text{nS}\,\text{m}^{-1}$$
$$L = 2.5\,\mu\text{H}\,\text{m}^{-1}$$
$$C = 8.1\,\text{pF}\,\text{m}^{-1}$$

Predict the characteristic impedance and the propagation constant if a 2 kHz test signal is transmitted down the line.

Taking the principal root to be the impedance, determine the current phasor if the test signal has a voltage of 5 $V_{\text{pk--pk}}$.

2. STARS AND DELTAS

In order to simplify network analysis, electrical engineers may use a technique called delta--star transformation to convert a delta circuit to its equivalent star circuit.

Figures 18.7 and 18.8 illustrate the general idea. Figure 18.7 shows three impedances connected in a delta configuration (so called because of its resemblance to the Greek letter) and Fig. 18.8 shows three impedances connected in star configuration.

In the star equivalent of the delta network the impedance between points A and B must appear the same as the impedance between points A and B of the delta

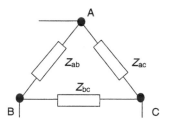

Figure 18.7 Delta configuration

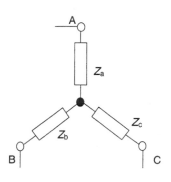

Figure 18.8 Star configuration

network. The same holds true for the impedances between points A and C and points B and C.

The following formulae can be used to determine the three impedances of the star network equivalent.

$$Z_a = \frac{Z_{ab}Z_{ac}}{Z_{ab} + Z_{ac} + Z_{bc}}$$

$$Z_b = \frac{Z_{ab}Z_{bc}}{Z_{ab} + Z_{ac} + Z_{bc}}$$

$$Z_c = \frac{Z_{ac}Z_{bc}}{Z_{ab} + Z_{ac} + Z_{bc}}$$

Given that in the original delta network the three impedances are:

$$Z_{ab} = 25.2 - j37$$

$$Z_{ac} = 41 + j14.6$$

$$Z_{bc} = 56.8 + j74.1$$

find, in Cartesian form, the three impedances of the star network and plot them as phasors on an Argand diagram.

3. POWER FACTOR CORRECTION

A reactive load such as a motor causes the load current to lag the supply voltage. This makes the distribution system work inefficiently because, for a given power, the motor draws more current than it needs. The cosine of the angle of phase between current and voltage is called the **power factor** and the power factor correction is the business of reducing the phase angle.

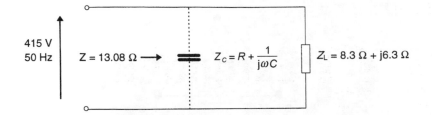

Figure 18.9 Z_L causes a lagging power factor

Figure 18.9 illustrates a situation where the inductive reactance of a load (Z_L) would cause a lagging power factor. One way of overcoming this is to connect enough capacitive reactance to overcome the inductive reactance. In Fig. 18.9 we strive for the ideal. We want just the right amount of capacitive reactance to cause the inductive reactance to cancel out. This means that, as far as the supply is concerned, the load impedance (Z) would be entirely resistive (real). A preliminary calculation shows that for maximum efficiency, the total impedance of the circuit should be 13.08 Ω.

Start with the product-over-sum formula for two parallel impedances and find out what value of capacitance (C) is required to bring this about.

4. THE POLAR QUOTIENT

Show that if, $\mathbf{z}_1 = r_1(\cos\theta_1 + j\sin\theta_1)$ and $\mathbf{z}_2 = r_2(\cos\theta_2 + j\sin\theta_2)$, then:

$$\frac{\mathbf{z}_1}{\mathbf{z}_2} = \frac{r_1}{r_2} \angle \theta_1 - \theta_2$$

19 Extended calculus

19.1 A brief review

Calculus is the Latin word for 'pebble'. The Romans used small stones or pebbles to do simple arithmetic or 'calculation'. Hence the general meaning of calculate and calculator.

The name given to the mathematical field known as 'calculus' you know, from Chapter 7, is a contraction of 'infinitesimal calculus' meaning calculation with infinitesimally small numbers. This gives a key to the whole thing.

Integral calculus goes back a long time, to the ancient Greeks, who calculated areas and volumes by treating them as wholes made up from an infinite number of infinitely small parts. For instance, the volume of a cuboid is treated as the integral (whole) of an infinite number of plane areas of infinitesimal thickness.

Newton gave us differential calculus which uses 'infinitesimals' for dealing with rates of change. He also provided the fundamental theorem of calculus which links integration and differentiation as inverse operations.

This flirting with infinity is a fascinating part of calculus but there is a very important practical side to it. For an engineer, calculus is a ready made tool for modelling dynamic systems, i.e. all those systems that change over time.

In this chapter and the next you will be extending your knowledge of this topic but, first some reinforcement of what you know so far.

We have a set of standard derivatives and integrals which allow us to deal with some of the more common mathematical functions by inspection. In other words there is no need to go back to first principles each time. There is also a set of rules for differentiation: the **chain rule**, the **product rule** and the **quotient rule**. All these appear in the summary on pages 213 and 214.

PRACTICE EXERCISE 19.1 (REVISION)

(a) Find the gradient of the curve of $y = 15x^3 - \dfrac{3}{\sqrt{x}} - \cos 3x + \ln x$ at the point where $x = 2$.

(b) Find the first derivative of $y = \ln 3x^2$.

(c) Differentiate $y = 3 \tan \theta$.

(d) Find the second derivative of $y = (2x^2 - e^{3x})^2$.

(e) Find the maximum and minimum values of y given that $y = \frac{1}{3}x^3 - \frac{1}{2}x^2 - 6x + 6$.

(f) Find $\int (3 \sin 2\theta + 5\theta - 2 e^{4\theta}) \, d\theta$.

(g) Find the distance travelled by an object whose speed is described by $v = 6 + 9.8t$ over the interval from $t = 5$ to $t = 10$ s.

(h) Find the average value of the voltage which varies according to $v = 200(1 - e^{-2t})$ V over the period of time from 0 to 2.5 s.

(i) Find the r.m.s. value of $v = 8 \sin 2\theta$.

Solutions

(a) 180.2; (b) $\dfrac{dy}{dx} = \dfrac{2}{x}$; (c) $\dfrac{dy}{d\theta} = 3 \sec^2 \theta$; (d) $\dfrac{d^2y}{dx^2} = 48x^2 - 4 e^{3x}(9x^2 + 12x + 2 - 9 e^{3x})$

(e) 13.3, −7.5; (f) $\frac{1}{2}(5\theta^2 - 3 \cos \theta - e^{4\theta} + D)$; (g) 398 m; (h) 160 V; (i) 5.66 V

19.2 Functions and their curves

A mathematical function is a rule which links two variables. If y is some function of x then y is the dependent variable and x is the independent variable.

A 'function of x' means any expression whose numerical value is determined by x.

The whole set of values which we allow the independent variable to take is called the **domain** of the function. For example the domain of the quadratic function illustrated below may be the whole set of all real numbers.

If,

$$y = f(x)$$

and,

$$f(x) = x^2$$

then,

$$y = x^2$$

where:

$$0 > x \geq 0$$

is the domain of the function.

Sometimes we may need to restrict the domain of a function. For example the volume of a sphere is a function of its radius. It would be logical to restrict the domain of this function to zero and all positive real numbers.

The volume of a sphere is a function of its radius:

$$V = \tfrac{4}{3}\pi r^3$$

where

$$0 \le r$$

In another situation it may be known that a relationship is only linear over a prescribed interval. So if speed as a function of time is known to be linear only over the first 5 seconds then the domain of the function is $0 \le t \le 5$.

If the domain of a function is not specified we take it to be the largest set of real numbers that make practical sense.

EXAMPLE 19.1

What is the largest domain of

(a) $f(x) = \sqrt{x}$?

(b) $f(x) = \dfrac{1}{x}$

(c) $f(x) = \arcsin x$

(d) $f(x) = \ln x$

Solutions

(a) $0 \le x$
(b) $x \ne 0$
(c) $-1 \le x \le 1$
(d) $0 < x$

A function may not necessarily be a single algebraic expression. An important example is where the function of x is the modulus of x. The modulus of x is equal to x if x is equal to or greater than zero *and* it is equal to $-x$ if x is less than 0.

$$f(x) = |x|$$

$$|x| = \begin{cases} x & \text{if } x \ge 0 \\ -x & \text{if } x < 0 \end{cases}$$

The whole set of values which the dependent variable can take is called the **codomain** of the function. You can think of the domains and codomains as the sets of inputs and outputs of a function.

$$\text{Domain} \to \boxed{f(x)} \to \text{Codomain}$$

A couple of other points which you may have forgotten.

For repeated calculations with a known function we can use functional notation with specified values. If,

$$f(x) = e^x$$

then,

$$f(0) = 1 \quad f(1) = 2.72 \quad f(2) = 7.39$$

Derivatives of functions are often denoted by **primes** (dashes). For example if the function of x is the sine of x then the first derivative, f-dashed-x, is the cosine of x and the second derivative, f-double-dashed-x, is minus the sine of x and so on. For obvious reasons we abandon the dashes with the fourth derivative and upwards. If,

$$f(x) = \sin x$$

then,

$$f'(x) = \cos x$$
$$f''(x) = -\sin x$$
$$f'''(x) = -\cos x$$
$$f^4(x) = \sin x$$

Finally, you know that we can represent different functions by different letters. So with the function of a function rule for differentiation, the 'inner' function, u, could be $g(x)$ and the 'outer' function, y, could be $f(u)$.

Function of a function rule:

$$\frac{d}{dx} f(g(x)) = f'(g(x)g'(x))$$

Standard functions

Some of these are well known to you but let's reiterate because familiarity with the curves of standard functions is a great aid to curve sketching of other less well known functions.

There are first, the linear, quadratic and cubic polynomials shown in generalized forms in (a), (b) and (c) of Fig. 19.1. Key points are illustrated.

Remember that the roots of quadratic and cubic functions may be fewer than shown. A quadratic may have no (real) roots. A cubic may only have one (real) root.

Likewise, the turning points of a cubic may just be a **point of inflexion**. At a point of inflexion the gradient goes to zero but does not reverse.

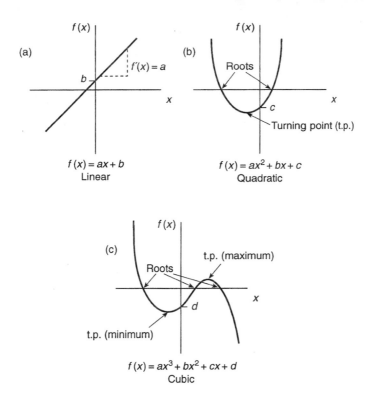

Figure 19.1 Curves of (a) linear, (b) quadratic and (c) cubic polynomials

$\dfrac{a}{x}$ CURVE

The curve of this function is called the **rectangular hyperbola** (see Fig. 19.2).

Key points occur when $|x| = a$.

The curve is **asymptotic**. When the modulus of x is very small the modulus of y is very large and vice-versa. The curve never crosses the x- or y-axis.

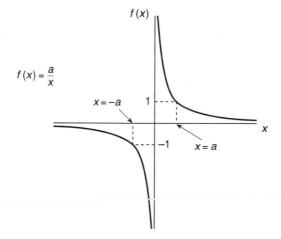

Figure 19.2 Curves of $y = a/x$

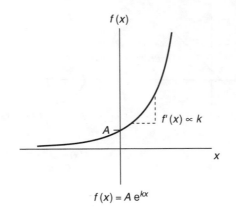

$f'(x) \propto k$

A

$f(x) = A e^{kx}$

Figure 19.3 Exponential growth curve

EXPONENTIAL

These are very special because they model the behaviour of many physical systems which vary over time.

Exponential curves have a gradient that is proportional to the value of the function.

The standard exponential growth curve is illustrated in Fig. 19.3. Note the key point where the curve crosses the y-axis.

LOGARITHMIC

The logarithmic function is the inverse of the exponential.

It is clear from Fig. 19.4 that the domain of this function is restricted to all positive values of x and *excludes* 0.

The key point occurs when $x = 1$. Below 1 the codomain becomes negative and above 1 it is positive. Note the asymptotic behaviour when x is very small.

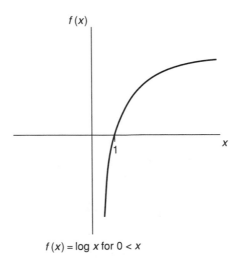

$f(x) = \log x$ for $0 < x$

Figure 19.4 Curve of a logarithmic function

TRIGONOMETRIC

These are important because they are periodic functions. They model systems which exhibit a regularly repeating pattern.

They are the basis for describing the behaviour of simple harmonic motion and phasors for example (see Fig. 19.5).

The key points on the curve of the sine and cosine functions occur at

$$x = 0, \quad |x| = \frac{\pi}{2}, \quad |x| = \pi$$

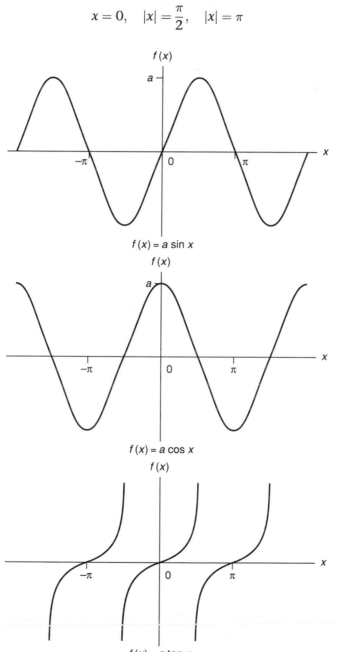

$f(x) = a \sin x$

$f(x) = a \cos x$

$f(x) = a \tan x$

Figure 19.5 Trigonometric curves

Remember also that the gradient of one is proportional to the corresponding value of the other.

It is interesting to note that with the sine and cosine functions it is the codomain that is restricted. If $a = 1$ then,

$$-1 \leq f(x) \geq 1$$

The tangent curve is altogether different, although it is derived from the sine and cosine functions. Remember the identity:

$$\tan x \equiv \frac{\sin x}{\cos x}$$

This leads to points on the curve that are **undefined**. Values of x which are odd-number multiples of $|\pi/2|$ give an output which is irrational.

CIRCULAR

Fundamentally, all trigonometric functions are circular functions (see Fig. 19.6). The simplest equation of a circle is for a circle that has its centre at the origin. The equation is Pythagorean:

$$x^2 + y^2 = r^2$$

This is the form in which it is normally written, with both the variables on the left.

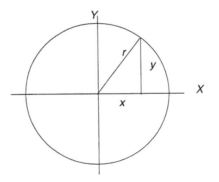

Figure 19.6 Circular function

EXAMPLE 19.2

Sketch the curve of the circle whose equation is:

$$(x - 4)^2 + (y - 3)^2 = 4$$

Transposing to make y the function of x gives:

$$y = 3 \pm \sqrt{4 - (x - 4)^2}$$

From the equation it is apparent that the circle has a radius of 2 units and it is **translated** 4 units in the direction of the x-axis and 3 units in the direction of the y-axis. Only four key points are needed. We all know the shape of a circle after all!

$$f(6) = 3, \quad f(4) = 5 \text{ and } 1, \quad f(2) = 3$$

Solution

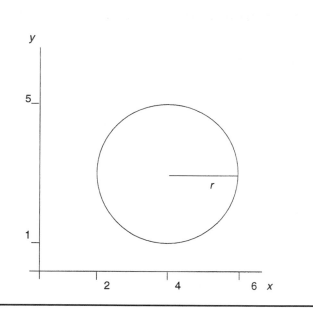

So the more general form of the equation of a circle is:

$$(x+a)^2 + (y+b)^2 = r^2$$

If a and b are positive they translate the centre of the circle from the origin to a point in the negative directions of the x- and y-axes.

ELLIPTICAL

The equation of this function is also Pythagorean

$$\frac{x^2}{a^2} + \frac{y^2}{b^2} = 1$$

where a and b are called the **semi-major** and **semi-minor axes** respectively (see Fig. 19.7).

Note, when a and b are equal, the ellipse becomes a circle.

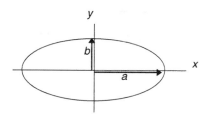

Figure 19.7 Ellipse

Transformations

From our knowledge of the curves of standard functions it is possible to deduce the curves of other functions. For example, if we know the curve of some function, $f(x)$, then it is possible for us to determine the curve of $af(x)$. $af(x)$ is called a **transformation** of $f(x)$. Let's take a look at some simple examples.

THE TRANSFORMATION $af(x)$

The effect of this transformation is to expand the curve of $f(x)$ parallel to the y-axis. For this reason a is called the scaling factor of y.

EXAMPLE 19.3

Given that $y = 2x - 1$ sketch the curve of the function and its transformation, $2f(x)$.

The curve of the function is a straight line which passes through:

$$f(0) = -1$$
$$f(2) = 3$$
$$f(-2) = -5$$

The $2f(x)$ transformation will pass through corresponding points at:

$$2f(0) = -2$$
$$2f(2) = 6$$
$$2f(-2) = -10$$

Solution

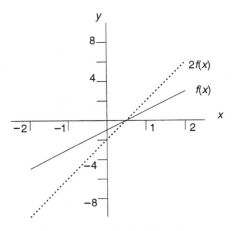

Dealing with a linear curve is straightforward enough. We only need two sets of coordinates; three were chosen in the last example because the third point provides a useful check. But what about a non-linear function? Let's see.

EXAMPLE 19.4

Sketch the curve of $y = x^3$ and deduce its transformation $-3f(x)$.

When x is very large and positive, y is very large and positive. When x is large and negative so is y. Key points are:

$$f(0) = 0 \quad \text{so} \quad -3f(0) = 0$$
$$f(1) = 3 \quad \text{so} \quad -3f(1) = -9$$
$$f(-1) = 1 \quad \text{so} \quad -3f(-1) = 9$$

Solution

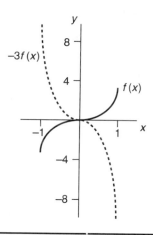

So the effect of this transformation is a 'y-expansion' by the scaling factor a and if a is negative, the expansion also involves a reflection about the x-axis. What about values of a which are less than 1?

The expansion becomes a contraction

THE TRANSFORMATION $f(x) + a$

The effect of this transformation is to translate the function in the positive or negative direction of the y-axis. It is quite straightforward. Think of it as a 'y-shift' on an oscilloscope.

EXAMPLE 19.5

Apply the transformation $f(\theta) + 3$ to:

$$y = \cos\theta$$

We simply add the value of 3 to key values. The average value of the curve of the function translates from 0 to 3, the peak value from 1 to 4 and the negative peak from −1 to 2.

Solution

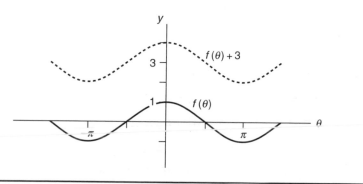

It should be clear that with this type of transformation a negative value of a would have the effect of translating the curve in a downward direction.

THE TRANSFORMATION $f(x+a)$

This involves a translation in the direction of the x-axis. So the curve of the function moves a units to the right or left.

EXAMPLE 19.6

Apply the transformation $f(x+2)$ to the curve of $f(x) = \ln x$

$$f(1) = 0 \qquad\qquad f(1+2) = 1.1$$
$$f(2) = 0.7 \qquad\qquad f(2+2) = 1.4$$
$$f(0) \text{ is undefined} \qquad f(0+2) = 0.7$$
$$f(-1) \text{ is undefined} \quad f(-1+2) = 0$$

This is a little more tricky but the general result is important.

Solution

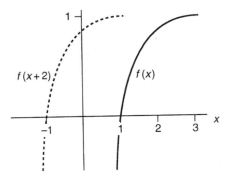

If a is positive it translates the curve in the negative direction of the x-axis. Conversely if a is negative it translates the curve in the positive direction of the x-axis.

EXAMPLE 19.7

Apply the transformation $f(x-1)$ to the curve of $y = x^2$

The key points here are:

$$f(0) = 0 \qquad f(0-1) = 1$$
$$f(4) = 16 \qquad f(4-1) = 9$$
$$f(-2) = 4 \qquad f(-2-1) = 9$$

Solution

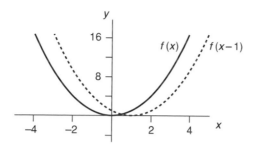

So there we have it. The transformation $f(x+a)$ brings about a translation in the positive direction of the x-axis if a is negative. You can think of this as the x-shift on an oscilloscope.

THE TRANSFORMATION $f(ax)$

This transformation expands the curve along the x-axis because a is a scaling factor of x. But there is a little more to it. Look at the following example.

EXAMPLE 19.8

Given that $y = x^2$ sketch the curve of this function and its transformation $y = (2x)^2$.

Key points:

$$f(0) = 0 \quad f[2(0)] = 0$$
$$f(4) = 16 \quad f[2(4)] = 64$$
$$f(-2) = 4 \quad f[2(-2)] = 16$$

Solution

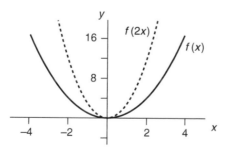

So the 'expansion' turns out to be a 'compression' which increases the gradient. Speculate. What do you think might cause an actual expansion with a corresponding decrease in gradient?

The curve would expand along the x-axis if $a < 1$

Let's see it happen with another function.

EXAMPLE 19.9

Apply the transformation $f(0.5x)$ to the function $f(e^x)$.

$$f(0) = 1 \qquad f[0.5(0)] = 1$$
$$f(-1) = 0.4 \quad f[0.5(-1)] = 0.6$$
$$f(1) = 2.7 \qquad f[0.5(1)] = 1.6$$

Solution

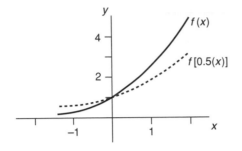

So the curve does indeed expand when a is a scaling factor with a value of less than 1. The expansion is apparent from the fall in the gradient of the curve.

EXAMPLE 19.10

Apply the transformation $f(2x)$ to the function, $f(x) = \sin x$

$$f(0) = 0 \qquad f[2(0)] = 0$$
$$f\left(\frac{\pi}{2}\right) = 1 \quad f\left[2\left(\frac{\pi}{2}\right)\right] = 0$$
$$f\left(\frac{\pi}{4}\right) = 0.7 \quad f\left[2\left(\frac{\pi}{4}\right)\right] = 1$$

Solution

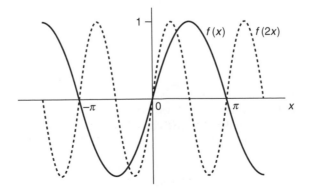

The transformation $f(2x)$ has the effect of doubling the frequency of a periodic function. Note again, the effect on the gradient of the curve, it becomes steeper.

Can you predict the effect of this transformation if a is less than 1?

Yes. It would reduce the gradient causing an x-expansion. With a periodic function it would reduce the frequency. See the solution to Example 19.10.

EXAMPLE 19.11

Apply the transformation $y = f(-x)$ to

$$f(x) = x^2 - 2x - 1$$

Key points:

$$f'(x) = 2x - 2$$

At t.p. $2x - 2 = 0, x = 1$

$$
\begin{aligned}
f(1) &= -2 & f[-1(1)] &= 2 \\
f(0) &= -1 & f[-1(0)] &= -1 \\
f(3) &= 2 & f[-1(3)] &= 16 \\
f(-2) &= 7 & f[-1(-2)] &= -1 \\
& & f[-1(-3)] &= 2
\end{aligned}
$$

Solution

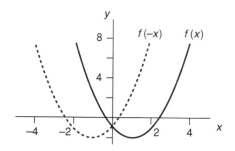

So with the transformation $f(ax)$ when a is negative the curve of the function undergoes a reflection about the y-axis.

Some transformations are not possible. Try the transformation $f(-x)$ on the function $\ln x$. It does not work because the transformation does not lie within the domain of the function.

This is a good point at which to summarize the transformations which we have considered (Table 19.1).

Table 19.1. Summary of simple transformations

Transformation	Function	Result	Comment
$af(x)$			y-expansion
$-f(x)$			Reflection about the x-axis
$f(x)+a$			Translation in the positive direction of the y-axis
$f(x+a)$			Translation in the negative direction of the x-axis
$f(ax)$			x-compression
$f(-x)$			Reflection about the y-axis

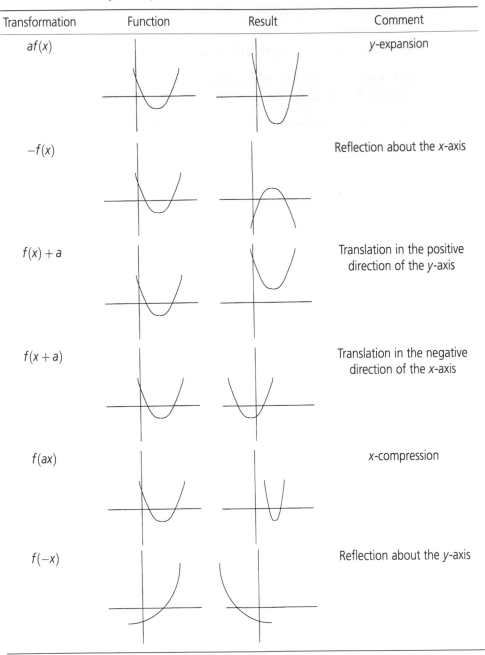

Classes of functions

One of these is the **periodic** class of functions. Periodic functions are those which are repetitive. Those that are familiar to you are the circular functions: equations of the circle and ellipse, and those which are derived from the circle, the sine, cosine and tangent. They all have a period of 2π.

There are other classifications of functions.

CONTINUOUS AND DISCONTINUOUS FUNCTIONS

A **continuous** function is defined as one for which a small change in x cannot produce a large change in $f(x)$. The graph of such a function has no breaks or jumps. Examples are the linear, and quadratic polynomials, and the sine and cosine functions. The derivative or integral of a continuous function can be evaluated over the entire range of the function.

A **discontinuous** function is one whose graph does have breaks or jumps.

One example is the tangent curve. Discontinuities occur when the modulus of x is an odd number multiple of $\pi/2$ (Fig. 19.8). The gradient of the curve cannot be found at this point. Nor is it possible to evaluate an integral over an interval which contains the discontinuity.

$$f(x) = \tan x$$

is discontinuous at:

$$x = \frac{\pi}{2}, \quad \frac{3\pi}{2}, \quad \frac{5\pi}{2} \dots$$

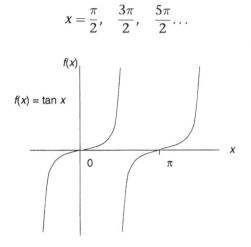

Figure 19.8 Tangent curve showing discontinuities

EXAMPLE 19.12

Find the area under the curve of $y = \tan x$

(a) between, $x = 0$ and $x = \pi$

(b) between, $x = 0$ and $x = \dfrac{\pi}{2}$

(c) between, $x = 0$ and $x = 1.570796326$

given that,

$$\int \tan x \, dx = -\ln(\cos x) + C$$

Solution

(a) Area $= \displaystyle\int_0^\pi \tan x \, dx = \Big[-\ln(\cos x) \Big]_0^\pi = [-\ln(\cos \pi)] - [-\ln(\cos 0)] = ?$

(b) Area $= \left[-\ln\left(\cos\dfrac{\pi}{2}\right) \right] - [-\ln(\cos 0)] = \ ?$

(c) Area $= [-\ln(\cos 1.570796326)] - [-\ln(\cos 0)] = 20.9 \text{ units}^2$

To understand the results look at the curve in Fig. 19.8. Any area which takes in the discontinuity at $\pi/2$ will become infinite and infinity is not mathematically defined.

The result that makes sense is the one in part (c). It is the area which comes very close to the discontinuity.

Here we came to within 10^{-10} of a radian from the discontinuity!

Visualize the height of the area found in the last example. My calculator gives me: $\tan 1.570796326 = 1.25 \times 10^9$.

Ponder the enormity of this result. Compare it say, with the tangent of 1.57. A slight twitch in x sends the function careering off toward infinity.

When you think that the tangent of $\pi/4$ is just 1, you realize how dramatically the gradient increases as the function approaches the discontinuity. Discontinuities are regions of chaos. An engineered system which is modelled by a discontinuous function may behave unpredictably.

Whilst the study of instability in systems is not new, for a long time scientists and engineers felt uncomfortable with the idea of potential chaos in systems which hitherto had appeared to be quite predictable. The study of chaos is a new development in mathematics and physics.

There are other examples of discontinuous functions. The rectangular hyperbola has a discontinuity at $x = 0$ and so does the graph of the log function.

When functions are specified by more than one expression their graphs may have 'corners'. Take

$$f(x) = \begin{cases} -2 & \text{if } -\pi < x < 0 \\ \ \ 3 & \text{if } 0 < x < \pi \end{cases}$$

which is the square wave sketched out in Fig. 19.9.

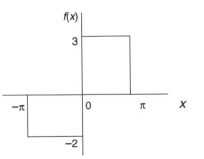

Figure 19.9 Square wave

Here, differentiation makes no sense. The gradient switches from zero to infinity at the corners.

Integration is possible if the intervals between discontinuities are treated separately.

EXAMPLE 19.13

Find the area under the curve of the function shown in Fig. 19.9 between $-\pi$ and π.

The total area is the sum of the two areas lying above and below $x = 0$.

Solution

$$\text{Area} = \int_{-\pi}^{\pi} f(x)\,dx$$

$$= \int_{-\pi}^{0} f(x)\,dx + \int_{0}^{\pi} f(x)\,dx$$

$$= \int_{-\pi}^{0} -2\,dx + \int_{0}^{\pi} 3\,dx$$

$$= \left[-2x \right]_{-\pi}^{0} + \left[3x \right]_{0}^{\pi}$$

$$= (0) - (2\pi) + (3\pi) - (0) = \pi$$

Studies of heat transfer show that, in the main, temperature is a linear function of heat. It is also a discontinuous function.

When water is supplied with heat energy its temperature rise is proportional until it undergoes a change in phase. When water changes from solid to liquid the graph of the function undergoes a discontinuity. This happens because, instead of raising the temperature, heat energy (latent heat) is used to bring about a change in state (see Fig. 19.10).

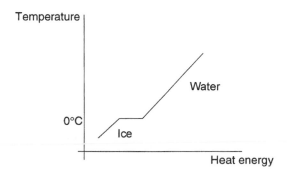

Figure 19.10 Discontinuity in the heat energy–temperature equation

ODD AND EVEN FUNCTIONS

A function is defined as **odd** if:

$$f(-x) = -f(x)$$

An odd function is easily recognized from its graph because it is always *symmetrical about the origin.*

A function is defined as **even** if:

$$f(-x) = f(x)$$

The graph of an even function is *symmetrical about the y-axis.*

PRACTICE EXERCISE 19.2

Identify the functions whose curves are sketched in Fig. 19.11 and state whether they are odd or even.

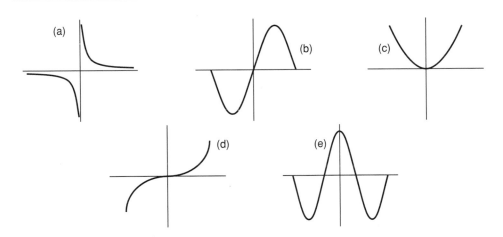

Figure 19.11 Curves for Practice Exercise 19.2

Solutions

(a) $f(x) = \dfrac{a}{x}$, odd;   (b) $f(x) = \sin x$, odd;   (c) $f(x) = x^2$, even;

(d) $f(x) = x^3$, odd;   (e) $f(x) \cos x$, even

Sketch the curve of the log function. Is it an odd or even function?

It is neither. So, there are functions which are neither odd or even; the exponential is another important example.

The terminology, *odd* and *even*, derives from the fact that odd number powers of a variable produce positive and negative values while even number powers are always positive. So functions which are linear, cubic, quintic, etc. are odd functions, while quadratic and quartic are even functions.

Inverse functions

It is sometimes useful to interchange the dependent and independent variables of a function. For example, a function that expresses distance as a function of time might be used to predict the time it takes an object to move a certain distance. Expressing time as a function of distance is the **inverse** of distance as a function of time.

The graphs of inverse functions are quite common in engineering, sometimes for dubious reasons. Consider the extension of a spring which obeys Hooke's law.

It is customary to plot the graph which links force and extension with force as the dependent variable and extension as the independent variable. This implies that force is a function of extension, which is nonsense.

The physical truth is that extension is a function of applied force but historically we have always used the inverse of the function (see Fig. 19.12).

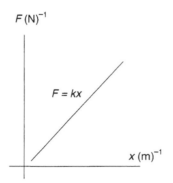

Figure 19.12 Graph of force and extension

The reason for this is, probably, that the gradient of the graph of force as a function of extension gives a numerically convenient measure of the constant of a spring: its **stiffness** (k) in Nm^{-1}. If we used the function rather than its inverse the gradient would give us the **compliance** of the spring which is the reciprocal of stiffness in mN^{-1}. For practical springs this would be a number which is inconveniently small.

The procedure for finding the inverse of a function is quite straightforward. We transpose the equation of the function and interchange variables.

EXAMPLE 19.14

Find the inverse function of $y = x + 3$.

Solution

$$f(x) = y = x + 3$$
$$x = y - 3$$
$$y = x - 3 = f^{-1}(x)$$

Note how the inverse is signified.

With some functions an inverse may not exist.

EXAMPLE 19.15

Find $f^{-1}(x)$ of $y = 2x^2 - 1$.

Solution

$$f(x) = y = 2x^2 - 1$$

$$x^2 = \frac{y+1}{2}, \quad x = \pm\sqrt{\frac{y+1}{2}}$$

$$y = \pm\sqrt{\frac{x+1}{2}} \neq f^{-1}(x)$$

If we look for the inverse of a quadratic function we obtain two possible solutions. This is not an inverse because for each value in its domain, the original function only generated one unique value in its codomain.

However, if the domain of the original function were restricted to $0 \le x$, then the inverse function would be valid.

Is there an inverse for a cubic or does it present the same problem as a quadratic? Let's see.

EXAMPLE 19.16

Find the inverse function of $y = x^3$.

Solution

$$f(x) = y = x^3$$

$$x = y^{\frac{1}{3}}, \quad x = \sqrt[3]{y}$$

$$y = \sqrt[3]{x} = f^{-1}(x)$$

An inverse for a cubic does exist because the cube root of a number is a single unique value.

We know of course that the exponential and logarithmic are inverse functions.

If,

$$f(x) = e^x$$

$$f^{-1}(x) = \ln e^x$$

$$f^{-1}(x) = \ln x$$

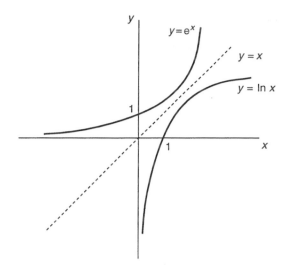

Figure 19.13 Exponential and logarithmic curves

Figure 19.13 shows that the exponential and logarithmic curves are reflections about the axis formed by:

$$y = x$$

PRACTICE EXERCISE 19.3

Sketch curves of the following functions and their inverses and decide in each case whether they are reflections about the axis formed by $y = x$.

(a) $y = 2x + 1$;   (b) $y = x^2 - 4$ for $x \geq 0$;   (c) $y = 2x^3 + 1$

(d) $y = \sin x$ for $-\dfrac{\pi}{2} \leq x \leq \dfrac{\pi}{2}$

Solutions

The curves that represent the solutions to this exercise appear at the end of the chapter. They show the inverses of all the functions to be reflections about the $y = x$ axis.

The curves of the inverse trigonometric functions are sketched out in Fig. 19.14.

Note the domains of these functions:

$$y = \arcsin x \quad \text{with} \quad -1 \leq x \geq 1$$
$$y = \arccos x \quad \text{with} \quad -1 \leq x \geq 1$$
$$y = \arctan x$$

The domain of the arctangent is not restricted.

However, the codomains of each of the inverse trig. functions are restricted to the **principal angles**. Figure 19.14 shows that the arcsin and arctan function

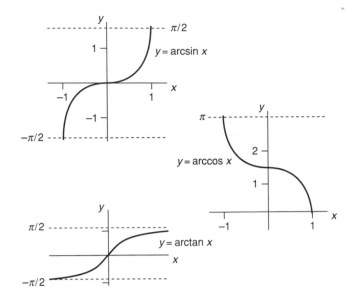

Figure 19.14 Curves of the inverse trigonometric functions

have a 'ceiling' at $\pi/2$ and a 'floor' at $-\pi/2$. Similarly, the codomain of the arccos function lies between 0 and π.

PRACTICE EXERCISE 19.4

State which of the inverse trigonometric functions sketched in Fig. 19.14 is even and which is odd.

Solution

$\arcsin x$ is odd, $\arccos x$ is neither even nor odd, $\arctan x$ is odd

19.3 Curve sketching

Sketching the curves of functions is a useful way of gaining an understanding of the relationship between two variables. Sometimes it is more than that. In Chapter 10 we found that it was an important initial step in applying the Newton–Raphson method to find the roots of a cubic equation. Curve sketching is also a skill which is useful in **curve fitting**. Curve fitting is the practice of formulating mathematical models from graphs of empirical data.

The art of curve sketching is not the business of graph plotting. The whole idea is to find the approximate coordinates of the key points of the curve in the region of interest, quickly. The following is a summary of the things that can help us.

- The curves of the standard functions and their domains.
- Transformations.
- The classes of functions: continuous, discontinuous, odd, even, periodic.
- The inverse functions.

A useful strategy for curve sketching

1. Consider if the given function is a standard function or a simple combination or transformation of a standard function.

2. How does the function behave when x is very large and positive and very large and negative? What is the domain of the function?

3. Try $x = 0$. This shows where the curve crosses the y-axis. Look for the point where the curve crosses the x-axis. Remember that the curve may not cross any axis.

4. Look for discontinuities. How does the function behave in the region of a discontinuity.

5. Find the coordinates of the turning points or points of inflexion, if they exist.

6. Sketch the curve by joining up the known coordinates with a smooth curve that follows the shape of the identified function. If necessary, choose other values of x to fill-in those parts of the curve of which you are not sure.

EXAMPLE 19.17

Sketch the curve of $y = (x - 3)^2$.

This is a quadratic whose curve we know to be a parabola. Expanding the brackets is simple because this is a perfect square so we obtain: $y = x^2 - 6x + 9$.

The quadratic term is positive so the parabola is the 'right-way-up' and has a local minimum. The linear term is negative so the parabola is right-shifted.

When x is very large (positive or negative) the x^2 term dominates and y is also very large and always positive.

When $x = 0$, $y = 9$.

When $y = 0$, we find there is a repeated root, $x = 3$. This is also the minimum turning point.

Solution

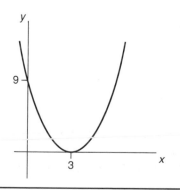

EXAMPLE 19.18

Sketch the curve of $y = -2x^3 + 2x^2 - 5$.

Consider the dominant term, which is the cubic. The transformation $-f(x)$ reflects about the x-axis. However, we also need to consider the quadratic and constant terms which introduce a transformation of the type $f(x) + a$. This would normally be a positive y-translation. But here, because of the reflection about the x-axis the curve undergoes a negative y-translation.

With x very large positive, y is very large negative.

With x very large negative, y is very large positive.

$$y(0) = -5$$

$$\frac{dy}{dx} = -6x^2 + 4x, \text{ at t.p. } -6x^2 + 4x = 0$$

$$x(-6x + 4) = 0 \quad x = 0 \quad \text{and} \quad x = \tfrac{2}{3}$$

$$\frac{d^2y}{dx^2} = -12x + 4 \text{ is positive for } x = 0 \text{ (minimum)}$$

$$\text{and negative for } x = \tfrac{2}{3} \text{ (maximum)}$$

$$y(\tfrac{2}{3}) = -4.7, \quad y(2) = -13, \quad y(-2) = 19$$

Here we have two turning points but they are not very distinct. They have almost combined to form a point of inflexion.

Solution

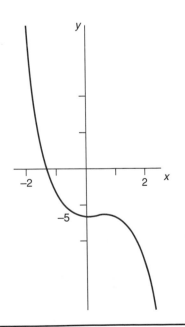

EXAMPLE 19.19

Sketch the curve of $y = 2 - e^x$.

The standard function is an exponential with two transformations: $-f(x)$ and $f(x) + a$; so we can expect the standard exponential growth curve to be reflected about the x-axis and translated in the positive direction of the y-axis.

When x is very large and positive, y is very large and negative. As, $\lim\limits_{x \to -\infty}$, $y = 2$.

$$y(0) = 1, \quad y(2) = -5.4, \quad y(-2) = -1.9$$

$$\frac{dy}{dx} = -e^x \neq 0 \text{ for any value of } x.$$

So there are no local minima or maxima.

Solution

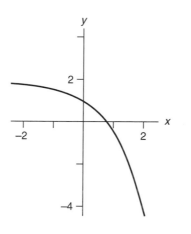

EXAMPLE 19.20

Sketch the curve of the function $y = \dfrac{1}{x+3}$.

This is the standard function of a rectangular hyperbola with a transformation of the kind $f(x + a)$ which brings about a translation in the negative direction of the x-axis by 3 units. So we could sketch the curve straightaway but let's see.

$$\text{As } \lim_{x \to \infty}, y = 0 \quad \text{and as } \lim_{x \to -\infty}, y = 0$$

$$y(0) = \tfrac{1}{3} \quad y(-3) \text{ is undefined}$$

$$y(-2.9) = 10 \quad y(-3.1) = -10 \quad y(-6) = -\tfrac{1}{3}$$

$$\frac{dy}{dx} = -\frac{1}{(x+3)^2} \neq 0 \quad \text{for any } x.$$

So, again we have no maximum or minimum. Note that $y = 0$ and $x = -3$ are asymptotes.

Solution

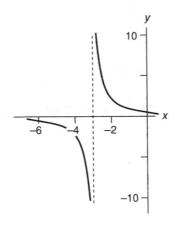

EXAMPLE 19.21

Sketch the curve of $y = \sqrt{25 - x^2}$.

It's not immediately clear but if we square both sides and rearrange we see that this is a standard function whose curve is a circle having a radius of 5 units,

$$y^2 + x^2 = 25$$

The variables have no components of translation so the circle is centred about the origin.

Solution

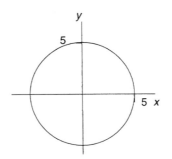

EXAMPLE 19.22

Sketch the curve of $y = 2 + e^{-x}\cos x$.

The first thing that is clear is that we are dealing with a standard periodic function with two transformations.

We have $af(x)$ where a is an exponential decay. So the amplitude of the cosine wave decays as x increases. Secondly there is a positive y-translation by two units.

The key points of a cosine wave are the points which coincide with its peak values. Peak values occur at $x = 0$ and at whole number multiples of π.

$$y(0) = 3, \quad y(\pi) = 1.3, \quad y(2\pi) = 2.5$$
$$y(3\pi) = 1.6, \quad y(4\pi) = 2.3, \quad y(5\pi) = 1.8$$

Solution

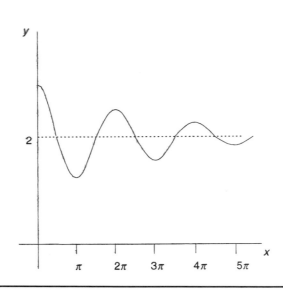

PRACTICE EXERCISE 19.5

Sketch the curves of the following functions. In each case: state its relationship to one of the standard functions; state whether it is odd, even, discontinuous or periodic; label the key points of the curve.

(a) $y = (x - 2)^2 + 3$; (b) $y = (x + 2)^3 - 3$; (c) $y = \dfrac{1}{x^2 + 2}$

(d) $y = f(x) = \begin{cases} -2 & \text{with } -\pi < x < 0 \\ 3 & \text{with } 0 < x < \pi \end{cases}$ of period 2π; (e) $y = 3\sin\left(x + \dfrac{\pi}{4}\right) - 2$

Solutions appear at the end of the chapter.

Solutions

PRACTICE EXERCISE 19.3

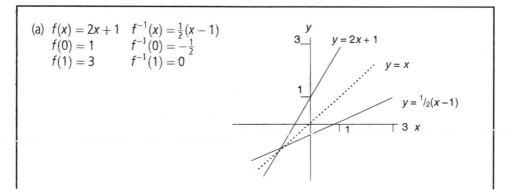

(a) $f(x) = 2x + 1$ $f^{-1}(x) = \frac{1}{2}(x - 1)$
    $f(0) = 1$ $f^{-1}(0) = -\frac{1}{2}$
    $f(1) = 3$ $f^{-1}(1) = 0$

$y = 2x + 1$

$y = x$

$y = \frac{1}{2}(x - 1)$

(b) $f(x) = x^2 - 4$ $f^{-1}(x) = \sqrt{x+4}$
$f(2) = 0$ $f^{-1}(0) = 2$
$f(3) = 5$ $f^{-1}(2) = 2.4$
$f(2.5) = 2.25$ $f^{-1}(3) = 2.6$

(c) $f(x) = 2x^3 + 1$ $f^{-1}(x) = \sqrt[3]{\dfrac{x-1}{2}}$

$f(0) = 1$ $f^{-1}(0) = -0.8$
$f(1) = 3$ $f^{-1}(1) = 0$
$f(2) = 17$ $f^{-1}(2) = 0.8$
$f(-1) = -1$ $f^{-1}(-1) = -1$
$f(-2) = 15$ $f^{-1}(-2) = -1.14$

(d) $f(x) = \sin x$ $f^{-1}(x) = \arcsin x$
$f(0) = 0$ $f^{-1}(0) = 0$

$f\left(\pm\dfrac{\pi}{4}\right) = \pm0.7$ $f^{-1}(\pm0.5) = \pm0.52$

$f\left(\pm\dfrac{\pi}{2}\right) = \pm1$ $f^{-1}(\pm1) = \pm1.57$

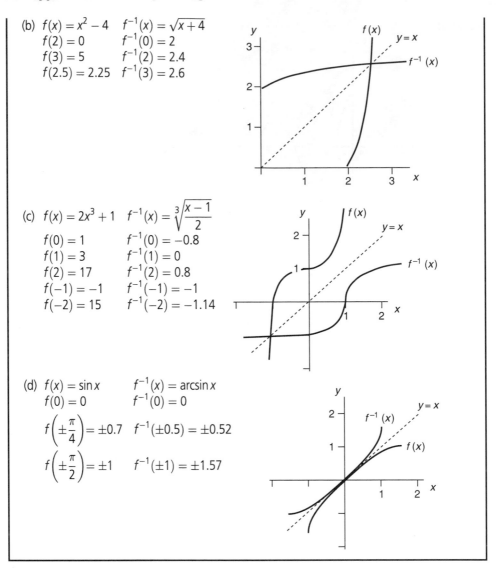

PRACTICE EXERCISE 19.5

(a) Quadratic, even function with x- and y-translations.

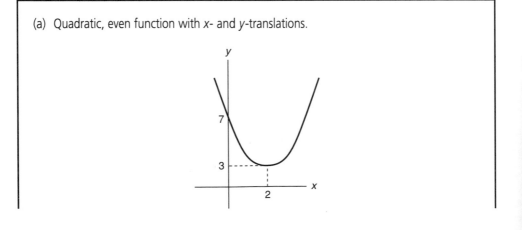

$y = x^2 - 4x + 7, \quad y(0) = 7$

$x = \dfrac{4 \pm \sqrt{16 - 28}}{2}$ has no real roots

$\dfrac{dy}{dx} = 2x - 4, \quad$ at t.p. $2x - 4 = 0, x = 2$

$\dfrac{d^2y}{dx^2} = 2,$ so $x = 2$ is a minimum

$y(2) = 3$

(b) Cubic, odd function with x- and y-translations.

$y = (x + 2)^3 - 3$

$f(x + 2)$ translates 2 units left of origin

$f(x) - 3$ translates 3 units below the origin

$\dfrac{dy}{dx} = 3(x + 2)^2 \quad$ at t.p. $3(x + 2)^2 = 0$

$x + 2 = 0, x = -2$

$\dfrac{d^2y}{dx^2} = 6(x + 2) = 0$ with $x = -2$

so point of inflexion.

$y(-2) = -3, \quad y(0) = 5, \quad y(-1) = 2, \quad y(-3) = -4$

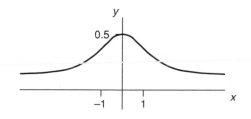

(c) Not a standard function.

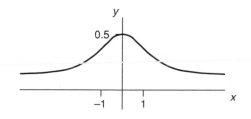

$$y = \frac{1}{x^2 + 2}$$

$\lim\limits_{x \to \infty}, y = 0$ and $\lim\limits_{x \to -\infty}, y = 0$

$y(0) = \frac{1}{2}, \quad y(1) = \frac{1}{3}, \quad y(-1) = \frac{1}{3}$

(d) Constant, periodic function with discontinuities at multiples of π.

$$y = \begin{cases} -2 & \text{with } -\pi < x < 0 \\ 3 & \text{with } 0 < x < \pi \end{cases}$$

$y\left(-\frac{\pi}{2}\right) = -2, \quad y\left(\frac{\pi}{2}\right) = 3, \quad y(0)$ is undefined

(e) Sine function with negative x-axis translation, y-expansion and negative y-axis translation.

$$y = 3\sin\left(x + \frac{\pi}{4}\right) - 2$$

$y\left(\frac{\pi}{4}\right) = 1, \quad y\left(\frac{3\pi}{4}\right) = -2$

$y\left(-\frac{\pi}{4}\right) = -2, \quad y\left(-\frac{3\pi}{4}\right) = 1$

chapter

20 More integration

The integrals that appear in Table 13.2 allow us to integrate by inspection. You recognize a function and write its integral. That's all. Now we will learn how to deal with some combinations of functions: products and quotients, both of which may involve a function of a function. There are three procedures which can help us:

- algebraic substitution
- trigonometric substitution
- integration by parts

20.1 Integration by algebraic substitution

You are familiar with the 'trick' of substituting u for a function when dealing with the derivative of a function of a function. The same idea can be applied to integration.

EXAMPLE 20.1

Evaluate: $\int x(x^2 + 1)^3 \, dx$

Here, we could start by expanding brackets and multiplying by x. Then we would have a straight-forward integral of a sum of terms.

But that's work. There is a more elegant method.

We substitute u for the more complicated function.

Then, despite the fact that we are trying to integrate, we differentiate u with respect to x.

That gives us an expression for x.

We can now substitute the u-functions for the x-functions in our integral and integrate with respect to u.

The last step is simple. Put everything in terms of x.

Solution

Let, $u = x^2 + 1$

and,

$$\frac{du}{dx} = 2x \quad \text{so,} \quad x = \frac{1}{2}\frac{du}{dx}$$

Now,

$$\int x(x^2 + 1)^3 \, dx = \int \left(\frac{1}{2}\frac{du}{dx}\right) u^3 \, dx$$

$$= \frac{1}{2}\int u^3 \frac{du}{dx} \, dx$$

but,

$$\frac{du}{dx} \, dx = du$$

so,

$$\frac{1}{2}\int u^3 \frac{du}{dx} \, dx = \frac{1}{2}\int u^3 \, du = \frac{1}{8}u^4 + C$$

and with, $u = x^2 + 1$,

$$\int x(x^2 + 1)^3 \, dx = \frac{1}{8}(x^2 + 1)^4 + C$$

Now let's deal with another example of a similar type of thing. One where different functions of x are linked as a quotient.

EXAMPLE 20.2

To evaluate:

$$\int \frac{2x + 3}{x^2 + 3x - 7} \, dx$$

Again, we try the substitution of u for the more complicated function and differentiate.

This time, we try a slightly different approach. Because there is no simple x term in the original integral we rearrange the differential coefficient for dx.

Now we are able to replace dx in the integral.

Note how conveniently the expression in the numerator cancels so we can restate the integral entirely in terms of u and integrate. Finally we put everything back in terms of x as before.

Solution

Let $u = x^2 + 3x - 7$

$$\frac{du}{dx} = 2x + 3, \quad du = (2x + 3)\,dx$$

so,

$$dx = \frac{du}{2x+3}$$

$$\int \frac{2x+3}{x^2+3x-7}\, dx = \int \frac{2x+3}{x^2+3x-7}\left(\frac{du}{2x+3}\right)$$

$$= \int \frac{1}{x^2+3x-7}\, du$$

$$= \int \frac{1}{u}\, du = \ln u + C$$

$$\int \frac{2x+3}{x^2+3x-7}\, dx = \ln(x^2+3x-7) + C$$

EXAMPLE 20.3

Find:

$$\int \cos\theta \sin^2\theta\, d\theta$$

We try u for $\sin\theta$. Differentiate and rearrange for $d\theta$. The rest falls out neatly.

Solution

Let $u = \sin\theta$, $\dfrac{du}{d\theta} = \cos\theta$

$$d\theta = \frac{du}{\cos\theta}$$

so,

$$\int \cos\theta \sin^2\theta\, d\theta = \int \cos\theta u^2 \frac{du}{\cos\theta}$$

$$= \int u^2\, du = \frac{1}{3}u^3 + C$$

$$\int \cos\theta \sin^2\theta\, d\theta = \frac{1}{3}\sin^3\theta + C$$

For comparison, try the approach we took in Example 20.1. Instead of rearranging for $d\theta$ and substituting. Substitute for $\cos\theta$.

EXAMPLE 20.4

Find:

$$\int \cos\theta \sin^2\theta\, d\theta$$

Here it is quicker to substitute the differential coefficient of u directly as we did in Example 20.1.

Solution

Let $u = \sin\theta$, $\dfrac{du}{d\theta} = \cos\theta$

so,

$$\int \cos\theta \sin^2\theta \, d\theta = \int \frac{du}{d\theta} u^2 \, d\theta$$

$$= \int u^2 \, du = \frac{1}{3}u^3 + C$$

$$\int \cos\theta \sin^2\theta \, d\theta = \frac{1}{3}\sin^3\theta + C$$

Look at the general form of the integrals we have been dealing with. Try to spot the reason why the substitution method works. Consider the form of the next integral carefully.

EXAMPLE 20.5

Evaluate:

$$\int x^2 \sqrt{x^3 - 2} \, dx$$

Here it pays to substitute for x^3.

Now, can you see that the integral we are dealing with is a product of some function of a function and its derivative?

We have a function which is cubic. Then we have the root of the cubic function – this is the function of a function. Then this is multiplied by the quadratic which is the first derivative of the cubic.

Solution

Try, $u = x^3 - 2$, $\dfrac{du}{dx} = 3x^2$

$$\frac{1}{3}\frac{du}{dx} = x^2$$

$$\int x^2 (x^3 - 2)^{\frac{1}{2}} \, dx = \int \left(\frac{1}{3}\frac{du}{dx}\right) u^{\frac{1}{2}} \, dx$$

$$= \frac{1}{3}\int u^{\frac{1}{2}} \, du = \frac{1}{3}\left(\frac{2}{3}u^{\frac{3}{2}}\right) + C$$

$$\int x^2 \sqrt{x^3 - 2} \, dx = \frac{2}{9}(x^3 - 2)^{\frac{3}{2}} + C$$

$$= \frac{2}{9}\sqrt{(x^3 - 2)^3} + C$$

The method works because we differentiate the cubic to obtain a quadratic which cancels in the substitution and leaves us with a simple standard function.

The procedure for integration by algebraic substitution works as follows.

Given:

$$\int kf'(x)(g(f(x))\,dx \quad \text{or} \quad \int \frac{kf'(x)}{g(f(x))}\,dx$$

where k is a constant,

try: $u = f(x)$, find $f'(x)$ then substitute for $f'(x)$ and use the rule

$$\frac{du}{dx}\,dx = du$$

to get:

$$\int kg(u)\,du$$

which is in the form of a standard integral.

This may seem complicated at first but it isn't once you get used to it. The thing to look for is to see if one factor of the product is some multiple of the differential coefficient of the other.

Work through the next example comparing each step to the procedure outline above.

EXAMPLE 20.6

Integrate with respect to x:

$$y = \frac{x}{1 - x^2}$$

Here we have $f(x) = 1 - x^2$.

The function of a function, $g(f(x)) = \frac{1}{1 - x^2}$.

And, $kf'(x) = x$ is a multiple of the derivative of the function.

So we have all the right ingredients which suit an algebraic substitution.

The procedure is as before and this time it is more convenient to rearrange the derivative for dx as we did in Example 20.2.

Solution

Try, $u = 1 - x^2$, so

$$\frac{du}{dx} = -2x$$

and

$$dx = -\frac{du}{2x}$$

$$\int y\,dx = \int \frac{x}{1 - x^2}\,dx$$

$$= \int \frac{x}{1 - x^2}\left(-\frac{du}{2x}\right)$$

$$= -\frac{1}{2}\int \frac{1}{u}\,du$$

$$= -\tfrac{1}{2}\ln u + C$$

$$\int y\,dx = C - \frac{1}{2}\ln(1 - x^2)$$

Sometimes it is not immediately apparent that substitution is possible.

EXAMPLE 20.7

Evaluate:

$$\int \frac{x^3}{x^2 + 1}\,dx$$

Here the numerator is not the 1st derivative of the more complicated function. But we can create a first derivative if we factorize the numerator in a clever way.

We must concentrate on the fact that we are to eliminate all the x-terms. We have:

$$
\begin{array}{ll}
u & \text{for } x^2 + 1 \\
u - 1 & \text{for } x^2 \\
du & \text{for } 2x\,dx
\end{array}
$$

After substitution we have a compound fraction that needs to be dealt with. We take u as the common denominator and express the integral as partial fractions that can be integrated.

Solution

$$\int \frac{x^3}{x^2 + 1}\,dx = \int \frac{\frac{1}{2}(2x)x^2}{x^2 + 1}\,dx$$

$$= \frac{1}{2}\int \frac{x^2}{x^2 + 1}\,2x\,dx$$

Let, $u = x^2 + 1$, so $u - 1 = x^2$ and

$$\frac{du}{dx} = 2x \quad \text{so} \quad du = 2x\,dx$$

$$\frac{1}{2}\int \frac{x^2}{x^2 + 1}\,2x\,dx = \frac{1}{2}\int \frac{u - 1}{u}\,du$$

$$= \frac{1}{2}\int \left(1 - \frac{1}{u}\right)du$$

$$= \tfrac{1}{2}(u - \ln u) + C$$

$$\int \frac{x^3}{x^2 + 1}\,dx = \frac{1}{2}[(x^2 + 1) - \ln(x^2 + 1)] + C$$

So, to integrate by algebraic substitution you need two things. First, you need a product or quotient of a function and its derivative. Second, you need to have the experience to spot it! But don't worry if you can't. Trial and error will take you in the right direction. Now practise.

PRACTICE EXERCISE 20.1

Evaluate the following integrals:

(a) $\int 5(x+2)^4 \, dx$; (b) $\int \frac{1}{\sqrt{3-2t}} \, dt$; (c) $\int \frac{\sin\theta}{\cos^2\theta} \, d\theta$; (d) $\int x \, e^{5-4x^2} \, dx$

(e) $\int 2x^2 \sqrt{6-5x^3} \, dx$; (f) $\int (3x+1)\sin(3x^2+2x) \, dx$; (g) $\int \tan\theta \, d\theta$

(h) $\int \frac{2x+1}{\sqrt{2x+4}} \, dx$; (i) $\int \cot\theta \, d\theta$; (j) $\int \frac{t^2}{(t-3)^6} \, dt$

Solutions

(a) $(x+2)^5 + C$; (b) $C - \sqrt{3-2t}$; (c) $\sec\theta + C$; (d) $C - \frac{1}{8} e^{5-4x^2}$

(e) $C - \frac{4}{45}\sqrt{(6-5x)^3}$; (f) $C - \frac{1}{2}\cos(3x^2+2x)$; (g) $C - \ln\cos\theta$

(h) $\frac{1}{3}\sqrt{(2x+4)^3} - 3\sqrt{2x+4} + C$; (i) $\ln\sin\theta + C$

(j) $C - \dfrac{1}{3(t-3)^3} - \dfrac{3}{2(t-3)^4} - \dfrac{9}{5(t-3)^5}$

Change of limits

When evaluating definite integrals using the substitution method it pays to change the limits and solve the whole thing in terms of u. Lets see what this means.

EXAMPLE 20.8

Find:

$$\int_2^4 x(x^2+1)^3 \, dx$$

The first solution is obtained in the way you would normally go about evaluating a definite integral.

After substitution it is necessary to declare that the limits are x-values.

After integration we put u in terms of x and then use the limits of x to find the value of the integral.

Solution 1

$$u = x^2 + 1, \quad \frac{du}{dx} = 2x, \quad x = \frac{1}{2}\frac{du}{dx}$$

$$\int_2^4 x(x^2+1)^3 = \int_{x=2}^{x=4} \left(\frac{1}{2}\frac{du}{dx}\right)(u)^3\,dx$$

$$= \frac{1}{2}\int_{x=2}^{x=4} u^3\,du = \frac{1}{2}\left[\frac{1}{4}u^4\right]_{x=2}^{x=4}$$

$$= \tfrac{1}{8}\left[(x^2+1)^4\right]_2^4$$

$$= \tfrac{1}{8}\left[((4)^2+1)^4 - ((2)^2+1)^4\right]$$

$$= 10\,362$$

Now look at an alternative approach shown in solution 2.

We find the substitution, as before but this time one extra step saves a bit of work later. We find the values of u which correspond with the limiting values of x.

Here, we find the upper and lower limits of u to be 17 and 5 respectively. Now there is no need to go back to x since the solution of a definite integral is just a numerical value. We get to it directly from the 'u-integral'.

Solution 2

$$u = x^2+1, \quad \frac{du}{dx}=2x, \quad x=\frac{1}{2}\frac{du}{dx}, \quad u(4)=17, \quad u(2)=5$$

$$\int_2^4 x(x^2+1)^3 = \int_5^{17}\left(\frac{1}{2}\frac{du}{dx}\right)(u)^3\,dx$$

$$= \frac{1}{2}\int_5^{17} u^3\,du = \frac{1}{2}\left[\frac{1}{4}u^4\right]_5^{17}$$

$$= \tfrac{1}{8}\left[(17)^4 - (5)^4\right] = 10\,362$$

PRACTICE EXERCISE 20.2

Evaluate the following integrals:

(a) $\displaystyle\int_0^2 x\sqrt{x^2+2}\,dx$; (b) $\displaystyle\int_0^1 \frac{2x^2}{4x^3+3}\,dx$; (c) $\displaystyle\int_{-0.1}^{0.3} \frac{x^2}{2\,e^{3x^3+2}}\,dx$

(d) $\displaystyle\int_{-\frac{\pi}{4}}^{\frac{\pi}{4}} \cos 2\theta(1+\sin 2\theta)\,d\theta$

Solutions

(a) 3.96; (b) 0.141; (c) 608×10^{-6}; (d) 1

20.2 Integration by trigonometric substitution

With some integrals for which an algebraic substitution may not work it is possible to substitute a trigonometric function. It may then be possible to evaluate the integral by applying our knowledge of trigonometric identities and double angle formulae.

Look at the triangles in Fig. 20.1 and see if you are satisfied that in each case the variable x can be equated with a trig. function.

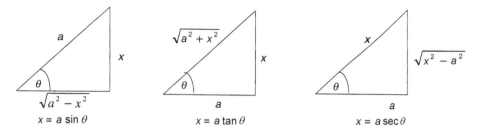

$$x = a\sin\theta \qquad\qquad x = a\tan\theta \qquad\qquad x = a\sec\theta$$

Figure 20.1 In these triangles the variable can be equated with a trig. function

Integrals which contain Pythagorean expressions may be simplified by applying one of the following substitutions (Table 20.1).

Table 20.1. Trigonometric substitutions

Expression	Substitution
$\sqrt{a^2 - x^2}$	$x = a\sin\theta$
$\sqrt{a^2 + x^2}$	$x = a\tan\theta$
$\sqrt{x^2 - a^2}$	$x = a\sec\theta$

We shall see the substitutions at work in a moment but first a reminder about some of the trigonometric identities and formulae which we worked with in Chapter 14.

Eight of them are listed below. Remember where they are. You will need to refer to them from time to time.

$$\tan A = \frac{\sin A}{\cos A}$$

$$\sin^2 A + \cos^2 A = 1$$

$$\sec^2 A = 1 + \tan^2 A$$

$$\text{cosec}^2 A = 1 + \cot^2 A$$

$$\sin 2A = 2\sin A\cos A$$

$$\cos 2A = \cos^2 A - \sin^2 A$$

$$\sin^2 A = \tfrac{1}{2}(1 - \cos 2A)$$

$$\cos^2 A = \tfrac{1}{2}(1 + \cos 2A)$$

EXAMPLE 20.9

Evaluate:

$$\int \sqrt{1 - x^2}\, dx$$

This is of the form $\sqrt{a^2 - x^2}$ where $a = 1$. A sketch helps

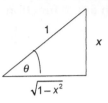

The initial steps are familiar. We determine the substitution and differentiate in order to restate the integral in terms of θ alone. Then it's trigonometry.

Let $x = \sin \theta$, so

$$\frac{dx}{d\theta} = \cos \theta$$

and

$$dx = \cos \theta\, d\theta$$

$$\int \sqrt{1 - x^2}\, dx = \int \sqrt{1 - \sin^2 \theta}\, \cos \theta\, d\theta$$

$$\therefore \quad \int \sqrt{1 - x^2}\, dx = \int \sqrt{\cos^2 \theta}\, \cos \theta\, d\theta$$

$$= \int \cos^2 \theta\, d\theta$$

Because

$$\sin^2 \theta + \cos^2 \theta = 1, \quad 1 - \sin^2 \theta = \cos^2 \theta$$

we cannot integrate yet but we know that,

$$\cos^2 \theta = \tfrac{1}{2}(1 + \cos 2\theta)$$

so we carry out the integration

$$\int \sqrt{1 - x^2}\, dx = \frac{1}{2}\int (1 + \cos 2\theta)\, d\theta$$

$$= \tfrac{1}{2}(\theta + \tfrac{1}{2}\sin 2\theta) + C$$

Now, because this is an indefinite integral we need to restate it in its original terms. First we sort out the double angle

$$\sin 2\theta = 2 \sin \theta \cos \theta$$

Then, if

$$x = \sin\theta, \quad \theta = \arcsin x$$

and from our sketch,

$$\sin\theta = x \quad \text{and} \quad \cos\theta = \sqrt{1-x^2}$$

Finally we can multiply out by the half if we wish.

$$\int \sqrt{1-x^2}\,dx = \frac{1}{2}\left(\theta + \frac{1}{2}(2\sin\theta\cos\theta)\right) + C$$

$$= \frac{1}{2}(\theta + \sin\theta\cos\theta) + C$$

so

$$\int \sqrt{1-x^2}\,dx = \frac{1}{2}\left(\arcsin x + x\sqrt{1-x^2}\right) + C$$

$$= \frac{1}{2}\arcsin x + \frac{1}{2}x\sqrt{1-x^2} + C$$

Sometimes for an integration to be successful it is necessary to combine a trigonometric substitution with an algebraic substitution. It takes patience and care, as you will see from the next example.

EXAMPLE 20.10

Evaluate:

$$\int \frac{1}{\sqrt{x^2-9}}\,dx$$

Here the following triangle is useful:

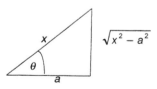

where $a = 3$.

We start with the trig. substitution for x and differentiate. This requires the function of a function rule. Then we can re-write the integral all in terms of θ.

Let, $x = 3\sec\theta = 3(\cos\theta)^{-1}$

$$\frac{dx}{d\theta} = -3(\cos\theta)^{-2}(-\sin\theta) = \frac{3\sin\theta}{\cos^2\theta}$$

$$\frac{dx}{d\theta} = 3\tan\theta\sec\theta$$

so

$$dx = 3\tan\theta\sec\theta\,d\theta$$

and

$$\int \frac{1}{\sqrt{x^2 - 9}}\, dx = \int \frac{3 \tan \theta \sec \theta}{\sqrt{9 \sec^2 \theta - 9}}\, d\theta$$

Using the identity $\sec^2 \theta = 1 + \tan^2 \theta$ allows us to simplify right down to what looks like a simple integral of a secant.

$$\int \frac{1}{\sqrt{x^2 - 9}}\, dx = \int \frac{3 \tan \theta \sec \theta}{\sqrt{9(\sec^2 \theta - 1)}}\, d\theta$$

$$= \int \frac{3 \tan \theta \sec \theta}{\sqrt{9 \tan^2 \theta}}\, d\theta$$

$$= \int \frac{3 \tan \theta \sec \theta}{3 \tan \theta}\, d\theta$$

$$= \int \sec \theta\, d\theta$$

However, this is not one of our standard integrals, we have some more work to do before we can integrate.

Multiplying by $(\sec \theta + \tan \theta)$ turns the numerator into the derivative of the denominator which is suitable for an algebraic substitution.

$$\int \frac{1}{\sqrt{x^2 - 9}}\, dx = \int \frac{\sec \theta (\sec \theta + \tan \theta)}{\sec \theta + \tan \theta}\, d\theta$$

$$= \int \frac{\sec^2 \theta + \sec \theta \tan \theta}{\sec \theta + \tan \theta}\, d\theta$$

By putting $u = \sec \theta + \tan \theta$ and differentiating we can rewrite the integral again. This time in terms of u which integrates very easily.

Let, $u = \sec \theta + \tan \theta$

$$\frac{du}{d\theta} = \sec \theta \tan \theta + \sec^2 \theta$$

$$d\theta = \frac{du}{\sec \theta \tan \theta + \sec^2 \theta}$$

The journey back requires two back substitutions. We substitute to get u in terms of θ.

Then we use our triangle to get θ in terms of x.

So,

$$\int \frac{\sec^2 \theta + \sec \theta \tan \theta}{\sec \theta + \tan \theta} \; \frac{du}{\sec \theta \tan \theta + \sec^2 \theta}$$

$$\int \frac{1}{u}\, du = \ln u + C$$

$$\int \sec \theta\, d\theta = \ln(\sec \theta + \tan \theta) + C$$

$$\int \frac{1}{\sqrt{x^2 - 9}}\, dx = \ln\left(\frac{x}{3} + \frac{\sqrt{x^2 - 9}}{3}\right) + C$$

So it can be difficult! It takes patience and a little bit of courage to try different substitutions to see if they work. In this case the main hurdle was
integrating a secant. Now you have seen how it can be done, next time
should be easier.

Sometimes, it seems at first that the appropriate trigonometric substitution will
not work. Again, there is nothing wrong with trial and error. The important
thing is to have a go.

EXAMPLE 20.11

Evaluate:

$$\int \frac{1}{(4+x^2)^2} \, dx$$

Here it seems that the tangent substitution might work so we give it a try.

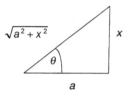

where $a = 2$.

Solution

We substitute the trig. terms for x and use the identity,

$$\sec^2 \theta \equiv 1 + \tan^2 \theta$$

to simplify the integral.

Let, $x = 2 \tan \theta$, $\dfrac{dx}{d\theta} = 2 \sec^2 \theta$, $dx = 2 \sec^2 \theta \, d\theta$

$$\int \frac{1}{(4+x^2)^2} \, dx = \int \frac{2 \sec^2 \theta}{(4 + 4 \tan^2 \theta)^2} \, d\theta$$

$$= \int \frac{2 \sec^2 \theta}{16(1 + \tan^2 \theta)^2} \, d\theta = \frac{1}{8} \int \frac{\sec^2 \theta}{\sec^4 \theta} \, d\theta$$

We get,

$$\cos^2 \theta = \tfrac{1}{2}(1 + \cos 2\theta)$$

which integrates easily.

$$= \frac{1}{8} \int \frac{1}{\sec^2 \theta} \, d\theta = \frac{1}{8} \int \cos^2 \theta \, d\theta$$

$$= \frac{1}{8} \int \frac{1}{2}(1 + \cos 2\theta) \, d\theta$$

$$= \tfrac{1}{16} \left(\theta + \tfrac{1}{2} \sin 2\theta\right) + C$$

$$= \tfrac{1}{16} \left(\theta + \sin \theta \cos \theta\right) + C$$

$$= \frac{1}{16}\left[\arctan\frac{x}{2} + \left(\frac{x}{\sqrt{4+x^2}}\right)\left(\frac{2}{\sqrt{4+x^2}}\right)\right] + C$$

$$= \frac{1}{16}\arctan\frac{x}{2} + \frac{x}{8(4+x^2)} + C$$

The final steps of putting θ in terms of x look a little awkward but it is quite straightforward if we put our faith in the relationships of our right-angled triangle.

EXAMPLE 20.12

Evaluate:

$$\int_1^2 \sqrt{4-x^2}\,dx$$

From the triangles in Fig. 20.1 it is suitable to try the sine substitution for x.

To save work we are going to change limits so we want the upper and lower limits of θ.

Putting the integral in terms of θ and using trig. identities to simplify is familiar to you now. Since this is a definite integral and we have changed limits we need not put the integral back in terms of x.

Solution

$$x = 2\sin\theta, \quad \frac{dx}{d\theta}2\cos\theta, \quad dx = 2\cos\theta\,d\theta, \quad \theta = \arcsin\frac{x}{2}, \quad \theta(2) = \frac{\pi}{2}, \quad \theta(1) = \frac{\pi}{6}$$

$$\int_1^2 \sqrt{4-x^2}\,dx = \int_{\frac{\pi}{6}}^{\frac{\pi}{2}} \sqrt{4-4\sin^2\theta}\,2\cos\theta\,d\theta$$

$$= \int_{\frac{\pi}{6}}^{\frac{\pi}{2}} \sqrt{4(1-\sin^2\theta)}\,2\cos\theta\,d\theta$$

$$\int_{\frac{\pi}{6}}^{\frac{\pi}{2}} \sqrt{4\cos^2\theta}\,2\cos\theta\,d\theta = \int_{\frac{\pi}{6}}^{\frac{\pi}{2}} 4\cos^2\theta\,d\theta$$

$$4\int_{\frac{\pi}{6}}^{\frac{\pi}{2}} \frac{1}{2}(1+\cos 2\theta)\,d\theta = 2\left[\theta + \frac{1}{2}\sin 2\theta\right]_{\frac{\pi}{6}}^{\frac{\pi}{2}}$$

$$= 2\left[\left(\frac{\pi}{2}+\frac{1}{2}\sin\pi\right) - \left(\frac{\pi}{6}+\frac{1}{2}\sin\frac{\pi}{3}\right)\right] = 1.23$$

So definite integrals are a little easier to deal with because we don't have the bother of substituting back the original terms.

PRACTICE EXERCISE 20.3

Evaluate the following integrals.

(a) $\int \sqrt{16-x^2}\,dx;$ (b) $\int \frac{1}{\sqrt{(1+x^2)^3}}\,dx;$ (c) $\int \frac{2}{\sqrt{x^2-2}}\,dx$

(d) $\displaystyle\int_0^1 \sqrt{1-x^2}\,dx$;   (e) $\displaystyle\int_2^3 \frac{1}{x^2\sqrt{9+x^2}}\,dx$

Solutions

(a) $8\arcsin\dfrac{x}{4}+\dfrac{x}{2}\sqrt{16-x^2}+C$;   (b) $\sin(\arctan x)+C$;   (c) $2\ln\left(\dfrac{x}{\sqrt{2}}+\dfrac{\sqrt{x^2-2}}{\sqrt{2}}\right)+C$

(d) $\dfrac{\pi}{4}$;   (e) 0.043

20.3 Integration by parts

Finally, we look at the integration of products where one factor is not the differential coefficient of the other so substitution methods don't work. Integration by parts is a method derived from the product rule for differentiation.

We take the formula for differentiating products and integrate with respect to x.

$$\frac{d}{dx}(uv) = u\frac{dv}{dx} + v\frac{du}{dx}$$

Integrating both sides:

$$uv = \int u\frac{dv}{dx}\,dx + \int v\frac{du}{dx}\,dx$$

Now there is a shorthand used because of the rules:

$$\frac{dv}{dx}\,dx = dv \quad \text{and} \quad \frac{du}{dx}\,dx = du$$

where dv is the differential coefficient of the v-function with respect to x and du is the differential coefficient of u with respect to x.

$$uv = \int u\,dv + \int v\,du$$

$$\int u\,dv = uv - \int v\,du$$

Rearranging provides the integral of the product of some function u and the differential coefficient of some other function v.

Integration by parts formula:

$$\int u\,dv = uv - \int v\,du$$

Integration by parts is quite easy to apply but it is important which of the two functions is selected to be the u-function and which the v-function, as you will see.

EXAMPLE 20.13

Find: $\int x \ln x \, dx$.

Now look closely at the 'by-parts' formula. On the left we have one factor which is just the u-function and the other which is dv, the derivative of some function, v. This is the more difficult one because we will need to integrate it to satisfy the need for the v-factor on the right.

If we choose $u = x$ and $dv = \ln x$ we create a problem for ourselves because there is no standard integral for a log function. The logical choice is to select the linear factor as dv and the log term as u.

Before we can proceed we need du, which is $\dfrac{du}{dx} dx$. And we also need v, the integral of dv with respect to x.

The rest is straightforward, we put the terms into the formula, integrate the second term and simplify.

Solution

$$\int u \, dv = uv - \int v \, du$$

Let,

$$u = \ln x \quad \text{and} \quad dv = x$$

so,

$$du = \frac{du}{dx} dx = \frac{1}{x} dx$$

and

$$v = \int dv = \int x \, dx = \frac{1}{2} x^2$$

Now,

$$\int \ln x (x) \, dx = \ln x \left(\frac{1}{2} x^2\right) - \int \frac{1}{2} x^2 \left(\frac{1}{x}\right) dx$$

$$= \frac{1}{2} x^2 \ln x - \frac{1}{2} \int x \, dx$$

$$= \tfrac{1}{2} x^2 \ln x - \tfrac{1}{4} x^2 + C$$

$$= \tfrac{1}{2} x^2 (\ln x - \tfrac{1}{2}) + C$$

Now let's stay with the log function for a while because it is interesting. Look back at Table 13.2, our list of standard integrals. The log function is conspicuous by its absence. Yet in the last example, integration by parts got us around the difficulty. So can we actually integrate a logarithmic or not? The answer is yes and no!

Let's apply integration by parts to a simple log function.

To find, $\int \ln x \, dx$

To apply the by-parts formula we need a product. Can $\ln x$ be considered a product? If so what are its factors?

Yes, it is the product of 1 and $\ln x$

So we can proceed. We nominate 1 as dv and $\ln x$ as u.

Let $u = \ln x$ and $dv = 1$, so

$$du = \frac{1}{x} \, dx \quad \text{and} \quad v = \int 1 \, dx = x$$

$$\int \ln x \, dx = \ln x (x) - \int x \frac{1}{x} \, dx$$

$$= x \ln x - \int 1 \, dx$$

$$= x \ln x - x + C$$

$$= x(\ln x - 1) + C$$

The result is quite astounding. We are able to find the integral of a function without actually integrating it!

So in general:

$$\int \ln x \, dx = x(\ln x - 1) + C$$

EXAMPLE 20.14

Evaluate: $\int x e^x \, dx$.

Let's start by selecting $u = x$. There is a good reason for this which you will see in a moment.

The rest is straightforward. Note, that we don't bother with the constant of integration until the end. There is no point. Every integration, no matter how many steps are involved, yields just the one constant.

Solution

Let $u = x$ and $dv = e^x$, so

$$du = 1 \, dx \quad \text{and} \quad v = \int e^x = e^x$$

Then,

$$\int x e^x \, dx = \int u \, dv = uv - \int v \, du$$

$$= x(e^x) - \int e^x \, dx$$

$$= x(e^x) - e^x + C$$

$$= e^x(x - 1) + C$$

Now look what might have happened had we selected $u = e^x$.

We now have to integrate x with respect to dx. Substituting into the formula we find ourselves with an integral which is more complicated than the one we started with.

Solution going wrong!

Let $u = e^x$ and $dv = x$, so

$$du = e^x \, dx \quad \text{and} \quad v = \int x \, dx = \frac{1}{2}x^2$$

Then,

$$\int x e^x \, dx = \int u \, dv = uv - \int v \, du$$

$$= (e^x)\frac{1}{2}x^2 - \int \frac{1}{2}x^2 \, e^x \, dx \ldots?$$

We could continue and get out of it by applying the by-parts formula again. But that would involve extra work. We would have to reverse the selection of the wrong factor as u in the first place.

So the *choice of u is important.* The general idea is to select the more complex of the factors as u so that differentiation (the du bit) makes it simpler. Let's look at an integral with a trig. factor.

EXAMPLE 20.15

Evaluate: $\int x \sin x \, dx$.

The integral of the sine function is simpler than the integral of the polynomial. So we make $u = x$.

Solution

Let $u = x$, $dv = \sin x$, so

$$du = 1 \, dx, \quad v = \int \sin x \, dx = -\cos x$$

Hence,

$$\int x \sin x \, dx = \int u \, dv = uv - \int v \, du$$

$$= x(-\cos x) - \int -\cos x \, dx$$

$$= -x \cos x + \sin x + C$$

$$= \sin x - x \cos x + C$$

Again the derivative of x is as simple as it can be and the integral of the sine function creates no difficulty either.

Can you see that if we had selected $u = \sin x$ we would have got into the same difficulty as before?

So is there a handy rule to remember what to select for u?

Yes there is. Follow the order of priority shown below. If there is a log factor, make that u. If there is no log then make the polynomial u. If there is no log or polynomial make the exponential u.

Priority for u-substitution

1. $\ln x$

2. x^n

3. e^x

If you forget – make u the term that would yield the more complicated integral.

Now, it may be that even though you select the factors correctly, you may have to use the by-parts formula more than once to arrive at a solution. The important thing is to keep your nerve and continue with the integration.

EXAMPLE 20.16

Evaluate: $\int x^2 e^x \, dx$.

The polynomial is the more complicated integral so make that u.

Let $u = x^2$ and $dv = e^x$, so

$$du = 2x \, dx \quad \text{and} \quad v = e^x$$

Then,

$$\int x^2 e^x \, dx = \int u \, dv = uv - \int v \, du$$

$$= x^2 e^x - \int e^x 2x \, dx$$

Substituting the appropriate terms into the by-parts formula leaves us with an integral which is still a product of two terms, e^x and $2x$. This cannot be integrated directly any more than the original integral could be.

So we try again, letting $u = 2x$ and use the by-parts formula to solve for the new integral.

Try again.

Let $u = 2x$ and $dv = e^x$, so

$$du = 2 \, dx \quad \text{and} \quad v = e^x$$

$$\int e^x 2x \, dx = uv - \int v \, du$$

$$= 2x e^x - \int e^x 2 \, dx$$

$$= 2x e^x - 2 e^x + C$$

This time we can integrate and obtain the solution:

$$2x e^x - 2 e^x + C$$

This can now replace $\int e^x 2x \, dx$.

$$\int x^2 e^x \, dx = x^2 e^x - (2x e^x - 2 e^x) + C$$

$$= x^2 e^x - 2x e^x + 2 e^x + C$$

$$= e^x (x^2 - 2x + 2) + C$$

Taking care of the negative sign and factorizing gets us what we wanted.

Or does it? How do we know this is right?

> The result of any integration
> can be checked by differentiation

Try it. We differentiate a product using the product rule:

$$\frac{dy}{dx} = u \frac{dv}{dx} + v \frac{du}{dx}$$

Solution check

$$\frac{d}{dx}(e^x(x^2 - 2x + 2) + C) = e^x(2x - 2) + (x^2 - 2x + 2) e^x$$

$$= 2x e^x - 2 e^x + x^2 e^x - 2x e^x + 2 e^x$$

$$= x^2 e^x$$

C is a constant so it goes to zero and the only term that survives the clear-out is the product in the original integral.

EXAMPLE 20.17

Evaluate: $\int e^x \cos x \, dx$.

Make u the exponential.

The first try gives us an integral which looks much like the original except we have a sine instead of a cosine.

Let $u = e^x$ and $dv = \cos x$, so

$$du = e^x \, dx \quad \text{and} \quad v = \sin x$$

Then,

$$\int e^x \cos x \, dx = uv - \int v \, du$$

$$= e^x \sin x - \int \sin x \, e^x \, dx$$

Integrating by parts again:

Let $u = e^x$ and $dv = \sin x$, so

$$du = e^x \, dx \quad \text{and} \quad v = -\cos x$$

Then,

$$\int \sin x \, e^x \, dx = e^x(-\cos x) - \int (-\cos x) \, e^x \, dx$$

$$= -e^x \cos x + \int e^x \cos x \, dx$$

$$\int e^x \cos x \, dx = e^x \sin x - \left(-e^x \cos x + \int e^x \cos x \, dx \right)$$

$$= e^x \sin x + e^x \cos x - \int e^x \cos x \, dx$$

Second time around, brings us back to where we started! Not only have we saddled ourselves with the original integral but there is baggage with it as well.

But there is a standard approach to this sort of thing. It's quite clever.

Let the original integral be I. Collect-up, divide by 2 and you have an expression which equates with the original integral.

Let,

$$I = \int e^x \cos x \, dx$$

$$I = e^x \sin x + e^x \cos x - I$$

$$2I = e^x \sin x + e^x \cos x$$

$$I = \tfrac{1}{2}(e^x \sin x + e^x \cos x)$$

Factorize the exponential and don't forget C.

$$\int e^x \cos x \, dx = \tfrac{1}{2} e^x (\sin x + \cos x) + C$$

The approach taken in the last example works whenever the integral is a product of an exponential and a sinusoid.

PRACTICE EXERCISE 20.4

Evaluate the following integrals. Some will require substitution methods as well as integration by parts.

(a) $\displaystyle\int x^2 \sin 3x \, dx$;　(b) $\displaystyle\int 2x^2 \, e^{-3x} \, dx$;　(c) $\displaystyle\int 3x^3 \ln 2x \, dx$;　(d) $\displaystyle\int x\sqrt{x+1} \, dx$

(e) $\displaystyle\int \ln(x^2 + 1) \, dx$;　(f) $\displaystyle\int e^x \sin 2x \, dx$;　(g) $\displaystyle\int_0^1 \frac{x}{e^x} \, dx$;　(h) $\displaystyle\int_0^{1.5} 4x^2 \, e^{2x} \, dx$

(i) $\displaystyle\int_{\frac{\pi}{6}}^{\frac{\pi}{3}} \frac{\cos \theta}{e^{3\theta}} \, d\theta$;　(j) $\displaystyle\int_1^2 \sqrt{x} \ln x \, dx$

Solutions

(a) $\frac{2}{9}(\frac{1}{3}\cos 3x + x\sin 3x) - \frac{1}{3}x^2\cos 3x + C$; (b) $e^{-3x}(-\frac{2}{3}x^2 - \frac{4}{9}x - \frac{4}{27}) + C$

(c) $\frac{3}{4}x^4(\ln 2x - \frac{1}{4}) + C$; (d) $\frac{2}{3}x(x+1)^{\frac{3}{2}} - \frac{4}{15}(x+1)^{\frac{5}{2}} + C$

(e) $x\ln(x^2+1) - 2x + 2\arctan x + C$; (f) $\frac{1}{5}e^x(\sin 2x - 2\cos 2x) + C$

(g) 0.264; (h) 49.2; (i) 0.0409; (j) 0.494

Assignment XVII

Calculus extended

In carrying out this assignment you will provide records of solutions to six engineering problems using calculus. The performance criteria for assessment are listed below.

- Engineering systems are represented using functions.
- Graphs of functions representing engineering systems are sketched.
- Transformations of functions of the form $y = f(x)$ are represented graphically.
- Combinations of standard functions are integrated.

Your solution to each problem should include the *sketch* of the curve of the relevant function. *You are not required to plot graphs*. What you are required to do is to demonstrate an understanding of how the function behaves at and around key points in its domain. Follow the procedure outlined under, 'A useful strategy for curve sketching' in Chapter 19. Each of your sketches should be supported by a brief justification:

- stating the standard function and its transformation(s),
- evaluating key points,
- applying differentiation to establish the position of turning points, if they exist,
- stating whether you think the function is odd, even, continuous, discontinuous, periodic,
- identifying the domain of the function.

You can use your sketches to help check your solutions of the definite integrals. To check the solutions of indefinite integrals you should differentiate where possible.

If you know what you are doing, it should not take you more than about one hour to complete each solution.

Be positive. Take control of the problems.

PROBLEM 1. A HELIUM BALLOON

A balloon is designed to have an elliptical cross-sectional area with a semi-major axis of 5 m and a semi-minor axis of 3 m. Sketch the curve of the standard function which describes the surface of the balloon. Then apply integration and the volume of revolution method to determine the total volume of helium required to fill the balloon.

PROBLEM 2. INVOLVING AN ALGEBRAIC SUBSTITUTION

Sketch the curve of: $f(x) = 2x\sqrt{9 - x^2}$. Then find an expression for the area under the curve of this function over its entire domain.

PROBLEM 3. WORK!

Work is done when a force moves a load some distance. In general, if the force, F, is constant the work done is simply the product of force and distance, s, i.e. $W = Fs$ joules. Frequently, the required force is dependent on distance in some way, the further a load is moved, the lighter or stronger the force required to move it. So we have $F = f(s)$ and work done becomes the integral:

$$\int_{s=a}^{s=b} f(s)\, ds$$

Given that for a particular load, force is a function of distance,

$$f(s) = \frac{2s}{s^2 + 1} \quad 0 \le s$$

sketch the curve of this function and find the work done in moving a load a distance of 2 m.

PROBLEM 4. INVOLVING TRIGONOMETRIC AND ALGEBRAIC SUBSTITUTION

Sketch the curve of:

$$f(x) = \frac{2}{\sqrt{x^2 - 2}}$$

then evaluate $\int f(x)\, dx$.

PROBLEM 5. TO BE DONE BY PARTS

Sketch the curve of $y = 3x^2 \ln 3x$, hence determine the area under the curve between $x = 1$ and $x = 2$.

PROBLEM 6. PARASITIC OSCILLATION

A voltage varies according to the function: $f(t) = 5\,e^{-0.4t} \sin t$, $0 \le t$.

Sketch a curve of this function, showing voltage, $v = f(t)$.

Then find the average value of voltage over the interval between $t = 0$ and $t = 4\pi$.

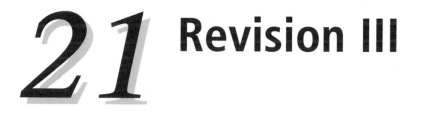

chapter

21 Revision III

Assignment XVIII

This is a summative exercise which requires you to use calculus, complex numbers and matrices to solve engineering problems. The following performance criteria apply.

- Engineering systems are represented using functions.
- The graphs of functions representing engineering systems are sketched.
- Transformations of functions of the form $y = f(x)$ are represented graphically.
- Combinations of standard functions are integrated.
- Numbers are represented on an Argand diagram.
- Engineering systems are represented using complex numbers.
- Calculations involving complex numbers are performed.
- Engineering systems are represented using matrix notation.
- Standard matrices are defined.
- Operations on matrices are performed.
- Engineering problems involving matrices are solved using appropriate techniques.

Read the criteria, revise the topics and, when ready, tackle the assignment. There are 21 multiple choice questions which you should answer in 2 hours, under normal examination conditions.

Good luck!

QUESTION 1

$$A = \begin{bmatrix} 2 & 13 & -6 & 0.22 \\ 17 & -1 & 7 & 10 \\ -5 & 2 & 0.7 & 0 \end{bmatrix}$$

The matrix **A** is

(a) of order four by three
(b) square order two
(c) of order three by four
(d) the transpose of an order three by four matrix

QUESTION 2

$$B = \begin{bmatrix} 3 & -5 \\ -1 & 8 \end{bmatrix} \qquad C = \begin{bmatrix} \pi & -4 \\ 1 & -11 \end{bmatrix}$$

The sum of the matrices shown, **B** + **C**, is

(a) $\begin{bmatrix} 3+\pi & -9 \\ 0 & -3 \end{bmatrix}$

(b) $\begin{bmatrix} 3\pi & 20 \\ -1 & -88 \end{bmatrix}$

(c) $\begin{bmatrix} 3-\pi & -1 \\ -2 & 29 \end{bmatrix}$

(d) $24 - 11\pi$

QUESTION 3

$$D = \begin{bmatrix} 2 & -3 \\ 5 & 7 \end{bmatrix} \qquad E = \begin{bmatrix} 1 & 2 \\ 3 & -4 \end{bmatrix}$$

The product of the matrices shown, **DE**, is

(a) $\begin{bmatrix} -7 & 16 \\ 26 & -18 \end{bmatrix}$

(b) $\begin{bmatrix} 2 & -6 \\ 15 & -28 \end{bmatrix}$

(c) $\begin{bmatrix} -7 \\ -18 \end{bmatrix}$

(d) $[12 \ -37]$

QUESTION 4

$$F = \begin{bmatrix} 2 & -3 \\ 5 & -1 \end{bmatrix}$$

The determinant of matrix **F** is

(a) 3
(b) 13
(c) −17
(d) $\begin{bmatrix} -1 & 3 \\ -5 & 2 \end{bmatrix}$

QUESTION 5

$$G = \begin{bmatrix} -1 & 8 \\ 2 & 0 \end{bmatrix}$$

The adjoint of matrix **G** is

(a) −16
(b) 16
(c) $\begin{bmatrix} 0 & -8 \\ -2 & -1 \end{bmatrix}$
(d) $\begin{bmatrix} 0 & 2 \\ 8 & -1 \end{bmatrix}$

QUESTION 6

$$H = \begin{bmatrix} 6 & -2 \\ 1 & -3 \end{bmatrix}$$

The inverse of matrix **H** is

(a) $\begin{bmatrix} -3 & 2 \\ -1 & 6 \end{bmatrix}$
(b) $\begin{bmatrix} \frac{3}{20} & -\frac{1}{10} \\ \frac{1}{20} & -\frac{3}{10} \end{bmatrix}$
(c) $\begin{bmatrix} -\frac{3}{8} & \frac{1}{8} \\ -\frac{1}{16} & \frac{3}{16} \end{bmatrix}$
(d) $\begin{bmatrix} \frac{3}{16} & -\frac{1}{8} \\ \frac{1}{16} & -\frac{3}{8} \end{bmatrix}$

QUESTION 7

Which of the following would *not* lead to a solution of a pair of simultaneous linear equations?

(a) $X = \left(\frac{1}{|A|} \text{adj} A \right) B$ (b) $X = \left(\frac{1}{\text{adj} A} \det A \right) B$

(c) $X = A^{-1} B$ (d) $IX = A^{-1} B$

QUESTION 8

Which one of the following options gives the correct j-operation?

(a) $j^3 = j$
(b) $j^4 = -1$
(c) $j^2 = 1$
(d) $j^5 = \sqrt{-1}$

QUESTION 9

$(3 + j5)(2 - j3)$ equals

(a) $21 + j$
(b) $-9 + j$
(c) $6 - j15$
(d) $21 - j$

QUESTION 10

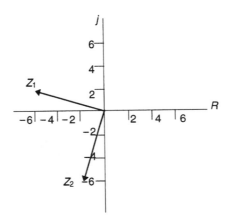

Figure 21.1 Questions 10 and 11

The phasor Z_1 shown in Fig. 21.1 is

(a) $2 - j6$
(b) $6 - j2$
(c) $6.32 \angle 162°$
(d) $6.32(\cos 162° - j\sin 162°)$

QUESTION 11

The phasor Z_2 shown in Fig. 21.1 is

(a) $Z_1 - j6$
(b) jZ_1
(c) $\sqrt{Z_1}$
(d) Z_1^2

QUESTION 12

For the two complex numbers shown,

$$Z_3 = 2.7 - j3.67 \qquad Z_4 = 1.5 + j4.20$$

$\dfrac{Z_3 Z_4}{Z_3 + Z_4}$ evaluates to

(a) $4.82 \angle 117°$
(b) $4.84 \angle 9.5°$
(c) $29.1 \angle -88°$
(d) $1.75 \angle -39°$

QUESTION 13

Taking the square root of the complex number Z_5,

$$Z_5 = 5 - j3$$

gives

(a) $2.41 \angle -15.5°$
(b) $2.41 \angle -5.6°$
(c) $2.82 \angle -37.8°$
(d) $2.92 \angle -5.6°$

QUESTION 14

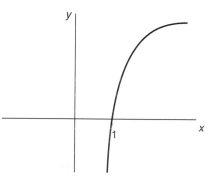

Figure 21.2 Question 14

The graph in Fig. 21.2 is of a standard function which

(a) is odd
(b) is periodic
(c) is discontinuous at $x = 0$
(d) has undergone the transformation $f(x + a)$.

QUESTION 15

Curve (i) of Fig. 21.3 is the curve of the standard function

(a) $f(x) = \ln x$
(b) $f(x) = e^x$
(c) $f(x) = \tan x$
(d) $f(x) = x^{-1}$

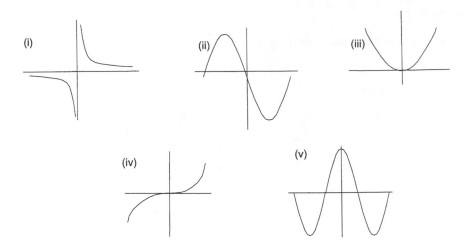

Figure 21.3 Questions 15–18

QUESTION 16

Curve (ii) of Fig. 21.3 represents the standard function, $f(x) = \sin x$, which has undergone the transformation

(a) $f(ax)$
(b) $f(-ax)$
(c) $-af(x)$
(d) $f(x-a)$

QUESTION 17

Select the one statement that is *false* about the standard function of curve (iii) in Fig. 21.3.

(a) The function is even.
(b) The function is continuous.
(c) As, $\lim_{x \to \infty}$, $\lim_{f(x) \to \infty}$.
(d) The function has undergone a reflection about the x-axis.

QUESTION 18

Select the one statement which is *false* regarding the curves in Fig. 21.3.

(a) Curve (v) represents a cosine function.
(b) Curve (iii) has a point of inflexion.
(c) Curve (iv) could represent the inverse sine function.
(d) Curve (iv) could represent a cubic function.

QUESTION 19

The integral $\int x(x^2 - 4)^3 \, dx$ evaluates to

(a) $\frac{1}{8}(x^2 - 4)^4 + C$
(b) $\frac{3}{2}(x^2 - 4)^2 + C$
(c) $2(x^2 - 4)^4 + C$
(d) $\frac{1}{4}(x^2 - 4)^4 + C$

QUESTION 20

The integral $\int \sqrt{9 - x^2} \, dx$ evaluates to

(a) $9\arcsin\dfrac{x}{3} + x\sqrt{9 - x^2} + C$

(b) $\dfrac{9}{2}\arcsin\dfrac{x}{3} + \dfrac{x}{3} + C$

(c) $\dfrac{9}{2}\arcsin\dfrac{x}{3} + \dfrac{x}{2}\sqrt{9 - x^2} + C$

(d) $\dfrac{9}{2}\arccos\dfrac{x}{3} + \dfrac{2\sqrt{9 - x^2}}{x} + C$

QUESTION 21

Evaluating the integral $\int x^2 \ln 2x \, dx$ gives

(a) $x^2(\ln x - 1) + C$
(b) $\frac{1}{3}x^3(\ln 2x - \frac{1}{3}) + C$
(c) $x^2(\ln x - \frac{1}{2}) + C$
(d) $x^2(e^{2x} - \frac{1}{2}) + C$

Appendix I

Système International d'Unités

The International System of Units (abbreviated SI) is commonly called the **metric** system because it is based upon the metre as the standard unit of length.

The origins of SI go back to Napoleonic France. It is said that Napoleon himself instigated the system annoyed at being cheated by the merchants who supplied cloth for his army's uniforms. Measurement up to that time was based on 'standards' that were variable to say the least. In England, the 'yard' was the distance between the nose and the end of an outstretched arm, the 'inch' was the width of a person's thumb and a 'foot', just that, the length of a foot. The wealthiest merchants must have all been the smallest people.

The present SI system was internationally adopted in 1960. It is based on seven **Fundamental Units** and two **Supplementary Units**. All other units are derived from these.

The fundamental units

Unit	Symbol	Dimension	Standard
metre	m	length	1 650 763.73 wavelengths of the orange–red spectral line of krypton-86.
kilogram	kg	mass	Cylinder of platinum–iridium alloy kept by the International Bureau of Weights and Measures in Paris. The only man-made standard remaining.
second	s	time	The time it takes to complete 9 192 631 770 cycles of radiation by a caesium-133 atom.
ampere	A	electric current	The current in each of two parallel conductors that produces an electromagnetic force of $2 \times 10^{-7}\,\text{N}$ between the conductors when they are 1 m long and spaced 1 m apart.

Unit	Symbol	Dimension	Standard
kelvin	K	temperature	$\frac{1}{273.16}$ of the thermodynamic temperature of the triple point of water. The 'triple point' of a substance is the temperature at the point where solid, liquid and vapour meet. 0 K is called 'absolute zero'.
mole	mol	amount of substance	The amount of atoms in 0.012 kg of carbon-12
candela	cd	luminous intensity	The luminous intensity of $\frac{1}{600\,000}$ m^2 of a blackbody (a perfect radiator) at the temperature of freezing platinum (2045 K)

The supplementary units

Unit	Symbol	Dimension	Standard
radian	rad	plane angle	The angle at the centre of a circle subtended by an arc equal to the radius of the circle.
steradian	sr	solid angle	The angle at the centre of a sphere subtended by an area of the surface of the sphere which is equal to the area of a square having sides equal to the radius.

Some of the derived units

Unit	Symbol	Dimension	Comment
square metre	m^2	area	The product of two lengths is the product of a metre and a metre.
cubic metre	m^3	volume	The product of three lengths. The litre is a unit of volume normally used with fluids. It is not an SI unit. $1\,\text{m}^3 = 1000\,\text{l}$
metre per second	m s^{-1}	velocity	Rate of change of distance over time.
metre per second squared	m s^{-2}	acceleration	Rate of change of velocity over time.
newton	N	force	$1\,\text{N} = 1\,\text{kg}\,\text{m}\,\text{s}^{-2}$. The force which gives a mass of 1 kg an acceleration of $1\,\text{m}\,\text{s}^{-2}$.
pascal	Pa	pressure, stress	$1\,\text{Pa} = 1\,\text{N}\,\text{m}^{-2}$. The force of 1 N distributed over an area of $1\,\text{m}^2$.

Unit	Symbol	Dimension	Comment
joule	J	work, energy	$1\,\text{J} = 1\,\text{N}\,\text{m} = 1\,\text{kg}\,\text{m}^2\,\text{s}^{-2}$. The amount of energy transferred when the point of application of a force moves a distance of 1 m.
watt	W	power	$1\,\text{W} = 1\,\text{J}\,\text{s}^{-1}$. The rate of energy conversion.
hertz	Hz	frequency	$1\,\text{Hz} = 1\,\text{s}^{-1}$. The reciprocal of 1 s.
coulomb	C	electric charge	$1\,\text{C} = 1\,\text{A}\,\text{s}$. The amount of charge moved by 1 A of current in 1 s.
volt	V	electric potential difference	$1\,\text{V} = 1\,\text{J}\,\text{A}^{-1}\,\text{s}^{-1}$. The potential which converts 1 J of energy in moving 1 C of charge. Potential difference is frequently and incorrectly called 'voltage'.
ohm	Ω	electric resistance	$1\,\Omega = 1\,\text{V}\,\text{A}^{-1}$. The amount of resistance present when 1 V produces 1 A of current.

There are a number of units called the **Additional Units**, which are not part of SI but because they are used so commonly by the scientific and engineering community they cannot be abandoned. They are tolerated within SI, for the present at least.

Some of the additional units

Unit	Symbol	Dimension	Comment
litre	l	volume (of a fluid)	$1\,\text{l} = 10^{-3}\,\text{m}^3$
gram	g	mass	$1\,\text{g} = 10^{-3}\,\text{kg}$
tonne	t	mass	$1\,\text{t} = 1000\,\text{kg}$
hour	h	time	$1\,\text{h} = 3600\,\text{s}$
minute	min	time	$1\,\text{min} = 60\,\text{s}$
degree celsius	°C	temperature	One interval on the celsius scale is equal to one interval on the kelvin scale.
degree	°	plane angle	$1° = \frac{\pi}{180}\,\text{rad}$
decibel	dB	power ratio	Obtained by taking $10\log_{10}$ of the ratio of two given powers. Used in electronics and audio engineering.

Since SI units form part of a coherent system of units whereby the derived units are all related to the fundamental units, we have to pay the price that some units are inconveniently large and some are inconveniently small. To cope with this, there is a set of prefixes which allows numbers to be expressed in a convenient form. In the **preferred standard form** the prefixes represent multiples or sub-multiples of 1000 (10^3).

Some of the prefixes of the preferred standard form

Prefix	Symbol	Exponent of 10	Value
exa	E	10^{18}	1 000 000 000 000 000 000
peta	P	10^{15}	1 000 000 000 000 000
tera	T	10^{12}	1 000 000 000 000
giga	G	10^{9}	1 000 000 000
mega	M	10^{6}	1 000 000
kilo	k	10^{3}	1 000
		10^{0}	1 (no prefix)
milli	m	10^{-3}	0.001
micro	μ	10^{-6}	0.000 001
nano	n	10^{-9}	0.000 000 001
pico	p	10^{-12}	0.000 000 000 001
femto	f	10^{-15}	0.000 000 000 000 001
atto	a	10^{-18}	0.000 000 000 000 000 001

Note the method of separating tens of thousands with a space. Commas should not be used because they can be confused with a decimal point. It is, in fact, accepted practice in some countries to represent the decimal place with a comma instead of a full stop.

The kilogram is a special case. Because it already contains a prefix and double prefixes are not used, we have to put the prefix in front of gram. For example 1000 kg would not be written as 1 k kg but either as 10^3 kg or 10^6 g.

Confusion sometimes arises with prefixes used with units^2 and units^3. An area expressed as $10 \, \text{km}^2$ is an area of:

$$10 \, (\text{km})^2 = 10 \times (10^3)^2 \, \text{m}^2 = 10 \times 10^6 \, \text{m}^2 = 10\,000\,000 \, \text{m}^2$$

Finally, the use of upper and lower case letters must obey the conventions of SI.

Appendix II

Over the years, the practitioners of mathematics, science and technology have developed a shorthand for communicating information and ideas in written form. How much easier to jot down the letter t when talking about time for instance.

However, our Roman alphabet is not sufficiently large to be used to represent the vast number of: dimensions, units, prefixes, variables and constants that exist. So, in addition to the Roman characters we also make extensive use of those Greek characters that differ from the Roman.

Letter	Upper case	Lower case	Some examples
alpha	A	α	α, β and γ are alternatives to x, y and z; the symbols used for variables in
beta	B	β	mathematics. They are also used to represent constants of transmission
gamma	Γ	γ	lines in telecommunications.
delta	Δ	δ	δ means a '*small* change in'. Δ is usually just 'change in'.
epsilon	E	ε	ε represents *strain*. Also *permittivity* in electrostatics.
zeta	Z	ζ	ζ is *damping constant* in control engineering.
eta	H	η	η is widely used to represent *efficiency*.
theta	Θ	θ	θ represents *plane angle* and also *temperature*.
iota	I	ι	
kappa	K	κ	
lambda	Λ	λ	λ represents *wavelength* of electromagnetic waves.
mu	M	μ	μ is much overused: prefix *micro*, *coefficient of friction* and *permeability* in magnetism.

Letter	Upper case	Lower case	Some examples
nu	N	ν	
xi	Ξ	ξ	
omicron	O	o	
pi	Π	π	No mistaking this one, $\pi \approx 3.142\ldots$, a constant.
rho	P	ρ	ρ represents *density* and also electrical *resistivity*.
sigma	Σ	σ	Σ represents 'sum of' in maths. σ is *stress*.
tau	T	τ	τ represents *time constant*.
upsilon	Y	υ	
phi	Φ	ϕ	Φ is *magnetic flux*. ϕ is *phase angle*.
chi	X	χ	
psi	Ψ	ψ	ψ *electric flux*.
omega	Ω	ω	Ω represents the unit of resistance, the *ohm*. ω is *angular frequency* which is often called *angular velocity*.

Appendix III

Various dictionary definitions of the mathematical meaning of **number** state it to be: a *particular value* in a count of things; an indication of *the position* of something in a series; an identification of *a member of* a series. Whatever the situation, we expect a number to be one particular value. However, there are numbers whose values are not fully defined. They are either arithmetically incomplete or simply do not belong to our **real number system**.

So there are: **real** numbers, **irrational** numbers and even **imaginary** numbers. The irrational and imaginary numbers happen to be some of the most important and interesting numbers that we deal with. They are interesting from a mathematical viewpoint and some are interesting from a physical viewpoint since they represent important constants such as the speed of light.

Number	Comment
0	An integer and therefore a rational number. It occupies the central position on the number line of the whole set of all real numbers. Some operations with 0 have dramatic effects, for example: $\frac{n}{0}$ and $0n$.
∞	The result of $\frac{n}{0}$, a number beyond our comprehension. We say that it is mathematically *undefined*.
1	*Unity.* Identifies the existence of a unit of something. If you have a kilogram of mass you have $1 \times$ kg. Some operations with 1 seem to have little effect, take: $\frac{n}{1}$, $1n$, 1^n, $1^{\frac{1}{n}}$.
$\sqrt{2}$	A *surd* meaning 'deaf'. A Latin mis-translation of the Greek word for mute; a number that cannot 'speak' its value. A calculator gives, $1.414213562\ldots$ a non-terminating decimal. This is what you get when you try to evaluate the hypotenuse of the triangle whose other two sides are 1 unit in length. Important in engineering because we use it to find the r.m.s. value of a sinusoid.

Number	Comment
$\sqrt{3}$	Another surd. It is the length of one side of a right-angled triangle whose other two sides are 1 and 2 units in length. In engineering it is the factor used to convert phase and line voltages in a three-phase distribution system. There are many other surds.
π	The number of times the diameter of a circle divides into its circumference. First discovered by Archimedes who established its value correct to 3 s.f. (3.14). It is another irrational number but vital for dealing with periodic functions.
e	A number with a unique property. When raised to any power its rate of change is proportional to its value. Essential in modelling dynamic systems which change exponentially.
$\log 2$	Taken to either base e or 10 the logarithm of 2 is irrational and also useful. $\log_{10} 2$ is used in audio engineering to find the 3 dB points, the points at which the human ear can detect a change in the level of sound.
μ_0	Permeability (distributed inductance) of free space. In SI, its value is $4\pi \times 10^{-7}\,\mathrm{H\,m^{-1}}$.
ε_0	Permittivity (distributed capacitance) of free space. Its SI value is $8.854 \times 10^{-12}\,\mathrm{F\,m^{-1}}$.
c	The speed of an electromagnetic wave in free space. A physical constant which has the approximate SI value of $300 \times 10^6\,\mathrm{m\,s^{-1}}$. It is a function of u_0 and ε_0. It is a point of interest that, very approximately, the speed of a sound wave in the atmosphere is around $300\,\mathrm{m\,s^{-1}}$. So in air, speed of electromagnetic waves (light, radio, etc) is around a million times greater than the speed of sound waves.
G	The *gravitational constant* as used in Newton's law for calculating force of gravity between two masses. In SI, $G = 6.670 \times 10^{-11}\,\mathrm{N\,m^2\,kg^{-2}}$ for any mass.
g	The averaged value of acceleration due to gravity on the surface of the Earth. In SI it is $9.80665\,\mathrm{m\,s^{-2}}$. Since this is an approximation in the first place, rarely do we bother taking it to an accuracy greater than 3 s.f.
i, **j** and **k**	Numbers with directional properties. These are the *unit vectors* which have a magnitude of 1 and lie in directions parallel to the axes of the three-dimensional system of Cartesian coordinates: x, y and z respectively.
i or j	One of the favourites, the imaginary $\sqrt{-1}$. Not a real number at all but used extensively in the solution of practical engineering problems.

Appendix IV

Some interesting derivations

Completing the square

A perfect square is a special case of a binomial that is easily factorized. Take the perfect square with terms x and α. Its structure consists of: the first term squared, the second term squared and in the middle is twice the product of the two terms.

$$x^2 + 2\alpha x + \alpha^2 = (x + \alpha)(x + \alpha) = (x + \alpha)^2$$

Now consider the general form of a quadratic that does not factorize. Let us subtract c and divide by a to make the LHS resemble the perfect square as closely as possible.

$$ax^2 + bx + c = 0$$

$$ax^2 + bx = -c$$

$$x^2 + \frac{b}{a}x = -\frac{c}{a}$$

Concentrating just on the expression on the LHS: we have the quadratic term and the linear term but not the α^2 term. So we add α^2 to make the expression equal to the perfect square.

$$x^2 + \frac{b}{a}x + \alpha^2 = x^2 + 2\alpha x + \alpha^2$$

We can subtract x^2 and α^2 from both sides, transpose and obtain an expression for α^2.

$$\frac{b}{a}x = 2\alpha x$$

$$\frac{b}{a} = 2\alpha \qquad \text{so} \qquad \alpha = \frac{b}{2a}$$

$$\alpha^2 = \left(\frac{b}{2a}\right)^2$$

Returning to our quadratic equation, we add the expression for a^2 to both sides. This is called **completing the square** because the LHS now has the structure of the perfect square: 1st term squared, 2nd term squared and in the middle twice the product of the two terms. So the LHS can now be factorized.

$$x^2 + \frac{b}{a}x + \left(\frac{b}{2a}\right)^2 = -\frac{c}{a} + \left(\frac{b}{2a}\right)^2$$

$$\left(x + \frac{b}{2a}\right)\left(x + \frac{b}{2a}\right) = -\frac{c}{a} + \left(\frac{b}{2a}\right)^2$$

$$\left(x + \frac{b}{2a}\right)^2 = \left(\frac{b}{2a}\right)^2 - \frac{c}{a}$$

We tidy up the LHS and the RHS.

The common denominator on the RHS is $4a^2$ so we use that.

$$\left(x + \frac{b}{2a}\right)^2 = \frac{b^2}{4a^2} - \frac{c}{a}$$

$$\left(x + \frac{b}{2a}\right)^2 = \frac{b^2 - 4ac}{4a^2}$$

Taking the square root of both sides breaks up the sum and allows us to subtract the b-over-$2a$ term.

$$x + \frac{b}{2a} = \pm\sqrt{\frac{b^2 - 4ac}{4a^2}}$$

$$x + \frac{b}{2a} = \frac{\pm\sqrt{b^2 - 4ac}}{2a}$$

$$x = -\frac{b}{2a} + \frac{\pm\sqrt{b^2 - 4ac}}{2a}$$

Now we get $2a$ as the common denominator on the RHS and tidy-up.

$$x = \frac{-b \pm \sqrt{b^2 - 4ac}}{2a}$$

The outcome is the well known formula for finding the roots of a quadratic equation. It turns out to be a solution-by-factors after all.

Differentiating a product

Given that y is the product of one function of x and some other function x. Let:

$$u = f(x) \qquad \text{and} \qquad v = g(x)$$

Then y is just the simple product of u and v.

Consider Fig. IV.1. If we take the coordinates of some point P on the curve to be (uv, y) then the coordinates of some point close to P, say Q, are:

$$((u + \delta u)(v + \delta v), (y + \delta y))$$

These have been brought about by the small change, δx.

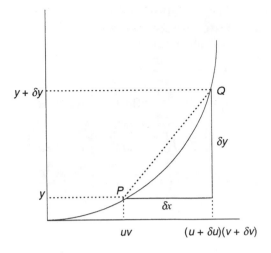

Figure IV.1 Differentiating a product

Now we substitute the coordinates of point Q into the equation, expand the brackets, subtract y from the left and uv from the right (for they are equal) and divide by δx.

$$y = f(x)g(x)$$

$$y = uv$$

$$y + \delta y = (u + \delta u)(v + \delta v)$$

$$y + \delta y = uv + u\delta v + v\delta u + \delta u\delta v$$

$$\delta y = u\delta v + v\delta u + \delta u\delta v$$

$$\frac{\delta y}{\delta x} = \frac{u\delta v + v\delta u + \delta u\delta v}{\delta x}$$

This gives an expression for the average rate of change between the points P and Q on the curve of $f(x)g(x)$.

$$\frac{\delta y}{\delta x} = u\frac{\delta v}{\delta x} + v\frac{\delta u}{\delta x} + \frac{\delta u\delta v}{\delta x}$$

As point Q is brought closer to P, δx will tend in the limit to zero and the average rate of change becomes the actual rate of change at point P.

$$\text{As} \quad \lim_{\delta x \to 0} \frac{\delta y}{\delta x} = \frac{dy}{dx}$$

The limiting value of δx affects the rate of change of u and the rate of change of v in the same way.

$$\frac{\delta v}{\delta x} = \frac{dv}{dx}$$

$$\frac{\delta u}{\delta x} = \frac{du}{dx}$$

$$\text{and} \quad \frac{\delta u\delta v}{\delta x} = 0$$

The last term becomes zero because δv and δu both depend on δx. If δx tends to zero so must δv and δu, and the term becomes the product of a differential coefficient and zero. It is either:

$$\frac{du}{dx} \times 0 \quad \text{or} \quad 0 \times \frac{dv}{dx}$$

This gives us the general rule for differentiating the product of two functions.

$$\frac{dy}{dx} = u\frac{dv}{dx} + v\frac{du}{dx}$$

Differentiating a quotient

Given that y is the quotient of two separate functions we let:

$$u = f(x) \quad \text{and} \quad v = g(x)$$

Now a small change in x will bring about a small change in u, a small change in v and a small change in y.

$$y = \frac{f(x)}{g(x)} \qquad y = \frac{u}{v}$$

$$y + \delta y = \frac{u + \delta u}{v + \delta v}$$

Subtracting y from the left is the same as subtracting u/v from the right.

$$\delta y = \frac{u + \delta u}{v + \delta v} - \frac{u}{v}$$

$$\delta y = \frac{v(u + \delta u)}{v(v + \delta v)} - \frac{u(v + \delta v)}{v(v + \delta v)}$$

Put the RHS over the common denominator:

$$v(v + \delta v)$$

$$\delta y = \frac{v(u + \delta u) - u(v + \delta v)}{v(v + \delta v)}$$

Expand brackets and collect-up.

$$\delta y = \frac{vu + v\delta u - uv - u\delta v}{v^2 + v\delta v}$$

$$\delta y = \frac{v\delta u - u\delta v}{v^2 + v\delta v}$$

We want the gradient so we divide both sides by δx.

$$\frac{\delta y}{\delta x} = \frac{v\delta u - u\delta v}{\delta x(v^2 + v\delta v)}$$

On the RHS we divide the numerator and denominator by δx because this clears it from the denominator.

$$\frac{\delta y}{\delta x} = \frac{\dfrac{v\delta u - u\delta v}{\delta x}}{v^2 + v\delta v}$$

Separate the numerator into partial fractions to get separate rates of change: u with respect to x and v with respect to x.

$$\frac{\delta y}{\delta x} = \frac{v\dfrac{\delta u}{\delta x} - u\dfrac{\delta v}{\delta x}}{v^2 + v\delta v}$$

Now as the change in x tends to zero, the average rates of change become the actual rates of change.

$$\text{As} \quad \delta x \lim_{\delta x \to 0} \frac{\delta y}{\delta x} = \frac{dy}{dx}$$

δv in the denominator goes to zero because v depends on x. So if δx tends to zero then δv tends to zero.

$$\frac{\delta u}{\delta x} = \frac{du}{dx} \qquad \frac{\delta v}{\delta x} = \frac{dv}{dx} \qquad \delta v = 0$$

and

$$\frac{dy}{dx} = \frac{v\dfrac{du}{dx} - u\dfrac{dv}{dx}}{v^2}$$

Glossary

Abscissa
The shortest distance of a point from the *y*-axis of a graph; the '*x*-value'. See *ordinate*.

Absolute error
The difference between the estimate, approximate or measured value and the true value of a quantity. See *relative error*.

Acceleration
Rate of change of velocity over time in m s^{-2}.

Additional units
Units which are not part of SI but are still used along with SI units.

Adjoint
Of a matrix: transpose of the matrix of its cofactors.

Algebra
Branch of mathematics that uses letters and symbols to represent numbers and quantities.

Ampere
Andre, French scientist, who worked with electrodynamic theory.
Ampere: fundamental SI unit of electric current.

Amplitude
The maximum, or peak value of an oscillation.

Analytical
An algebraic approach to problem solving.

Argand
Jean, Swiss mathematician. His diagram is used to represent the position of complex numbers in the real–imaginary plane.

Argument
The angular displacement of a point in a plane in the polar coordinate system. The phase angle of a phasor or direction of a vector.

Associative
Rule that determines whether operations involving grouped quantities gives the same result whatever the order of grouping.

Asymptote
Line that continually approaches a curve but does not meet it at a finite distance.

Average
The most common or representative value of a sample. Usually the arithmetic mean.

Axis
Line about which rotation takes place. Also, one of the reference lines in the system of Cartesian coordinates.

Base
The whole number on which a counting system is based.

Binary
The counting system which has the base of 2. The two symbols are 0 and 1.

Binomial
The sum (or difference) of two distinct algebraic terms such as $a + b$.

Binomial approximation	Method of finding approximate errors using the first two terms of the binomial series.
Binomial theorem	Rule for generating a binomial series.
Boolean	An algebra concerned with logical process. For instance, the statement $A.B = C$ means that C is true only if A *and* B are true. Applied to the analysis of digital devices, usually electronic. After George Boole, British mathematician.
By-parts	Formula based on the product rule for differentiation which helps us to integrate two functions when one is not related as a derivative of the other.
Calculus	From 'infinitesimal calculus'; calculation with infinitesimally small numbers. According to Newton's *Fundamental Theorem*, differential and integral calculus are inverse operations.
Candela	Fundamental SI unit of luminescence.
Capacitance	The ability to store electric charge in units of the Farad (F).
Cardano	Gerolamo, 16th c. Italian mathematician, philosopher, astrologer and professional gambler. Together with Bombelli he proposed the acceptance of $\sqrt{-1}$ as a solution to a quadratic equation that has no 'real' roots.
Cartesian	After Rene Descatres, 17th c. French philosopher, mathematician and intellectual giant. He was the one who said 'Cogito, ergo sum' (*I think, therefore I am*). The coordinate system for representing a point in a plane (graphs) was one of his many contributions.
Celsius	Anders, Swedish astronomer, founder of the $100°$ scale for temperature. His name makes the 'centigrade' scale obsolete.
Centroid	The point in a plane or solid body that is the centre of mass. It is the focus of internal and external forces and can be considered as the centre of gravity.
Chain rule	See, *function-of-a-function-rule*.
Charge	The basis of electrical energy. The property of an electron and a proton to attract (or repel) other particles that carry charge. In SI the charge in one electron is known to be -1.602×10^{-19} coulomb.
Codomain	The set that includes all the possible outputs of a function.
Coefficient	Mathematically, a factor that accompanies a variable. In engineering this factor is a constant such as the coefficient of friction or electrical resistance.

Cofactor	The value obtained when the minor of a matrix is multiplied by -1 raised to a power determined by a certain rule.
Common	A factor or denominator is common if it operates on more than one quantity.
Commutative	Rule which determines whether an operation can be carried out in reverse.
Complex number	A number that has two parts: real and imaginary.
Complexor	Rotating line which provides a graphical model of sinusoids. The basis of phasor representation.
Compliance	The reciprocal of *stiffness* (of a spring) in $m\,N^{-1}$.
Component	Part of a whole. Contributing to the composition of a whole. Vectors and phasors are made up from Cartesian components. A resultant vector or phasor is made up from component vectors or phasors.
Conformability	Suitable, similar, consistent. Two matrices must be conformable for addition and multiplication, for example.
Congruent	In geometry: coinciding exactly when superimposed.
Conjugate	A value or expression so related to another that the product of the two is rational. A complex conjugate, it is found by multiplying the imaginary part by -1.
Conservation laws	Of energy: within a system the total amount of energy is constant. Of momentum: within a group of objects total momentum is constant providing the objects are not subject to external forces.
Constant	The part of a relationship between variables whose value does not change. May be a distinct term or coefficient of a variable.
Continuous	Of a variable: one that can be measured. Of the curve of a function: one that has no breaks or jumps so that a small change in x will only bring about a small change in y.
Convergent	Coming together. Results of iterative methods that come ever closer to some true or focal value. Convergent series lead to a solution.
Cosecant	The reciprocal value of the sine ratio.
Cosine	Complementary sine. Value assigned to an angle in a right-angled triangle. The ratio of the adjacent side to the hypotenuse.
Cosine rule	For any triangle: $a^2 = b^2 + c^2 - 2bc\cos A$
Cotangent	The reciprocal of the tangent ratio.
Coulomb	Derived SI unit of electric charge. Named after Charles Coulomb, French physicist.
Cross-product	The vector product of two vectors.

Cubic	An equation of the third degree. One in which the highest power of the variable is 3.
Decibel	Non-SI unit of a power ratio. See Appendix I.
Definite integral	An integration between specified limits which produces a numerical value such as an area in units^2 or a volume in units^3.
Degree	Non SI unit of angular measure. Obtained by dividing the angle subtended by the circumference of circle into 360 equal parts. Of a differential equation: the highest power of the highest order term.
Denary	The counting system that has the base of 10. The ten symbols are: 0 to 9.
Denominator	The number below the line of a fraction. Also called the divisor.
Density	Of a substance: the ratio of mass over volume in $kg\,m^{-3}$.
Derivative	The rate of change of a *primitive*. Where the primitive is some mathematical function. Also called the *differential coefficient* of one variable with respect to another.
Derived units	All those SI units that are not fundamental. See Appendix I.
Determinant	The value obtained when products of the elements of a rectangular array are added according to a certain rule.
Difference	The result of a subtraction.
Differentiation	The process of finding a rate of change.
Dimension	A measurable quantity. Also the number of factors in an algebraic term: xyz and $2x^2$ are three-dimensional terms.
Discontinuous	The curve of a discontinuous functions has breaks or jumps. Differentiation (and therefore integration) is not feasible at the discontinuity.
Discrete	Separate, distinct, discontinuous. A discrete variable can be counted.
Dispersion	In statistics: a measure of the spread of data. Range and standard deviation are examples.
Distinct	Two or more roots of an equation that are different values; as opposed to roots that are *repeated*.
Distributive	Law that determines whether an operation on a group of quantities can also be carried out on individual members of the group and still produce the same result.
Divergent	A series which is not *convergent* so it does not lead to a solution.
Domain	The whole set of numbers that can be assigned to the independent variable of a function.
Dot product	The scalar product of two vectors.

Dynamic	A thing or system that changes over time.
e.m.f.	Electro-motive force measured in the unit of the volt. Often and incorrectly called *voltage*.
Einstein	Albert, 20th c. mathematician. Founder of quantum mechanics, the study of waves and the dynamics of sub-atomic particles. His theories have superseded Newtonian mechanics but not replaced them.
Elasticity	The property of a material to become deformed by a force and return to its original dimensions when the force is removed.
Electromagnetic waves	Waves composed of oscillating electric and magnetic fields. They are transverse waves, which means that they travel in a direction perpendicular to the fields. The speed of all electromagnetic waves in free space is the physical constant $c \approx 300 \times 10^6 \,\mathrm{m\,s^{-1}}$. EM waves exist at all frequencies of the spectrum, at the low end they are called radio waves; higher frequencies are light waves, X-rays, and gamma rays.
Element	Of a matrix: one of the values or expressions that make up a matrix.
Empirical	Based on observation or experiment, not theory.
Energy	The ability of particles or waves to do work.
Equality	The relationship that exists between the two sides of an equation. Another name for an equation.
Equilateral	A triangle having all sides equal.
Equilibrium	State of balance. Of forces: a state of equilibrium exists when the combination of all external forces on a mass produces zero acceleration.
Error	See *absolute error*.
Evaluate	Generally: to appraise or assess. Mathematically: to find the number or expression that represents a quantity.
Expansion	The operation of 'clearing brackets' by multiplying or dividing all the terms inside the brackets by the term or terms outside.
Exponent	Symbol indicating the power of a factor.
Exponential	Relating to an exponent. A function that changes at a rate which is proportional to the variable. e is sometimes called *the exponential*.
Expression	Collection of mathematical terms linked as a sum and/or difference.

Extrapolate	Extend the curve of a graph beyond available data.
Factor	A multiplier. Or one of the numbers or symbols that make up a term or expression by multiplication.
Factorization	Resolving factors by separating and writing them as a product.
Farad	SI unit of capacitance. Named after Michael Farady, famous for his work with electromagnetism.
Force	The external agent that is capable of accelerating an object. According to Newton's second law it is given by $F = ma$. The SI unit of force is the newton.
Frequency	The rate at which a periodic function changes in cycles per second. The SI unit of frequency is the hertz. See Appendix I.
Frustum	The part of a cone or pyramid that remains after a portion of its upper part has been removed.
Function	The rule that specifies how one variable depends on another.
Function-of-a-function	A rule for differentiation, also called the *chain rule*. It allows us to differentiate an expression in which one function is nested inside another function.
General	General mathematical statements are those in which variables and constants are represented by symbols. Of differential equations: a general solution is one for which the constant(s) of integration have not been evaluated.
Gram	10^{-3} kg. Non-SI unit of mass. See Appendix l.
Graphical	An approach to problem solving that makes use of graphs or diagrams.
Harmonic	Simple harmonic motion is sinusoidal: the extension of a vibrating spring or velocity and position of a pendulum.
Heat	The form of energy that is transferred when two objects at different temperatures are brought together.
Henry	SI unit of inductance. Named after Joseph Henry, American physicist.
Hertz	SI unit of frequency. Named after Gustav Hertz, the German physicist.
Hexadecimal	The counting system that has the base of 16. The 16 symbols are: 0, 1, 2, 3, 4, 5, 6, 7, 8, 9, A, B, C, D, E, F.
Histogram	Form of graph that presents a frequency distribution in vertical columns.

Hyperbola	The curve of one of the sections through a cone. Also the curve traced by a point that moves so that the difference between its distances from two fixed points remains constant. A rectangular hyperbola is the curve of $f(x) = 1/x$.
Hyperbolic	Non-circular functions, sinh, cosh and tanh are hyperbolic equivalents of sine, cosine and tangent. Base e logs are also called hyperbolic logs.
Hypotenuse	The side opposite the right angle, in a right-angled triangle.
Ill-conditioning	In general, a problem is ill-conditioned if a small change in one variable brings about a large change in the result so that errors are amplified. Simultaneous equations are ill-conditioned if their graphs are almost parallel.
Inconsistency	Of simultaneous equations: equations are inconsistent when no simultaneous solution is possible. The graphs of such equations are parallel.
Identity	Holds true for any value of a variable, therefore not an equality. Of matrices: the matrix equivalent of 1 in ordinary numbers.
Imaginary	A number with $\sqrt{-1}$ as a factor and belonging to the set that is not real.
Indefinite integral	An integration between unspecified limits. The evaluation of an indefinite integral leads to an algebraic expression, not a numerical value.
Index	Another name for *exponent* or *power* of a factor.
Inductance	The electromagnetic property that resists changes in electric current. Its SI unit is the henry.
Inequality	A mathematical statement in which terms are not balanced as in an equation (equality). An example of an inequality is $y > 2x$.
Inertia	The property of an object to remain at rest or at constant velocity. Mass is a measure of inertia. It indicates how difficult it is to accelerate an object.
Inflexion	On a graph the point at which the gradient of the curve becomes zero but does not reverse.
Instantaneous	Of periodic quantities: the value at an instant (not an interval) in time.
Integer	Whole positive or negative number including zero.
Integral	Whole, complete. In maths, the result of an integration.

Integration	The process of finding the whole from a number of parts.
Intercept	The point where the curve of a graph crosses the y-axis. The value of y when $x = 0$.
Interpolate	Join up the points of a curve between available data.
Inverse	An operation that is the reverse of another. An inverse function takes a value from the codomain of the original function and finds the corresponding value in its domain.
	Of a matrix: the matrix equivalent of a reciprocal in ordinary numbers. Found by dividing the adjoint of a matrix by its determinant.
Irrational	In general: not logical. In maths: not a natural number such as a non-terminating decimal or a number not defined in the real number system such as ∞ or $\sqrt{-1}$.
Isosceles	A triangle having two sides equal in length.
Iteration	A repeated process or operation which, generally, leads closer and closer to a solution.
Joule	James, British physicist: verified the conservation of energy laws and worked with thermodynamics.
	Joule: SI unit of work or energy. See Appendix I.
Kelvin	William Thomson, British physicist: involved in developing the science of thermodynamics. Kelvin: SI unit of temperature. See Appendix I.
Kinetic energy	Mechanical energy an object possesses due to its motion $E = \frac{1}{2}mv^2$.
Limit	A quantity that a function (or series) may be made to approach as closely as desired.
Limiting values	The numerical values of the independent variable which specify the upper and lower limits of integration.
Linear	Involving one dimension; of a straight line; an equation of the 1st degree.
Litre	10^{-3} m^3. Non-SI unit of volume. See Appendix I.
Logarithm	The logarithm of a number is the power to which a chosen base must be raised to give that number.
m.m.f.	Magneto-motive force. The force that magnetizes a substance. Its SI unit is the ampere but because the force is also proportional to the number of turns of a conductor the non-SI unit 'ampere-turn' is often used.

Maclaurin's series — A general series based on Maclaurin's theorem. It can be used to generate other series.

Magnitude — Order of: value quoted to the nearest power of 10.
The magnitude of a vector is the quantity of the vector without specified direction. The modulus of a two-dimensional quantity.

Mass — The property of *inertia* of an object. The SI unit of mass is the kilogram.

Matrix — A rectangular array of values or expressions that must be treated as a single quantity.

Mean — The representative value (an average) of a sample. Found by taking the sum of the values and dividing by the number of values.

Median — The middle value in a sample.

Mid-ordinate rule — Method for evaluating the approximate area of an irregular plane by considering it to be made up of a collection of rectangular strips.

Minor — Of the element of a matrix: value obtained according to a certain rule.

Mode — The value in a sample that occurs most often.

Model — A mathematical (or other) construct designed to represent or simulate the properties and behaviour of a real system.

Modulus — Magnitude of a vector or phasor. Also, constant multiplier or coefficient that gives the ratio between variables (e.g. Young's modulus).

Moment — Product of force and the distance of the point of application of the force from an axis of rotation. Turning force or *torque*.

Naperian — Logs to the base of e, also called *natural* and *hyperbolic* logs. Named after John Napier, Scottish mathematician.

Natural — Of numbers: whole positive numbers including zero.

Newton — Isaac, Sir. Founder of Classical Mechanics; the dynamics of large bodies and inventor of differential calculus.
newton: SI unit of force.

Newton's laws — *1st law*: the motion of an object remains at constant velocity unless subject to a force.
2nd law: leads to $F = ma$.
3rd law: each application of a force leads to a force of reaction which is equal and opposite.
Law of gravitation: the force of attraction between two masses is proportional to the product of the masses and inversely proportional to the square of the distance between the centres of the masses.

Newton–Raphson	Numerical method for finding the roots of an equation.
Normal	A particular type of frequency distribution that gives rise to a 'bell-shaped' frequency distribution curve, which is symmetrical about the mean. Also see *perpendicular*.
Number	A particular value in a count of things; the position of something in a series; identification of a member of a series.
Numerator	The number above the line of a fraction. Also called the dividend.
Numerical	An approach to problem solving that makes use of numerical values.
Octal	The counting system which has the base of 8. The eight symbols used are, 0 to 7.
Ohm	Georg, Simon. German physicist stated his famous law in 1828. ohm: SI unit of electrical resistance.
Order	The order of a differential equation is the value of the highest derivative. See also *Magnitude, order of*.
Ordinate	The shortest distance of a point from the *x*-axis of a graph; 'the *y*-value'. See *abscissa*.
Origin	The point having coordinates $(0, 0)$ on a Cartesian graph.
Pappus	Theorem, of: allows us to calculate volumes of circular objects by taking the product of the cross-sectional area and the path of its centroid.
Parabola	The curve of one of the sections through a cone. Also: curve of a quadratic function and the path taken by a projectile under the influence of the force of gravity.
Parallelogram rule	Rule for the addition of two vectors or phasors.
Particular	A particular mathematical statement is one in which variables are symbols and the constants are numbers. Of differential equations: a particular solution is one for which the constant(s) of integration have been evaluated.
Pascal	SI unit of pressure and stress. See Appendix I.
Pascal's triangle	Triangle formed by the coefficients of a binomial expansion.
Percentile	One of 99 values of a variable dividing a statistical sample into 100 equal groups. The median is the 50th percentile. Quartiles are the 25th and 75th percentiles.
Periodicity	The property of recurring at equal intervals in time. The angular displacement of a period is

	always 2π rad and the time taken to complete a cycle is the *period* or *periodic time*.
Perpendicular	At right angles to a line or plane. Also called *normal*.
Phase	The angular displacement of a phasor from a chosen reference.
Phasor	An electrical quantity having amplitude and phase.
Physics	Branch of science concerned with the behaviour of energy and matter.
Polar	A coordinate system that represents points in a plane by modulus and argument.
Pole	The point on a polar graph that has the coordinates, $0 \angle 0$.
Polygon rule	An extension of the triangle rule for addition of vectors. It allows more than two vectors to be added.
Polynomial	An expression containing more than two terms.
Potential difference	The difference of electric potential between two points. See Appendix I.
Potential energy	Mechanical energy an object possesses due to its position $E = mgh$.
Power	Rate of energy transfer/conversion. SI unit is the watt.
Pressure	Perpendicular force per unit area. SI unit is the pascal.
Prism	A solid with congruent and parallel end planes and sides which are parallelograms.
Product	The result of a multiplication.
Product rule	Rule for differentiating an expression that contains the product of two different functions of the same variable.
Proportionality	Two quantities are said to be proportional to each other if the change by a factor in one brings about a change by the same factor in the other. A proportion is written in the form $y \propto x$. It is not an equality.
Pythagoras' theorem	States that the square of the hypotenuse is equal to the sum of the squares of the other two sides, i.e. $a^2 = b^2 + c^2$.
Quadratic	The common name given to an equation of the second degree.
Qualitative	An analysis or treatment that is concerned with the qualities of a system; usually descriptive.
Quantitative	An analysis or treatment that is concerned with the quantities of a system: usually numerical.
Quartic	An equation of the 4th degree.

Quartile	One of three values of a variable dividing a statistical sample into four equal groups. They are the 25th and 75th percentiles and the median.
Quintic	An equation of the 5th degree.
Quotient	The result of a division.
Quotient rule	Rule for differentiating an expression that contains two different functions of the same variable linked as a quotient.
r.m.s.	Root-mean-square value. In electrical engineering it is the a.c. equivalent of d.c.
Radian	Dimensionless SI unit of plane angle. See Appendix I.
Rational	Numbers that are arithmetically complete. They include the integers.
Real	Real numbers are those that are members of the whole set that makes up the number line. They include all the rational and irrational numbers.
Reciprocal	A value, expression or function so related to another that the product of the two is 1. To obtain a reciprocal, exchange the denominator with the numerator.
Recurrence relation	A fixed rule for generating successive terms of a *sequence*.
Relative error	Absolute error divided by the true value. Often expressed as a percentage. See *absolute error*.
Repeated	Of roots; we say a solution of an equation has repeated roots if it consists of only one distinct value. It occurs when a function has a turning point that coincides with the *x*-axis.
Right-handed system	The convention of representing the directions of the axes of the three-dimensionl coordinate system. The directions of the extended thumb, index finger and second finger represent the directions of the *x*-, *y*- and *z*- axes respectively.
Right-hand screw rule	The convention of representing the direction of a vector product. A clockwise rotation in the *x*–*y* plane gives a product vector that is in the positive direction of the *z*-axis.
Scalar	A quantity which can be fully specified by its magnitude.
Scalene	Non-equal sided triangles.
Secant	The reciprocal of the cosine ratio.
Sequence	A set of terms stated in a definite order. Each term of a sequence is formed according to some rule that is called a *recurrence relation*.
Series	The sum of the terms of a *sequence*.
Significant figures	One way of expressing the accuracy of a number.

Similar	Two triangles are said to be similar if they have the same angles but different sides.
Simpson's rule	Method for evaluating the approximate area of an irregular plane by considering it to be made up of a collection of rectangular strips with tops that are paraboloids.
Sine	Value assigned to an angle in a right-angled triangle. The ratio of the side opposite to the hypotenuse.
Sine rule	For any triangle: $\dfrac{a}{\sin A} = \dfrac{b}{\sin B} = \dfrac{c}{\sin C}$.
Singular matrix	A matrix that has no inverse because its determinant is zero.
Sinusoid	A periodic function that varies in proportion with the sine (or cosine) of angular displacement.
Speed	The magnitude of velocity. A scalar.
Static	Systems in *equilibrium* that do not (normally) change over time. See *dynamic*.
Stationary point	See *turning point*.
Steradian	Dimensionless SI unit of solid angle. See Appendix I.
Stiffness	The constant of proportionality between applied force and extension (or compression) of a spring. A measure of the ability of a spring to resist extension or compression in units of $\mathrm{N\,m^{-1}}$. See *compliance*.
Strain	The ratio of the change in length over the original length of a sample subjected to *stress*.
Stress	The perpendicular force per unit area applied to a material. It is not pressure because it may be tensile as well as compressive. Nevertheless, the SI unit of stress is the pascal.
Sum	The result of an addition.
Tangent	Value assigned to an angle in a right-angled triangle. The ratio of the side opposite to the side adjacent.
Tangent to a curve	A line (or surface) that meets the curve at a point without intersecting the curve at any point.
Temperature	A measure of the amount of heat energy present. The SI unit is the kelvin.
Term	A value or quantity that forms part of an *expression, series* or *sequence*. A term may be a product of factors such as $2\pi f$.
Tonne	1000 kg. Non-SI unit of mass.
Torque	Turning force. See *moment*.
Transformation	Of a function: change from one function to another function belonging to the same class.

Translation
Of the curve of a function: a shift of the curve in the x–y plane brought about by a certain kind of transformation.

Trapezoidal rule
Method for evaluating the approximate area of an irregular plane by considering it to be made up of a collection of trapezium-shaped strips.

Triangle rule
The triangle rule for the addition of two vectors. Can be extended to the *polygon rule* for the addition of more than two vectors.

Trigonometry
Branch of mathematics that deals with the relationships between the angles and sides of triangles.

Turning point
The point on the curve of a function where the gradient changes from positive to negative and vice-versa. A turning point corresponds to a maximum or minimum value of the function.

Unit vector
One of the **i**, **j** and **k** vectors that have a magnitude of 1 and respectively lie in directions parallel to the x-, y- and z-axes of the three-dimensional Cartesian coordinate system.

Unity
The number 1. The unity factor leaves the number on which it operates unchanged.

Variable
An independent variable is one that can take any value from the domain of the function. A dependent variable takes a corresponding value from the *codomain* of the function.

Vector
A quantity that must be specified by two components, magnitude and direction.

Velocity
The rate of change of position over time. A vector quantity: it is speed in $m\,s^{-1}$ in a specified direction.

Volt
The SI unit of potential difference. Named after Alessandro Volta.

Watt
James, Scottish mechanical engineer, inventor of the steam engine.
watt: SI unit of power.

Weight
The gravitational force exerted on an object by the Earth (or any other massive body). The SI unit of weight is the newton.

Work
Energy: transferred or expended.